EVOLUTION:
Process and Product

REINHOLD BOOKS IN THE BIOLOGICAL SCIENCES

Consulting Editor:
PROFESSOR PETER GRAY
Department of Biological Sciences
University of Pittsburgh
Pittsburgh, Pennsylvania

EVOLUTION:
Process and Product

EDWARD O. DODSON
Professor of Biology, University of Ottawa

Revised Edition of *A Textbook of Evolution*

ILLUSTRATED

REINHOLD PUBLISHING CORPORATION
NEW YORK
Chapman and Hall, Ltd., London

To the memory of

PROFESSOR ERNEST BROWN BABCOCK

and

PROFESSOR RICHARD BENEDICT GOLDSCHMIDT

*for many years distinguished proponents
of the Modern Synthesis of evolutionary research*

this book is respectfully
and affectionately dedicated

Consulting Editor's Statement

I AM HAPPY to welcome Dodson's *Evolution: Process and Product* to the Reinhold books in the Biological Sciences. It is the kind of book we want because it is exciting without being fanciful and solid without being dull. Evolution is, in last analysis, the philosophical thread that holds contemporary biology together. However, there are some books on evolution written with wide-eyed enthusiasm as though they were dealing with a new concept. Others plod stolidly along so preoccupied with detailing the facts that they miss the beauty of the concept. Dodson avoids either extreme.

This is one of those textbooks that can be read with pleasure by the teacher and with profit by the student. It belongs also in the library of every thinking layman who wants a well-balanced account of the facts, the concept, and the result of Darwin's great contribution to our contemporary thought.

PETER GRAY
Consulting Editor

Pittsburgh, Pennsylvania
April, 1960

Preface

THERE ARE FASHIONS in science as in all other things, and the study of evolution is at present highly fashionable among biologists. This is reflected not only by the immense number of technical papers on various aspects of evolution which are published annually, but also by the fact that courses in evolution are presented in many colleges and universities, while most courses in biological departments deal with evolution to some extent.

Our generation has witnessed a complete reversal of the character of evolutionary thinking. During the early decades of the present century, after the great enthusiasm of the immediate post-Darwinian era had spent itself, widespread pessimism prevailed regarding the very possibility of gaining any real insight into the mechanics of evolution. This pessimism was based upon many things, including a psychological reaction against the unbridled and uncritical enthusiasm of post-Darwinian biologists; a misconstruction of the significance of the new science of genetics; the disrepute into which taxonomy had fallen; and the mutation theory of DeVries, which seemed to make Darwinian variation and selection unnecessary.

Even while this pessimism prevailed, however, its bases were being destroyed by research in many apparently unrelated fields. In 1937, Dobzhansky published *Genetics and the Origin of Species,* in which he brought together many lines of research, and demonstrated that the prospects were bright indeed for understanding the mechanics of evolution in terms of the genetics of natural populations. This stimulated a reassessment of the relationship of many biological sciences (and some of the physical sciences) to evolution, and the result has been a *modern synthesis* in which all biological sciences seem to converge fruitfully upon evolution. This modern synthesis has been formalized in a series of books of such importance that any one of them would have required the revision of existing texts and justified the publications of new ones. Dobzhansky's book, which is now in its third edition, was the first of these. It was followed in 1940 by Goldschmidt's *Material Basis of Evolution,* in 1943 by Mayr's *Systematics and the Origin of Species,* in 1945 by Simpson's *Tempo and*

Mode in Evolution, in 1950 by Stebbins' *Variation and Evolution in Plants,* in 1953 by Simpson's *Major Features of Evolution,* and in 1957 by Darlington's *Zoogeography.* In addition to these books, an enormous amount of valuable evolutionary research has been published in a host of technical journals.

The present book was first published in 1952 in an attempt to make this flood of evolutionary scholarship available to students. I am indebted to those many professors and students who used the book and encouraged me to believe that it served its purpose well. In the eight years which have intervened since its publication, however, the flood of new research on evolution has continued and expanded. Many new data must be presented. Old theories and interpretations must be reassessed, and new ones must be tentatively put forward. Hence this second edition is mandatory, and I offer it with the hope that it will prove even more useful than its predecessor.

The general organization of the first edition has been retained. Part I includes seven chapters, summarizing the contents of the older books on evolution, and serving primarily to define the subject. These chapters present the major evidences for evolution. A new chapter, on evidence from comparative physiology and comparative biochemistry, has been added. Part II, Phylogeny, deals with the evolution of the higher categories, and attempts to trace the main lines of evolution in the Plant and Animal Kingdoms, including the probable lineage of man. Part III, the Origin of Variation, deals with the mode of origin of those hereditary variations which form the necessary substrate for the action of natural selection. Part IV, the Origin of Species, deals with those factors which sort out the varying arrays of organisms into species, genera, and higher groups. A new chapter on quantitative aspects of evolution has been added here. Lastly, Part V, Retrospect and Prospect, summarizes briefly what has gone before, attempts to put it in perspective, and brings together some predictions of bolder scientists as to what future evolution may bring.

The title of Huxley's book, *Evolution: the Modern Synthesis,* is particularly apt, for the modern study of evolution requires a synthesis of material from all fields of biology, together with some material from a good many other sciences. It has now been just over a century since Johannes Muller allegedly committed suicide because of despondency resulting from the realization that it would not be possible to master the whole domain of science. The difficulty of mastering all phases of modern evolutionary science is quite comparable, and the present work no doubt shows many deficiencies. Yet the task of writing a text which can introduce students to this important and fascinating field must be undertaken. The kind reception which was accorded the first edition of this book leads me to believe that the present edition, which has been improved in many respects, should fill even more successfully the urgent need for such a book.

I am indebted to many persons for the successful completion of this revision, and I am happy to acknowledge their help. The illustrations of

Mr. Frederick S. Beckman remain a valuable asset of the book. The entire manuscript was read critically by Dr. L. N. Garlough, Dr. Peter Gray, and Dr. Theodora N. Salmon, while the chapter on Man and the Primates was read by Dr. William L. Straus, Jr. Their criticisms have resulted in significant improvements. I am also indebted to Dr. Alfred S. Romer and Dr. Warren P. Spencer, whose criticisms of the first edition were still in many respects helpful in the revision. I am also indebted to those many users of the first edition who have submitted suggestions and corrections, many of them most helpful. Finally, I am indebted to the many publishers who have generously granted permission to use quotations and figures from their books. Each of these is separately acknowledged at the appropriate place.

This book is based upon a course of lectures which was first presented in 1947 at the Dominican College of San Rafael, California, was then developed over a period of years at the University of Notre Dame, and which is currently being given at the University of Ottawa.

EDWARD O. DODSON

April, 1960
Ottawa, Ontario

Contents

PART ONE

A Definition of Evolution

CHAPTER ONE

Evolution by Natural Selection: Darwin's Conception

EVOLUTION MAY BE DEFINED as "descent with modification" (Darwin), closely related species resembling one another because of their common inheritance, and differing from one another because of the hereditary differences accumulated since the separation of their ancestors. Stated in another way, evolution is the derivation of new species (or higher groups) of plant or animal from previously existing ones.

These definitions, as well as others commonly given, are general in their terms. They do not imply a particular line of descent for any specified organism. The descent of man from "monkeys" is not a point of definition for evolution. Nor does any competent student of evolution regard man as derived from any other organism now living. But both man and the great apes are regarded as coming from a common ancestor, an unknown primate. Man does not even play a very useful role in the study of evolution, because he is not available for laboratory experimentation to nearly the extent that other animals are, and because primitive man was rarely fossilized, though apparently more frequently fossilized than other primitive Primates. However, the student of evolution belongs to an egocentric species, and so a chapter on the evolution of man will be included in this book.

DARWINIAN PREMISES

Evolution, as conceived by Darwin, can be briefly summarized. All plants and animals reproduce in excess of the numbers which can actually survive, yet adult populations remain relatively constant. Hence, there must be a struggle for survival. Now the members of any species vary one from another. Some of the variations may be neutral, but others help or hinder the organism in its struggle for survival. As a consequence, the "survival of the fittest" (Spencer) variants will be expected, with the less fit being destroyed either by their physical or biotic environments (natural selection). Thus species will be gradually modified in the direction of the most advantageous variants.

3

The Prodigality of Nature. The prodigality of nature with respect to reproduction is well known. A single salmon produces 28,000,000 eggs in a season; an oyster may pass as many as 114,000,000 eggs at a single spawning; and *Ascaris lumbricoides* var. *suum,* a common parasite of hogs, has been observed to pass as many as 700,000 eggs in a single 24-hour-period under laboratory conditions. That such immense numbers of individuals should survive and themselves reproduce in similar numbers is simply unimaginable. For example, a thorough study of a small sector of the Pacific coast just north of San Francisco revealed about one hundred starfish (mainly *Pisaster ocraceus,* but a few other species were included). Assuming that half of these were females, and that each produced one million eggs (a modest estimate), the population in the next year would be about 50,000,000. These would include about 25,000,000 females, all of which would again produce about a million eggs each. It is obvious that, if the ordinary rate of reproduction were to continue for even a few generations with 100 per cent survival of all offspring, soon the starfish would fill the seas and be pushed out across the lands by sheer pressure of reproduction. Indeed, at the rate of reproduction here described, it would take only fifteen generations for the number of starfish to exceed the estimated number of electrons in the visible universe (10^{79})!

The animals discussed above are intentionally chosen from among the more prolific members of the Animal Kingdom. Essentially the same situation applies, however, to even the most slowly breeding animals. Frogs, while generally regarded as quite prolific, produce at the most 20,000 eggs annually (the bullfrog *Rana catesbeiana*). Most species of frogs produce fewer than 1,000 eggs annually, while a few (robber frogs, family Leptodactylidae) may lay as few as six eggs annually. Perhaps the most slowly breeding of organisms is the elephant. Darwin calculated the results of a minimal rate of reproduction for this animal. Elephants have a life span of about 100 years, with active breeding life from about thirty to about ninety years of age. During this period a single female will probably bear no fewer than six young. If all of these young survived and continued to reproduce at the same rate, then after only 750 years, the descendants of a single pair would number about 19,000,000.

Thus, regardless of the rate of reproduction of a species, it is clear that its numbers would soon become impossibly large if all survived and reproduced, simply because the rate of increase is geometric and the abundant young try desperately to survive. Because of this, there must be a struggle for existence, with the majority of the participants losing. This struggle may take many forms, as the struggle of the individual to overcome adverse environmental conditions such as cold or drought, or to escape predators, or to obtain an adequate share of a limited food supply for which there are many competitors. Darwin thought of the struggle as being most intense between members of the same species which must compete for identical requirements of life.

The constancy of adult populations is not true to the degree that Darwin thought, for populations of wild species may vary tremendously from year to year. However, they never approach the size calculated from the reproductive rate, and so it is clear that a severe struggle for existence must account for the difference.

Organic Variation. The fact of variation among living things is so obvious that it need be proved to no one. Even the proverbial peas in a pod vary from one another visibly. With the exception of identical twins, any two individuals of a given species generally show easily recognizable differences, differences which can often be measured. Not infrequently, a whole population may show a definite pattern of variation which differentiates it from the rest of its species. Such a population may be called a subspecies (frequently called a variety by breeders and fanciers). Such subspecies Darwin regarded as "incipient species," that is, species in process of formation. Many of these natural variations are completely neutral, conferring on their bearers neither advantage nor disadvantage in the struggle for existence. Others, however, may influence the chances of survival of their bearers. Thus, any variation which tends to reduce water loss will favor a desert plant, one which increases the speed of an ungulate will aid it in escaping predators, and one which increases the sensitivity of sense organs will aid a predator in detecting its prey.

Natural Selection. The outcome of this struggle for existence among varying plants and animals can be only one, as was very effectively stated by Darwin in Chapter 3 of the "Origin of Species"; . . . "how is it that varieties, which I have called incipient species, become ultimately converted into good and distinct species, which in most cases obviously differ from each other far more than do the varieties in the same species? How do those groups of species, which constitute what are called distinct genera and which differ from each other more than do the species of the same genus, arise? All these results . . . follow from the struggle for life. Owing to this struggle, variations, however slight and from whatever cause proceeding, if they be in any degree profitable to the individuals of a species, in their infinitely complex relations to other organic beings and to their physical conditions of life, will tend to the preservation of such individuals, and will generally be inherited by the offspring. The offspring, also, will thus have a better chance of surviving, for, of the many individuals of any species which are periodically born, but a small number can survive. I have called this principle, by which each slight variation, if useful, is preserved, by the term natural selection. But the expression often used by Mr. Spencer, of the Survival of the Fittest, is more accurate, and is sometimes equally convenient. We have seen that man by selection can certainly produce great results, and can adapt organic beings to his own uses, through the accumulation of slight but useful variations, given him by the hand of Nature. But Natural Selection, we shall hereafter see, is a power incessantly ready for action, and is as immeasurably superior to man's feeble efforts as the works of Nature are to those of Art."

BIOGRAPHICAL SKETCH OF DARWIN

Some idea of the comprehensive basis upon which Darwin proposed his theory may be obtained from a review of his life, taking the autobiography which he wrote for his children as a guide. Charles Robert Darwin was born on February 12, 1809, the fifth of six children born to Robert Waring Darwin and Susannah Wedgwood Darwin. Most of Darwin's elementary schooling was obtained at a boarding school in Shrewsbury, England, where his father practised medicine with notable success. The curriculum was almost entirely classical, and Darwin professed to have found it exceedingly dull and profitless. While he did not distinguish himself scholastically, he developed a great love for dogs, for collecting all manner of things, and for hunting birds. His father once said to him, "You care for nothing but shooting, dogs, and rat-catching, and you will be a disgrace to yourself and all of your family." But Darwin adds, ". . . my father, who was the kindest man I ever knew and whose memory I love with all my heart, must have been angry and somewhat unjust when he used such words."

While Darwin found his formal schooling rather fruitless, he did enjoy some cultural avocations during these years. He was fond of poetry, particularly of the historical plays of Shakespeare. He collected minerals and insects with great zeal, but, he says, rather unscientifically. He took much pleasure in watching the habits of birds, and he took some notes on his observations. One of his greatest pleasures was to assist his older brother in his chemical experiments, yet his schoolmaster publicly rebuked him for this, on the grounds that it was a useless pursuit.

In the fall of 1825, Darwin was sent to the medical school at Edinburgh. His account of the two years at Edinburgh make them seem utterly futile. Instruction was exclusively by means of lectures, which he described as "incredibly dull." He felt very little motivation to come to grips with his medical studies, for "I became convinced from various small circumstances that my father would leave me property enough to subsist on with some comfort, though I never imagined that I should be so rich a man as I am; but my belief was sufficient to check any strenuous efforts to learn medicine." But his achievements at Edinburgh could not have been so mediocre as he himself indicates, for he gained the friendship and respect of well-established scientists, such as Dr. Ainsworth, a geologist; Dr. Coldstream and Dr. Grant, zoologists; and Mr. Macgillivray, an ornithologist who was also curator of the museum. While he took no courses under these men, he enjoyed their company and learned much natural history from them. Also, he joined a students' scientific society before which he read papers on some small research problems which he had undertaken.

In any event, Darwin did not complete his medical education because his father learned that he did not want to be a physician. So the elder Darwin then suggested that he prepare himself to be a clergyman of the Church of England. Darwin says that the life of a country clergyman appealed to him, and after some study he was convinced of the truth of the Creed of the Church of England. In order to achieve this goal, it was necessary to have a degree from an English university, and so Darwin enrolled at Cambridge in January, 1828, and was graduated in January, 1831. Of his Cambridge years, Darwin says, ". . . my time was wasted, as far as the academical studies were concerned, as completely as at Edinburgh and at school." He says that the only things he enjoyed in his studies at Cambridge were geometry, and the works of Paley, a distinguished eighteenth century theologian whose beautiful logic and clear expression he admired. He felt that these were the only things in his formal education which contributed to the development of his mind.

Again, Darwin's own estimate of his achievements at Cambridge must have been unduly harsh. He says that he wasted his time with a crowd of sporting men, including some dissipated, low-minded men. But he also developed a taste for the fine arts, and made friends among the more cultured students at Cambridge. And as at Edinburgh, he attracted the friendship and respect of distinguished men of science, who must have seen in this youth something far better than the dilettante which he pictures himself to have been. Most important among these was the botanist, Dr. Henslow, through whom young Darwin met many of the most distinguished men of that time. But the major interest of Darwin's Cambridge years was in collecting beetles, a study which he pursued with great energy and with some distinction.

The Voyage of the Beagle. It was through Henslow that Darwin got the major opportunity of his life. The British Admiralty planned a voyage of exploration on H.M.S. Beagle, with the surveying of Tierra del Fuego as the primary objective. Henslow was asked to nominate a young naturalist for the voyage, and he urged Darwin to accept this appointment. Darwin's father objected, because he felt that this would simply delay the establishment of his son in the clergy. But he added, "If you can find any man of common sense who advises you to go, I will give my consent." Darwin's uncle, Josiah Wedgwood (of the "China" family), whom Dr. Darwin had always regarded as one of the most sensible men in the world, kindly fulfilled this condition. The Beagle was originally scheduled to sail in September, 1831, but did not actually get underway until December 27, 1831. The ship visited some of the islands of the Atlantic Ocean, many points on the coast of South America, and some of the islands of the South Pacific, of which the Galapagos Islands were much the most important for the development of Darwin's ideas on the mutability of species. During this voyage he took voluminous notes on the geology, botany, and zoology of the regions visited. These notes, together with the many specimens collected, formed the basis of several books which he later published, and made valuable contributions to his major work. The Beagle finally returned to England on October 2, 1836, after a voyage of nearly five years.

Darwin's Publications. Back in England again, Darwin at once set to work upon his "Journal of Researches," which was based upon the journal which he had kept during the voyage of the Beagle. This was published in 1839, and was an immediate success. Darwin said that the success of this, his first literary child, always pleased him more than that of any of his other books. Also, in 1839, he was married to his cousin, Anna Wedgwood. Two daughters and five sons were born to them. They lived in London until September of 1842. During this time, Darwin was active in scientific society, being secretary of the Geological Society from 1838 to

1841. His closest associate during this time was Lyell, who perhaps has contributed more than any other man to the modernization of geology. But his health became progressively worse, and he was unable to bear much excitement, and so the Darwins moved to Down, a country residence, in 1842. It was here that most of his life's work was done. As his health forced him to remain in seclusion, the remainder of his biography becomes largely a catalogue of his books. Much evidence indicates that his problems were largely psychosomatic.

In 1842, Darwin published the first of his major geological works resulting from the voyage of the Beagle, "The Structure and Distribution of Coral Reefs." In this book, he presented a theory of the structure and mode of formation of coral reefs which was very different from the one then generally accepted. But Darwin's keen observations and accurate thinking on the subject won support, and his theory is even now generally accepted among geologists. This was followed in 1844 by "Geological Observations on Volcanic Islands" and in 1846 by "Geological Observations on South America."

In 1846, Darwin began work on a study of the Cirripedia, or barnacles. This began with the study of an aberrant barnacle, which burrows into the shells of other species, collected when the Beagle visited the coast of Chile. In order to understand the structure of this new species, it was necessary to dissect more typical forms. Gradually the scope of the study broadened until it included descriptions of all known species of barnacles, living and fossil. This great Monograph on the Cirripedia was published in four volumes. The Ray Society published two volumes on the living Cirripedia in 1851 and 1854, respectively, while the Palaeontological Society published the two volumes on fossil species in the same years. Of this work, Darwin said, "I do not doubt that Sir E. Lytton Bulwer had me in his mind when he introduced in one of his novels a Professor Long, who had written two huge volumes on limpets. . . . My work on the Cirripedia possesses, I think, considerable value, as besides describing several new . . . forms, I made out the homologies of the various parts . . . and proved the existence in certain genera of minute males. . . . The Cirripedia form a highly varying and difficult group of species to class; and my work was of considerable use to me, when I had to discuss in the 'Origin of Species' the principles of natural classification. Nevertheless, I doubt whether the work was worth the consumption of so much time." Yet Sir Joseph Hooker, a distinguished botanist, wrote to one of Darwin's sons that "Your father recognized three stages in his career as a biologist: the mere collector at Cambridge; the collector and observer in the Beagle, and for some years afterwards; and the trained naturalist after, and only after, the Cirripede work." T. H. Huxley seems to have concurred in this opinion.

The Origin of Species. No further books followed until the "Origin of Species" in 1859. Yet this had really been in the making for more than twenty years. During the voyage of the Beagle, various facts of paleontology and biogeography which Darwin observed had suggested to him the possibility that species might not be immutable. But he had no theory to work upon. Lyell had attacked geological problems by accumulating all applicable data in the absence of a working theory, in the hope that the sheer weight of facts might throw some light upon his problems. As Darwin greatly admired the geological work of Lyell, he determined to apply the same method to the species problem. Accordingly, he began in July, 1837, his first notebook on variation in plants and animals, both under domestication and in nature. No possible source of information was overlooked: personal observations and experiments, published papers of other biologists, conversations with breeders and gardeners, correspondence with biologists at home and abroad, all are represented. As a result of this, Darwin soon saw that man's success in producing useful varieties of plants and animals depended upon selection of desired variations for breeding stock. But he did not see how selection could be applicable to nature.

In October, 1838, Darwin happened to read for pleasure "Malthus on Population," and it struck him at once that the struggle for existence among plants and animals offered a basis for *natural selection* of those variants which were best fitted to compete. But it was only in 1842, four years later and after the collection of a great deal more data, that he wrote out the first outline of his theory, a pencil draft of thirty-five pages. In 1844, this outline was enlarged to 230 pages. From the time of the completion of the Cirripede work in 1854, Darwin devoted all of his time to the study and organization of his notes, and to further experiments on transmutation of species.

Early in 1856, Lyell advised Darwin to write out a full account of his ideas on the origin of species, and he began this work on a much larger scale than that which finally appeared in the "Origin of Species." Then early in the summer of 1858, when this work was perhaps half completed, Alfred Russel Wallace, a young and little-known English naturalist then working at Ternate in the Dutch East Indies, sent Darwin a short essay "On the Tendency of Varieties to Depart Indefinitely from the Original Type." Wallace asked Darwin, if he should think well of this essay, to send it to Lyell for his criticism. Darwin thought very well of it, for he recognized his own theory, and he felt that he ought to withhold his own publication in favor of Wallace. However, Lyell and Hooker had for years been familiar with Darwin's work on the transmutation of species, and Lyell had read Darwin's outline of 1842. These men therefore suggested that Darwin write a short abstract of his theory, and that it be published jointly with Wallace's paper in the Journal of the Proceedings of the Linnean Society. These papers appeared in that journal in 1859, together with portions of a letter which Darwin had written to Asa Gray, the great American botanist, in September, 1857, in which he set forth his views on natural selection and the origin of species.

Following this, Lyell and Hooker urged Darwin to prepare for early publication a book on transmutation of species. Accordingly, he condensed the manuscript which he had begun in 1856 to about one third or even one fourth its original size, and then completed the work on the same reduced scale. The "Origin of Species," thus produced, was finally

FIGURE 1. CHARLES ROBERT DARWIN. (From Fuller and Tippo, "College Botany," Revised Ed., Henry Holt & Co., Inc., New York, N.Y., 1954.)

published in November, 1859. With regard to the great success of this work, Darwin wrote that "The success of the 'Origin' may, I think, be attributed in large part to my having long before written two condensed sketches, and to my having finally abstracted a much larger manuscript, which was itself an abstract. By this means I was enabled to select the more striking facts and conclusions. I had, also, during many years followed a golden rule, namely, that whenever a published fact, a new observation or thought came across me, which was opposed to my general results, to make a memorandum of it without fail and at once; for I had found by experience that such facts and thoughts were far more apt to escape from the memory than favourable ones. Owing to this habit, very few objections were raised against my views which I had not at least noticed and attempted to answer."

Most of the succeeding books of Darwin presented more fully data and viewpoints which were summarized tersely in the "Origin," or were otherwise supplementary to his great work. These include "The Fertilization of Orchids," 1862; "The Variation of Plants and Animals under Domestication," 1868; "The Descent of Man," 1871; "The Expression of the Emotions in Men and Animals," 1872; "Insectivorous Plants," 1875; "The Effects of Cross- and Self-Fertilization in the Vegetable Kingdom," 1876; "Different Forms of Flowers on Plants of the Same Species," 1877; "The Power of Movement in Plants," 1880; and finally, "The Formation of Vegetable Mould through the Action of Worms," 1881. In addition to this immense program of publication, he also brought out revised editions of many of his books, including five revisions of the "Origin."

This, then, is the scientific background of the man who wrote the "Origin of Species." It can be equalled by very few either for breadth or for depth.

Darwin's Mental Qualities. Before concluding these biographical notes, it may be of interest to review Darwin's estimate of his own mental qualities. His writing is remarkably clear and persuasive, and his style has a charm seldom found in scientific works. Yet he says that "There seems to be a sort of fatality in my mind leading me to put at first my statement or proposition in a wrong or awkward form." Again, it is difficult to escape the feeling that he is himself his harshest critic, for his letters are also very effectively written, and it seems unlikely that these were carefully planned and revised, as were his books. The general manner in which the "Origin" was developed, through a series of outlines based upon a large series of notes, has already been described in detail. This was the general plan of work for all of his larger books, though none other was done quite so thoroughly and over so long a period of years as the "Origin."

As a young man, Darwin enjoyed poetry, particularly Shakespeare, and such other poets as Milton, Byron, Wordsworth, and Shelley. While at Cambridge, he developed a taste for fine paintings and music. But in later years this taste for the fine arts was lost. "I have tried lately to read Shakespeare, and found it so intolerably dull that it nauseated me." The only artistic taste which remained was for novels. He says, "I often bless all novelists. . . . A novel, according to my taste, does not come into the first class unless it contains some person whom one can thoroughly love, and if a pretty woman all the better." He regarded his loss of taste for the arts in general as a personal defect, and said that he would cultivate such tastes every week if he had his life to live over. "My mind seems to have become a kind of machine for grinding general laws out of large collections of facts, but why this should have caused the atrophy of that part of the brain alone, on which the higher tastes depend, I cannot conceive."

Darwin regarded himself as rather slow of apprehension, and as being incapable of following for long a purely abstract train of thought. But against the charge of some of his critics that he had no powers of reasoning, he defended himself. For he pointed out that the "Origin" is one long argument, and that it convinced many able men, and he felt justified in saying that this could not have been done by a man without some powers of reasoning. But he felt that he did not exceed in this respect the average successful doctor or lawyer. However he believed that his powers of observation and his love of natural science were superior. His mind was kept open, and indeed he exercised unusual care in recording any data contrary to his hypotheses. ". . . with the single exception of the Coral Reefs,

I cannot remember a single first-formed hypothesis which had not after a time to be given up or greatly modified."

He concluded his autobiography with the statement that "... my success as a man of science, whatever this may have amounted to, has been determined ... by complex and diversified mental qualities ... the love of science—unbounded patience in long reflecting over any subject—industry in observing and collecting facts—and a fair share of invention as well as of common sense. With such moderate abilities as I possess, it is truly surprising that I should have influenced to a considerable extent the belief of scientific men on some important points."

Darwin died on April 19, 1882, at the age of seventy-three, and was buried in Westminster Abbey near the grave of Newton.

ALFRED RUSSEL WALLACE—CO-DISCOVERER

Alfred Russel Wallace was born on January 8, 1823. As a young man, he made a journey of explorations into the Amazon Valley with H. W. Bates, a distinguished entomologist. This served as the basis for a book, "Travels on the Amazon and Rio Negro," which he published in 1853. In 1854, he began a zoological exploration of the Malay Archipelago, a work which occupied him until 1862, and which resulted in a book entitled "The Malay Archipelago," which he published in 1869. While on the island of Ternate in February, 1858, he was stricken with intermittent fever. During an attack of the fever, he happened to think of Malthus' "Essay on Population," and "suddenly there flashed upon me the idea of the survival of the fittest." The theory was thought out during the rest of the ague fit, written out roughly the same evening, and then written out in full in the two succeeding evenings. The resulting paper he sent to Darwin, with whom he was somewhat acquainted. The rest of this story has been told above.

Wallace also was active in the further development of evolutionary literature. His major contribution was in the field of biogeography. His most important work in this field was "Geographical Distribution of Animals," which was published in 1876. He expressed the hope that this book might do for the biogeographical chapters of the "Origin" what Darwin's book on "The Variation of Plants and Animals under Domestication" had done for the corresponding chapters of the "Origin." This hope was realized, for this was perhaps the most outstanding of the classical treatments of biogeography. A second book relating to geographical problems of evolution, "Island Life," appeared in 1880. He also published "Contributions to the Theory of Natural Selection" in 1870, "Tropical Nature and Other Essays" in 1878, and "Darwinism" in 1889. While Wallace shares with Darwin the honor of first publication upon the theory of the origin of species by natural selection, he always very generously (and properly, for it was the "Origin" which convinced scientists) gave Darwin full credit for the theory, as indicated by the title of the last book mentioned above. Wallace died on November 7, 1913.

REFERENCES

DARWIN, CHARLES, 1897. "The Life and Letters of Charles Darwin. Including an Autobiographical Chapter." Edited by his son, Francis Darwin. D. Appleton & Co., New York, N.Y.

DARWIN, CHARLES, 1958. "The Autobiography of Charles Darwin." Edited with appendix and notes by his granddaughter, Lady Nora Barlow. Collins, London.

DARWIN, CHARLES, 1859. "On the Origin of Species by Means of Natural Selection, or the Preservation of Favoured Races in the Struggle for Life," Modern Library (Giants Series), New York, N.Y. This is still the most basic book on evolution, and should be read by every serious student of this subject.

SEARS, P. B., 1950. "Charles Darwin, the Naturalist as a Cultural Force," Charles Scribner's Sons, New York, N.Y. This book demonstrates the great influence of Darwin upon the culture of his time, general as well as scientific.

WALLACE, ALFRED RUSSEL, 1859. "On the Tendency of Varieties to Depart Indefinitely from the Original Type," Linnean Society, *Journal of the Proceedings,* 3, 53–62.

CHAPTER TWO

Evidences of Evolution
I: Biogeography

BIOGEOGRAPHY, taxonomy, physiology, comparative anatomy, embryology, and paleontology have usually been presented as the fields from which the evidence of evolution is primarily derived. Some of the more important evidence from each of these fields will be summarized in this and the following chapters. However, genetics must be regarded as a seventh field of evidence for evolution, for a large part of the current literature on evolution is drawn from this field.

The study of biogeography, or the geographical distribution of plants and animals, is of particular interest, because this is the field which first directed Darwin's attention while on the Beagle to the possibility of the origin of species by means of evolution. He regarded the voyage of the Beagle as the most important event in his life. In his autobiography he states: "During the voyage ... I had been deeply impressed by discovering in the Pampean formation great fossil animals covered with armour like that on the existing armadillos; secondly, by the manner in which closely allied animals replace one another in proceeding southwards over the continent; and thirdly, by the South American character of most of the productions of the Galapagos Archipelago, and more especially by the manner in which they differ slightly on each island of the group; none of the islands appearing to be very ancient in a geological sense.

"It was evident that such facts as these, as well as many others, could only be explained on the supposition that species become modified; and the subject haunted me. But it was equally evident that neither the action of the surrounding conditions, nor the will of the organisms (especially in the case of plants) could account for the innumerable cases in which organisms of every kind are beautifully adapted to their habits of life—for instance, a woodpecker or a tree frog for climbing trees, or a seed for dispersal by hooks or plumes. I had always been much struck by such adaptations, and until these could be explained it seemed to me almost useless to endeavour to prove by indirect evidence that species have been modified....

"In October 1838, that is, fifteen months after I had begun my systematic enquiry, I happened to read for amusement 'Malthus on Population,' and being well prepared to appreciate the struggle for existence which everywhere goes on from long-continued observation of the habits of animals and plants, it at once struck me that under these circumstances favourable variations would tend to be preserved, and unfavourable ones to be destroyed. The result of this would be the formation of new species. Here then I at last got a theory by which to work; but I was so anxious to avoid prejudice, that I determined not for some time to write even the briefest sketch of it."

DISCONTINUOUS DISTRIBUTION

The actual distribution of many organisms presents problems which are very difficult to understand if it be assumed that, in the words of Linnaeus, "There are just so many species as in the beginning the Infinite Being created," and if their present distributions correspond to their places of origin. Thus, the same or closely similar species sometimes exist in widely separated places, with no representatives in the intermediate territory. Alpine species are frequently identical with, or closely similar to, species much farther north. Organisms separated by great physical barriers are usually quite different, even though their physical surroundings may be much the same. Yet the fossil organisms of a particular area are usually similar to those now living in the same area. Finally, the inhabitants of

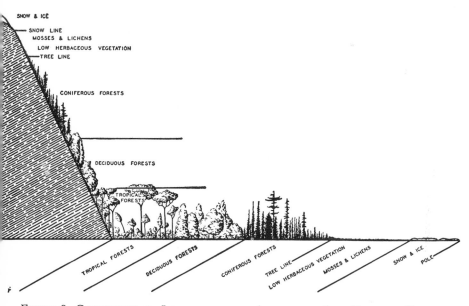

FIGURE 2. COMPARISON OF LATITUDINAL AND ALTITUDINAL LIFE ZONES IN NORTH AMERICA. (After Wolcott, from Allee, *et al.*, "Principles of Animal Ecology," W. B. Saunders Co., Philadelphia, Pa., 1949.)

15

oceanic islands are usually few in number of species, but a large proportion of these are peculiar to each island, and are similar to the inhabitants of the nearest continent. Also, amphibians and terrestrial mammals are usually not found on oceanic islands.

Thus, the magnolias at present occur naturally only in the southeastern portion of the United States and of Asia. The physical and biotic environments of these areas are similar. In the vast expanse which separates these two populations, magnolias do not occur naturally, although there are areas, such as southern California, in which they flourish when introduced. The predecessors of Darwin interpreted such facts as indicating that one and the same species had been created independently in more than one place. Many species, however, have very wide ranges, and Darwin pointed out that, if a wide-ranging species were to become extinct in the intermediate portions of its range, the result would be widely separated populations of the same species. Actually, fossil evidence indicates that, during a much warmer age, magnolias (and the associated flora and fauna) were distributed continuously over the vast area which now separates the extant populations. During the glacial ages, the climate became too severe for these subtropical plants over most of this vast range, with the result that they became extinct except for the mildest parts of their total range.

One of the most striking examples of widely separated populations of the same or closely related species is the common case of the inhabitants of high mountains, which may be identical even though they are separated by great expanses of lowland in which alpine species could not possibly survive. Or again, such mountains may have a flora and fauna closely similar to that of lowlands far to the north (Figure 2). Thus Darwin pointed out that the plants of the White Mountains of New England are the same as those of Labrador, and similar to those of the highest peaks of Europe. It was not difficult to explain this on the basis of a single origin for each plant species, with subsequent migration and modification. For, as the glacial ages advanced, arctic plants would progress ever further southward, replacing the temperate plants which, in turn, would migrate closer to the tropics. At the height of the glaciation, an essentially arctic flora would prevail over the entire northern United States, lowlands as well as mountains. As the glaciers retreated, arctic plants would again move northward in the lowlands and upward on the mountains, while the temperate plants moved out from their southern refuges to recapture their former territories. Thus, mountain plants would be expected to be the same as those of the lowlands to the north. And the close similarity of the alpine plants of Europe and America is understandable because of the observed fact that circumpolar plants are quite uniform in all places. Similar phenomena are known for animals. Thus the ptarmigan and the varying hare are found in the higher mountains of western United States and in the arctic and subarctic lowlands of Canada and Alaska. In Europe, the mountain hare is found in mountains from the west coast eastward to the Caucasus and Ural Mountains, and also in the lowlands of arctic Europe, but it does not inhabit the intervening lowlands.

FIGURE 3. THE BIOGEOGRAPHICAL REGIONS OF THE WORLD.

BIOGEOGRAPHICAL REGIONS

One of the important aspects of biogeography is the division of the world into six very distinct biogeographical regions (Figure 3), so that a biological explorer may well feel that he is entering an entirely different world when he goes from one region to another. These were originally defined on the basis of the avian faunas of the various parts of the world, but their validity is general. Thus, the *Holarctic Region* includes all of Europe, Asia north of the Himalaya and Nan Ling (mountains), Africa north of the Sahara Desert, and North America north of the Mexican Plateau. Typical mammals of this great region include the caribou and the elk, foxes of the genus *Vulpes,* bears, and the marmot tribe. This Holarctic Region is often broken up into the Palearctic Region (Old World) and the Nearctic Region (North America), because the two do present characteristic differences, though commonly only on the specific or generic levels.

The *Ethiopian Region* comprises Africa south of the Sahara Desert, and is marked by such mammals as the gorilla, giraffe, lion, and hippopotamus. The *Oriental Region* includes the portions of Asia south of the Himalayas and the Nan Lings and is marked by tarsiers, the orang-utan, the Indian elephant, and flying "foxes" (frugivorous bats). The *Neotropical Region* includes South and Central America, and is marked by tapirs, sloths, prehensile-tailed monkeys, and vampire bats. Finally, the *Australian Region* includes Australia and (like all of the above regions) the associated islands. As is well known, it is marked by a predominantly marsupial mammalian fauna, together with many other relic forms, and by the complete absence of any native placental mammals other than bats (and a few rodents which may have been introduced by primitive man).

All of these biogeographical regions are separated from one another by very nearly impassable barriers of sea, desert, or mountain, or by climatic zones, and these barriers are very ancient geologically. Thus the Palearctic and Nearctic Regions are separated by the Atlantic and Pacific Oceans. But because the north Pacific is quite shallow in the vicinity of Alaska, the two regions have been connected in the past, and hence the desirability, for some purposes, of describing them as a single region, the Holarctic. The southern continents are widely separated from each other by the great ocean basins of the world. The Ethiopian Region is separated from the Palearctic by the Sahara Desert, a very formidable barrier indeed to any organism adapted to temperate conditions, or to cold. South America is at present connected to North America by the Isthmus of Panama. During great eras of the earth's history, however, this tenuous connection has been submerged, so that South America was completely isolated from all other land masses. And even at present, climatic factors prevent most plants and animals from using this connection between the regions. The Oriental Region is separated from the Palearctic by the most lofty mountain chains in the world—the Himalayas and the Nan Lings. A glance at the map shows that the Malay Archipelago extends down from southern Asia and approaches quite near to Australia. Thus there is a broken link between the two regions, and one might imagine that it may have been complete at one time. But the channels between some of the islands are very deep, a fact which favors permanence, or at least very long duration. Nonetheless, the physical conditions within one region may be so closely similar to those of another as to be indistinguishable, so long as biotic factors are neglected. Thus the climate and physiography are much alike over large areas of South America and Africa. It can scarcely be doubted that each presents habitats quite suitable for the plants and animals of the other. Yet they have few organisms in common, and those they do have may often be explained as isolated survivors from once world-wide groups. Thus the Dipnoi (lungfishes) are now represented by only three living genera, *Neoceratodus* in Australia, *Protopterus* in Africa, and *Lepidosiren* in South America. The African and South American forms belong to the same family, while the Australian form is the sole member of its family. If only the living species be considered, it appears that the fishes of the southern continents have an especial relationship, despite the great ocean barriers which separate them. However, the fossil record shows that the lungfishes were once of world-wide distribution. They have long since become extinct in the face of competition with better adapted forms in most parts of the world, but the southern continents have been a last refuge of survival for these and so many other primitive forms.

That the flora and fauna within any one region show a certain consistency is to be expected on the basis of any theory of origin. Increasing populations, with their subsequent dispersal throughout the available territory, should achieve this. But the fact that many plants and animals are *excluded* from lands for which they are eminently well suited is difficult to explain on any theory other than the evolutionary one. This is exemplified by the problem of widely separated populations of the same or closely similar species, which has already been discussed above. It is difficult to understand why, if these represent independent creations of the same species, they should not be placed in similar parts of *different* regions. Why, for example, should both populations of magnolias be located in the Holarctic Region, when both the Oriental and the Neotropical Regions present eminently satisfactory habitats? And why should generally similar organisms be grouped together in such regions when suitable habitats for almost any organism can be found outside its own region?

ECOLOGICAL ZONES IN THE OCEAN

Distinct biotic regions are not limited to the land masses of the world. Although there is a degree of physical continuity between the oceans of the world, the habitats presented in different parts are quite different, and so ecological factors produce barriers within the oceans (Figure 4). On every shore, there is a narrow strand which is alternately covered and exposed by the tides. Beyond this *intertidal* or *littoral* zone, there is the broad, gently sloping continental shelf, the higher portions of which form the continental islands. The seas over the continental shelves are generally shallow, not over 100 fathoms (600 feet) deep, and they comprise the

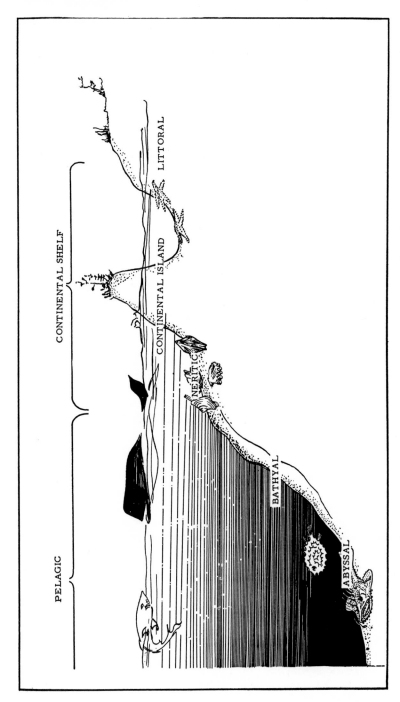

Figure 4. Ecological Zones in the Ocean. (Redrawn from Storer, "General Zoology," McGraw-Hill Book Co., Inc., New York, N.Y.)

neritic zone. But at the edge of the continental shelf, the ocean floor drops off rapidly to great depths. In this great expanse of open sea, there are several depth zones. The surface waters, to a depth of 100 fathoms, comprise the *pelagic* zone, which is inhabited by widely ranging fishes. The water here is subject to wave action, and is well oxygenated and lighted. The deeper water, down to 1000 fathoms, comprises the *bathyal* zone. Here the water is always quiet and poorly lighted. It grows progressively colder with increasing depth, and is scantily inhabited. Below this is the *abyssal* zone, into which the sunlight never penetrates. Here the water is always cold and quiet. The living forms to be found there are profoundly modified for life in the abyss.

The littoral and neritic zones are by far the most richly inhabited. The deep ocean basins form a barrier to the dispersal of these inhabitants of the continental shelves, with the result that different marine floras and faunas may be quite as isolated from one another as are those of the different continents. Thus Darwin pointed out that the organisms of the east and west coasts of the Americas are quite different, because they are separated by a great land mass. Yet about 30 per cent of the fishes on opposite coasts of Panama are identical. Correlated with this is the known geological fact that the Isthmus of Panama was submerged during much of the Tertiary Period. But westward of the continental shelf of the west coast there lies a great expanse of open sea until the islands of the Orient and the south Pacific are reached. The organisms found here are utterly different from those of the American continental shelf, because the open sea has been a formidable barrier to the littoral and neritic flora and fauna. But from the islands of the Orient to Africa, a much greater expanse, there is an almost continuous chain of islands or of continental coast, and the flora and fauna are rather uniform throughout this great region.

DARWIN'S EXPLANATION

It appeared to Darwin that these and other puzzling problems of the distribution of plants and animals could be understood if it be assumed that all organisms of a particular group (species or higher group) had migrated from a common place of origin, with subsequent modification. On this basis, one would expect the floras and faunas of those areas which have been isolated from one another longest (the biogeographical regions) would be most sharply differentiated. The inhabitants of different parts of the same region (for example, the mountains and plains of South America) should resemble one another more closely than the inhabitants of similar parts of different regions (for example, the mountains of South America and of Africa). That is, when the mountains in any region are elevated during geological ages, the new mountains will be colonized by the inhabitants of the surrounding lowlands. Some of these will be altogether unfit for the mountain environment; others will be adaptable to the lower but not to the higher altitudes; while a few can invade the highest ranges; and so the proportions of the various organisms will be different from those which characterize the surrounding lowlands. As a result, not only the physical surroundings but also the biotic environment of the mountain colonizers will be different, and so natural selection will favor modification of the colonists. But these should bear within their structure and habits evidences of their close relationship to their lowland progenitors. On the other hand the mountain colonists of different regions will be different because of the long isolation of their ancestors. Similar considerations apply to the colonization of any new territory whatever. And finally, the fossil remains in any region should resemble the living organisms of the same region, more closely in the case of recent fossils, less closely in the case of ancient fossils. This is required by the rather obvious fact that the present inhabitants of any region must have been descended from the past inhabitants. Great migrations of the past may modify the truth of this proposition in some instances, but they cannot take away its general validity. Darwin was quoted above for one example, namely the fossil armored mammals of South America which closely resemble the present armadillos of the same continent. Further discussion of this important topic may be deferred to the chapter dealing with paleontological evidences of evolution. All of the above facts follow logically from the Darwinian hypothesis, yet each is anomalous if it be assumed that each species has been independently created in its present range.

DISTRIBUTION OF FRESH-WATER ORGANISMS

Darwin regarded fresh-water organisms as the most noteworthy exception to the principle that organisms separated by a barrier are quite different. River systems and lakes are, of course, separated from one another by barriers of land. While many fresh-water systems frequently empty into the same ocean, salt water is a barrier to most fresh-water organisms which is no less formidable than land. Hence one might expect an unusual degree of differentiation in fresh-water floras and faunas. But the opposite is the case. There is great similarity between fresh-water organisms throughout the world, and many individual species are world-wide in distribution. Darwin believed that this could be accounted for by the fact that most fresh-water organisms must, in order to survive, be adapted for frequent short migrations from pond to pond or from stream to stream within a limited locality. But these migrations will inevitably lead to longer ones occasionally. Given time on a geological scale, this should result in very widespread species.

Darwin gave much attention to what might be called accidental means of transport of fresh-water organisms. A very common phenomenon is the joining of different rivers and lakes by their floodwaters in the spring. This should permit a very extensive exchange of their inhabitants. More selective and certainly less common is the transport of fishes and other small organisms by whirlwinds and tornados. A whirlwind, when passing over water, may pick up the surface waters together with any small organisms which happen to be near the surface. Later, when the force of the wind abates, the water and its contents will be dropped. If it should happen to drop over another body of water, the organisms so transported might then multiply and become established in the new locality. This is the basis of the rains of fishes which are occasionally reported. While such reports are usually received with much well-justified skepticism, nonetheless Gudger has examined all such reports critically, and he believes that at least seventy-eight recorded rains of fishes are valid. Shorebirds and waterfowl may also act as agents of dispersal for fresh-water organisms. For as the birds arise from the water, small organisms, eggs, larvae, and mud containing seeds are likely to cling to the feet of the birds. As the flight is likely to terminate in a comparable body of fresh water, and as these birds are among the most wide-ranging, it seems probable that much dispersal of small organisms occurs in this way. Also, the seeds of many plants retain their viability after passing through the digestive tract of a bird. Thus seeds which are eaten in one pond may be discharged in a quite distant pond, there to germinate.

All of this is not to say that there is a single, world-wide fresh-water flora and fauna. Discontinuities do exist among the inhabitants of fresh-water systems. But they are less marked than might at first be expected, and they correspond to the most ancient and imposing geographical barriers.

ISLAND LIFE

The final category of geographical evidence, and the one which had the greatest effect upon the thinking of Darwin, is that of oceanic islands and their living inhabitants. Darwin observed that such islands are typically poor in numbers of species present, although the success of animals and plants introduced by man has proven that these islands are well adapted to support a much greater variety of organisms than existed upon them aboriginally. He reasoned that, if all organisms had been created in their present localities, there is no reason why oceanic islands (islands, that is, which are located beyond the continental shelf) should not be as richly inhabited as comparable areas of the continents. Yet this is readily understandable upon his theory of migration from a common place of origin for all members of any group, with subsequent modification. For relatively few species could cross the great water barrier separating oceanic islands from the continental centers of origin.

Of the few species found on oceanic islands, a large number are endemic, that is, found nowhere else. Darwin found twenty-six species of land birds in the Galapagos Islands (Figure 5). Of these, twenty-one and possibly twenty-three are endemic. But of the eleven species of marine birds, only two are endemic. This is again just what one would expect in accordance with Darwin's theory. For, the occasional immigrants from the distant mainland (South America) would, upon arrival in their new environment, compete with quite different species from their cousins on the mainland, and so would be modified, eventually reaching the status of new and distinct species. But the great water barrier would greatly reduce the probability of these new species spreading to other localities. Yet for the marine birds the barrier ought to be less formidable, and hence the

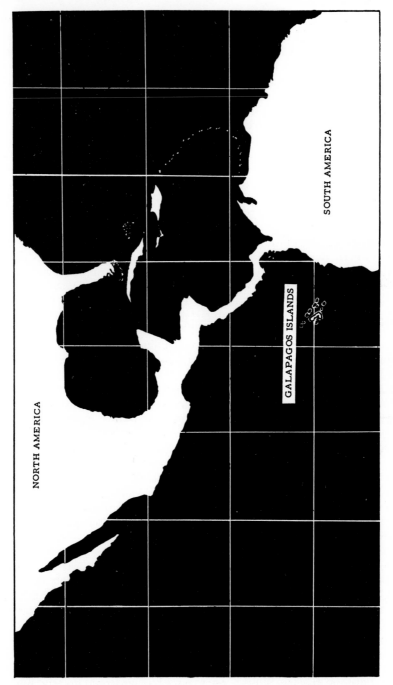

FIGURE 5. THE GALAPAGOS ISLANDS.

smaller proportion of endemics is not surprising. Lest thirty-seven species of birds for a small group of islands sound like a large number, the number of species on a restricted continental area may be given for comparison. The 1944 checklist of birds on the campus of the University of California at Berkeley lists 105 regular residents or seasonal migrants and forty species which have been recorded as occasional visitors.

The Amphibia and terrestrial mammals, though not the bats, are usually entirely absent from oceanic islands. When they have been introduced by man, they frequently have multiplied so greatly as to become a nuisance. The west coast toad, *Bufo marinus,* for example, was introduced into Hawaii in the hope that it would aid in the control of insects; but the toads themselves have now become a nuisance in the islands. Yet these groups are unable to cross large water barriers (or salt water barriers in the case of the Amphibia, which are quickly killed by salt water). But a barrier across which a mouse, for example, could not swim, might be easily flown by a bat. Had all species been created in the places where they now exist, then Amphibia and terrestrial mammals should be as frequent on oceanic islands as on comparable continental areas. Certainly terrestrial mammals should have been created on these islands as frequently as were bats. But bats are the very mammals which should reach the islands most readily if all mammals arose first on the continental land masses and then subsequently invaded such territories as they could.

Finally, the inhabitants of the several islands of an archipelago are commonly specifically distinct, yet plainly closely related; all of them, however, show a less close relationship to the inhabitants of the nearest mainland. Thus, when the Beagle visited the Galapagos Archipelago, located between 500 and 600 miles west of South America, Darwin felt that he was stepping upon American soil because of the obvious similarity of the plants and animals of these islands to those of the South American continent. The Galapagos Islands include 332 species of flowering plants. Of these, 172 species, more than half of the total, are endemic, and many species are restricted to one or a few islands in the archipelago. Yet all of these plants show close relationship to South American plants. But the climate and the geological character of the islands are utterly different from those of South America, hence the relationship of their plants cannot be understood on the basis of creation of similar plants for similar lands, but only on the basis of migration of plants from the continent to the outlying islands, with subsequent differentiation. The Bermuda Islands are located about 700 miles off of the coast of North Carolina, and their inhabitants are all North American in character. Many terrestrial vertebrates have been successfully introduced into the islands, but only one, a lizard, is native there. It belongs to a North American genus, but the species is endemic. Land birds are represented by many species, but none are endemic, for Bermuda is on one of the major migration routes for North American birds; hence it is not at all isolated from the viewpoint of the birds. Bats are also common to the mainland and the islands because these flying mammals can readily cross the water barrier. But the land molluscs

25

include a high proportion of endemics, no doubt because of the rarity of a successful crossing of the water barrier.

The only understandable basis for these facts is Darwin's hypothesis, that the islands were colonized from the mainland, with the colonists becoming modified subsequently. As they spread to the various islands of the archipelago, each isolated population was modified independently, with the result that groups of closely related, endemic species were formed. The connection, then, between the various similar species of an archipelago and of the nearest continent is simply heredity.

Very tersely then, the actual geographical distribution of organisms is readily understandable on the assumption that each group has originated in one of the major regions of the world, then spread to occupy as much space as it could in the face of physical and climatic barriers, and of competition from other organisms. The result has been continuous pressure of natural selection, leading to adaptation of the group to a wide variety of circumstances, that is, to evolution. Much of the data of distribution would be anomalous on any other basis. It is little wonder that first-hand experience with so impressive and persuasive a series of facts should have suggested to Darwin the possibility that species are mutable.

REFERENCES

BARLOW, LADY NORA (Ed.), 1946. "Charles Darwin and the Voyage of the Beagle," Philosophical Library, New York, N.Y. Darwin's granddaughter here presents selections from his letters to his family and from his notebooks on the voyage.

DARLINGTON, P. J., 1957. "Zoogeography: the Geographical Distribution of Animals," John Wiley & Sons, Inc., New York, N.Y. A readable, thoughtful, and thought-provoking rethinking of the entire field—the first such since Wallace's.

DARWIN, CHARLES, 1845. "Journal of Researches," 2nd Ed., Appleton & Co., New York, N.Y., and London. The original report of the voyage of the Beagle, and Darwin's "favorite literary child."

DE BEAUFORT, L. F., 1951. "Zoogeography of the Land and Inland Waters," Sidgwick & Jackson, London. A useful summary.

EKMAN, S. P., 1953. "Zoogeography of the Sea," Sidgwick & Jackson, London. Brief but excellent.

WALLACE, ALFRED RUSSEL, 1876. "The Geographical Distribution of Animals," The Macmillan Co., New York, N.Y. After more than 80 years, this is still the most fundamental work in its field.

CHAPTER THREE

Evidences of Evolution II: Taxonomy, Comparative Anatomy, and Embryology

THE SECOND MAJOR CATEGORY of evidence for evolution is taxonomy, the science of the classification of organisms. Classification would be absolutely necessary because of the sheer numbers of species, even if no other purpose beyond that of facilitating study were to be served. There are something on the order of 1,000,000 species of animals and 250,000 species of plants described in biological literature. Large numbers of these may live even in very restricted localities. Thus, in Lake Maxinkukee, Indiana, Jordan identified sixty-four species of fish, eighteen species of amphibians, and 130 species of molluscs. This does not infer that other groups may not be as liberally represented, but simply that they are not cited. It is evident that no extensive and orderly study of the living world would be possible unless it be divided into categories about which generalizations could be made.

LINNAEUS AND BIOLOGICAL NOMENCLATURE

Modern classification is based upon the work of Carolus Linnaeus, a Swedish botanist (1707–1778) who undertook the classification of the entire living world. Previously, scientific names of organisms had been short (or not so short!) descriptions written in Latin. Thus, Mark Catesby in 1754 referred to the red-headed woodpecker as *Picus capite toto rubro* and to the red-winged blackbird as *Sturnus niger alis superne rubentibus*. Linnaeus introduced the practice of giving each animal a binomial name, the first member of which is the generic name, and is shared with other closely similar species; while the second member, the specific name, differentiates the species from other members of the same genus. This binomial system of nomenclature is now universally accepted. Thus, the red-headed woodpecker becomes *Melanerpes erythrocephalus* and the red-winged blackbird becomes simply *Agelaius phoeniceus*.

The Species Concept. Basic to this is the species concept, the idea that there are definite kinds of plants and animals, the individuals of any one kind differing from each other only in minor traits, except sex; sharply separated in some traits from all other species; and mutually fertile, but at least partially sterile when crossed to other species. In the viewpoint of Linnaeus, such species were absolute. "There are just so many species as in the beginning the Infinite Being created." But many present-day biologists feel that the species is to some extent artificial, that is, the boundaries between closely related species are arbitrary rather than natural.

Taxonomic Categories. Linnaeus recognized that several species may have so much in common that they must be grouped together as a genus, distinct from other species-clusters (genera). Thus, the current edition of the American Ornithologists' Union Checklist of North American Birds lists two species of *Melanerpes* and three of *Agelaius*. But genera can be much larger. Thus, the Checklist gives twelve species of the genus *Puffinus* (shearwaters). And there are a few very large genera, for example the weed *Crepis,* of which Babcock recognizes 196 species!

The genera, also, Linnaeus found to fall naturally into larger clusters, based on similarity in rather fundamental characters. These groups of like genera he called orders. The genus *Melanerpes* belongs to the order Piciformes, along with ten other genera of woodpeckers and allied birds. *Agelaius* belongs to the great order Passeriformes, including the great majority of song birds. Finally, these orders were grouped into classes, the diverse members of which share only very fundamental characteristics. Thus, all of the orders of birds comprise the single class Aves.

Linnaeus grouped the classes into the two kingdoms, Plantae and Animalia, but did not feel the need of any category intermediate between class and kingdom. Ernst Haeckel, in the immediate post-Darwinian era, introduced the term *phylum* to include related classes. Phylum means, etymologically, a line of descent, and the term was chosen especially for its appropriateness to the new study of evolution. For a similar reason, Haeckel introduced the *family* as a category intermediate between genus and order.

Thus the full hierarchy of essential taxonomic categories is species, genus, family, order, class, phylum, and kingdom. Every organism which is described must be assigned to each of these categories, either overtly or implicitly. Returning to the two birds discussed above, their complete classification is as follows:

Kingdom	Animalia	Animalia
Phylum	Chordata	Chordata
Class	Aves	Aves
Order	Piciformes	Passeriformes
Family	Picidae	Icteridae
Genus	*Melanerpes*	*Agelaius*
Species	*erythrocephalus*	*phoeniceus*

All of the taxonomic categories are commonly subdivided for purposes of detailed study. The prefixes sub- and super- may be added to any of the standard categories to indicate subdivisions or larger assemblages. Other intermediate categories are sometimes used, but do not have official status in zoology (as determined by the International Congress of Zoology).

T. H. Huxley has said, "That it is possible to arrange all the varied forms of animals into groups, having this sort of singular subordination one to the other, is a most remarkable circumstance." It is indeed. Linnaeus accounted for this by the theory of archetypes, which assumes that the Creator worked from a series of plans, the archetypes, which was limited in number. These archetypes, like the plans in an architect's folio, were not all equally distinct, but fell into definite, classifiable categories. Thus each class would correspond to a major archetype, the various orders within a class would be lesser archetypes, and so on down the hierarchy. Thus, Linnaeus attributed the similarity between species of a genus not to descent from a common ancestor, but to the supposed fact that each is a more or less imperfect copy of rather similar archetypes, plans of the Creator.

SIGNIFICANCE OF THE TAXONOMIC HIERARCHY

Darwin's explanation of this "remarkable circumstance" is quite different. The taxonomic categories simply represent degrees of blood relationship. Thus, all members of the phylum Chordata have common ancestors, but they are exceedingly remote, and hence only the most fundamental chordate characters are held in common by extreme members of the phylum. Within any class, however, the degree of relationship is much closer, and hence more numerous and less fundamental characters are held in common by diverse members of a class. All birds, for example, share many characters in common. As one goes down the taxonomic scale this trend becomes stronger until finally members of a single species differ only in minor characters, and this because of their common inheritance. It is difficult to study any group of organisms in detail without feeling that this argument is a cogent one.

The Tree of Life. Taxonomists have always tried to summarize their studies with diagrams. A *tree* is the most successful type of diagram, but this has not always been evident. Linnaeus experimented with map-like diagrams, but it was found that no arrangement was possible which would always place similar forms together and always separate dissimilar forms. Later, Lamarck and others tried to arrange living forms on a ladder-like diagram, on the principle that the adaptation of living forms is progressive, so that any particular animal should be preceded by one somewhat lower in the scale of life and followed by another one somewhat higher. In a general way, this can be done. It may be easily conceded that amphibians are more advanced than fishes, and that reptiles are more advanced than amphibians. But one cannot continue this series—birds, then mammals. For, while one mammal has advanced so far beyond all other animals that he alone studies the world in which he lives, nonetheless the majority of birds are in every respect quite as "high" as are the majority of mammals. And so it appears that two rungs are required at the same level on the ladder. This type of dilemma is not only repeated frequently among the major groups of organisms, but it becomes very much more frequent when the study is extended to the lower levels of classification. As soon as the need for parallel rungs at many levels became apparent, with parallel series above them, it immediately was evident that a tree would make a much better diagram, and this was generally accepted long before Darwin.

Now everyone understands that the parts of a real tree are related to one another in a branching fashion because the whole organism is the product of growth from a single seed, growth accompanied by branching and differentiation. Independent creation and secondary union of the many parts would be unthinkable. The taxonomic tree is not strictly comparable to an actual tree, for here the processes of branching and differentiation are not generally amenable to direct observation. But the analogy cannot be avoided: the fact that no other type of diagram can symbolize the data of taxonomy so readily as a tree strongly suggests that, like a real tree, the tree of life owes its branching character to organic growth and differentiation—in other words, to evolution. Pre-Darwinian biologists were unable to understand why it was that classification seemed to fall into a tree-like pattern, yet they agreed that it was so. Long awareness of this fact undoubtedly helped to prepare the way for the ultimate acceptance of Darwinism.

A final characteristic of the tree of life deserves special consideration. In any group, there are likely to be some members which are simpler and

FIGURE 6. TEN SPECIMENS OF WARBLERS REPRESENTING FIVE SPECIES OF *Dendroica*. Note the gradual deepening of color from left to right. Members of the same species are not always adjacent in this series. The transition appears even more gradually in color. (Specimens loaned by the Zoological Museum of the University of Notre Dame.)

less specialized than the others. These may be known only as fossils, but many such archaic or primitive species are still living. In this case, the archaic species generally has its closest affinities with fossil rather than with living members of its group. Now it is a general rule that such primitive species resemble members of *other* groups more than do the more specialized members of the same group. To paraphrase this, species which are placed near the points of branching in the tree of life show especial resemblances to other species in both branches. If each species were created independently of all others, this fact would be inexplicable. Common characters should then be distributed among various groups without regard to their level of specialization. Yet, if the evolutionary theory be correct, then the most primitive members of related groups are those which have diverged the least from their common ancestry, and so primitive species should illuminate the relationships between groups. This is exactly what is found in nature.

Another suggestive aspect of taxonomy is the difficulty of distinguishing closely related species. In not a few cases, the ranges of variation of related species overlap. For example, in the genus *Dendroica*, the wood warblers, it is possible to set up a series ranging from the yellow warbler through magnolia, myrtle, and palm warblers to the chestnut-sided warbler in which the color pattern gradually shifts from predominantly yellow to predominantly black. The gradations are so fine that expert knowledge is needed to separate the species (Figure 6). Again, the fruit flies *Drosophila pseudoobscura* and *D. persimilis* are so similar that statistical analysis of populations is needed to distinguish them. Such situations strongly suggest descent from rather recent common ancestors.

EVIDENCE FROM COMPARATIVE ANATOMY

As comparative anatomy is the field from which inferences of relationship among animals are most commonly drawn, it is an especially important source of evidence for evolution. If any particular organ system is studied in diverse representatives of a single phylum, one gets the impression that the system is based upon a prototype which is simply varied from class to class (with finer variations within each class). Examples from the comparative anatomy of the vertebrates will be discussed.

The Vertebral Column. The vertebral columns of all vertebrates originate from similar embryonic rudiments, four pairs of condensed mesenchymal masses in each somite, the arcualia, with associated looser mesenchyme (Figure 7). From these simple beginnings, the typical vertebra with its centrum, neural arch, spinous and transverse processes, and its various articular processes, is formed. Among the Cyclostomata, the most primitive of living vertebrates, only the two dorsal pairs of arcualia (Figure 8) are formed, and these do not advance much beyond the embryonic condition, for they merely form two pairs of cartilage spikes flanking the spinal cord.

In the Chondrichthyes (sharks and their allies), a much more extensive development occurs (Figure 9). The two dorsal pairs of arcualia (= arch-

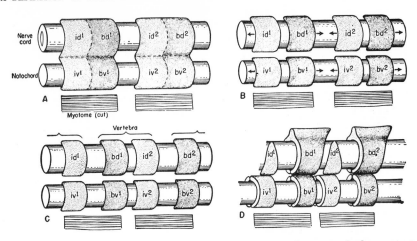

FIGURE 7. THE ARCUALIA WHICH FORM THE VERTEBRAE. *A* shows two body segments, each with four arcualia (actually four pairs, but only the left side is shown). *B* and *C* illustrate the separation of the anterior from the posterior arcualia in each segment. *D* shows how a vertebra comes to span two muscle segments because it is made up of arcualia derived from two adjacent body segments. The arcualia are named interdorsal, id; interventral, iv; basidorsal, bd; and basiventral, bv. (From Romer, "The Vertebrate Body," 2nd Ed., W. B. Saunders Co., 1955.)

formers) form neural arches and so-called intercalary arches over the spinal cord. The two ventral pairs fail to form arches in the trunk region, but in the caudal region the anterior ventral pair of arcualia forms a haemal arch in each segment, enclosing the caudal artery and vein. A centrum is formed around the notochord partly by the notochordal sheath, and partly by the bases of the arcualia and the surrounding mesenchyme. The articulation is strengthened by the notochord, which is still continuous in adult sharks. Ribs are formed in conjunction with these vertebrae. The entire skeleton of the adult is cartilaginous, a fact which is probably properly regarded as the retention of a basically embryonic condition.

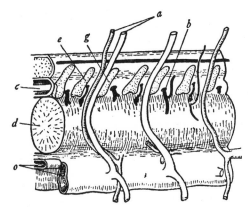

FIGURE 8. THE "VERTEBRAE" OF THE LAMPREY, CONSISTING ONLY OF THE TWO PAIRS OF DORSAL ARCUALIA IN EACH SEGMENT. a shows intersegmental blood vessels; b, segmental nerve; c, neural tube; d, notochord; e and g, dorsal arcualia; and o, longitudinal blood vessels. (From Hyman, "Comparative Vertebrate Anatomy," 2nd Ed., University of Chicago Press, 1942.)

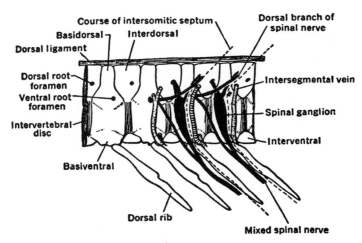

FIGURE 9. SHARK VERTEBRAE (*Squalus acanthias*). (From Goodrich, "Studies on the Structure and Development of Vertebrates," 1909, by permission of The Macmillan Co., publishers.)

The vertebrae of Osteichthyes (bony fishes) are not substantially different from those of the Chondrichthyes, except that they are ossified in whole or in part. In many fishes, *two* centra are formed for every body segment, an anterior *hypocentrum* based upon the anterior ventral pair of arcualia, and a posterior *pleurocentrum* based upon the posterior dorsal pair of arcualia. This is exemplified by the Rhipidistia, an extinct group of lung-breathing fishes which is believed to have given rise to the Amphibia.

The most primitive fossil amphibians have vertebrae much like those of the Rhipidistia, but they change in two different lines of descent (Fig-

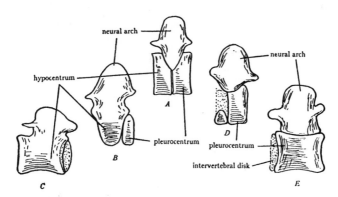

FIGURE 10. DERIVATION OF AMPHIBIAN AND AMNIOTE CENTRA. *A* is the primitive amphibian type. *A, B,* and *C* illustrate the transition to the type of vertebrae found in the modern amphibia, while *A, D,* and *E* illustrate the transition to the type found in amniotes. (From Hyman, "Comparative Vertebrate Anatomy," 2nd Ed., University of Chicago Press, 1942.)

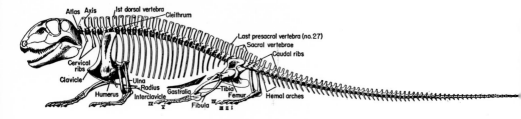

FIGURE 11. SKELETON OF A PRIMITIVE REPTILE, HAPTODUS. (From Romer, "The Vertebrate Body," 2nd Ed., W. B. Saunders Co., Philadelphia, Pa., 1955.)

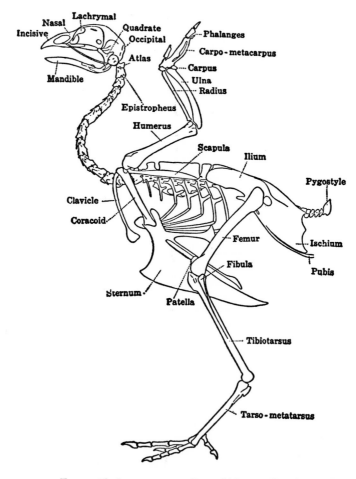

FIGURE 12. SKELETON OF A BIRD. (After Bradley.)

ure 10). In one, leading toward the modern Amphibia, the hypocentrum gradually replaces the pleurocentrum. In the other, leading to the reptiles, the pleurocentrum gradually replaces the hypocentrum. This replacement is not yet complete in the Rhyncocephalia, the most primitive order of living reptiles.

The pre- and post-zygopophyses, articular processes on the neural arches, first appear among the Amphibia. They strengthen the joints between successive vertebrae. Another innovation is the addition of two more differentiated types of vertebrae in addition to the trunk and caudal vertebrae. One or two *cervical* vertebrae serve to make a freely movable joint between the skull and the vertebral column, a character of adaptive value to land-dwellers, but not to the strictly aquatic fishes. Between the trunk and caudal regions is a single sacral vertebra specialized by fusion with its ribs to attach the pelvic girdle to the axial skeleton.

The vertebrae of the Reptilia differ from those of the Amphibia mainly in that the centrum is a pleurocentrum rather than a hypocentrum (in extant forms). Also, the number of cervical and sacral vertebrae is increased. In the cervical region, the ribs are now fused to the vertebrae (Figure 11). The vertebral column of the birds differs from that of the reptiles chiefly in that a highly flexible articulation has been developed between the cervical vertebrae, and those of the body are largely fused. The caudal vertebrae are much reduced (Figure 12). The vertebral column of the mammals differs from that of the reptiles mainly by the development of curvatures associated with the mechanics of locomotion (Figure 13).

Thus, a survey of the vertebral columns of the extant classes of vertebrates leads back to the statement with which this discussion was opened, that, within a single phylum, any particular system seems to be based upon a prototype which is simply varied from class to class.

Serial Homology in the Crustacea. Structures, which are similar in different species because of common inheritance, irrespective of the diverse uses to which they may be put, are called *homologous* structures. There is no group of living things which does not present homologies in all of its structural systems, but perhaps no set of homologous structures

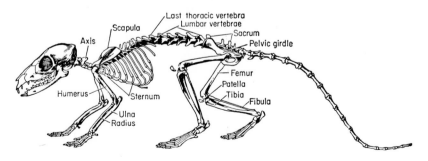

FIGURE 13. THE SKELETON OF A PRIMITIVE MAMMAL, THE TREE SHREW, *Tupaia.* (After Gregory, from Romer, "The Vertebrate Body," 2nd Ed., W. B. Saunders Co., 1955.)

35

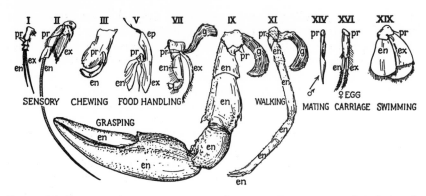

FIGURE 14. TYPES OF APPENDAGE OF THE CRAYFISH. Roman numerals indicate the body segment from which each was taken. *pr,* protopodite; *ex,* exopodite; *en,* endopodite; *ep,* epipodite; *g,* gill. (From Storer and Usinger, "General Zoology," 3rd Ed., McGraw-Hill Book Co., Inc., 1957.)

is more striking than the appendages of the Arthropoda, especially those of the Crustacea. In the typical crustacean, there is one pair of appendages borne by each segment of the body. These are all referable to a single structural plan, and so they are considered to be serially homologous. Serial homology is typical of all animals whose body structure is based upon a series of essentially similar segments. The major examples are in the phyla Annelida, Arthropoda, and Chordata. In the crustacean appendage, there is typically a basal *protopodite* of two segments. Distal to this are two parallel structures, a medial *endopodite* and a lateral exopodite, each consisting of a series of segments. The appendages of various body segments have been modified in the most diverse ways, adaptive to widely different uses (Figure 14). Their condition in the crayfish is a familiar example. The first two pairs of appendages have been greatly modified to form antennae. Whether the eyestalk also represents a greatly reduced appendage is still subject to debate among specialists. The heavy, biting mandibles are formed by the shortened protopodite and endopodite of the third segment, while the exopodite has been eliminated altogether. The fourth pair of appendages forms the first maxilla, in which the protopodite and the endopodite form a flattened plate which is used for handling food. The fifth pair are the second maxillae, but the exopodite is present here, and together with a dorsal outgrowth called the epipodite, it forms a vane to pass a current of water over the gills. This is the last of the appendages of the head region. The first three pairs of thoracic appendages are maxillipeds, the protopodites of which are flattened and serve for the handling of food, just as do the maxillae. The endopodite and exopodite are present on these appendages, but they are not very large, and they may be sensory in function. The first maxilliped bears an epipodite, while the second and third bear gills which extend dorsally under the carapace. The remaining five pairs of thoracic appendages are specialized as walking legs, with the exopodite absent and the endopodite

36

terminating in a pincer. The first pair of walking legs forms the major pincers (or chelae) of the organism, serving as defensive and food-procuring organs. All but the last pair of walking legs bear gills. There are six pairs of abdominal appendages. The first is often much reduced or even absent in females, while in males it is modified together with the second abdominal appendage to form a copulatory organ. In females, the second appendage forms a swimmeret or pleopod in which there is a short protopodite followed by a more or less equally developed, filamentous exopodite and endopodite. In both sexes, the third, fourth, and fifth abdominal appendages follow this pattern, which appears to be the primitive appendage pattern. The final appendage consists of a short protopodite and a very broad, flat endopodite and exopodite, forming the telson, or terminal fan of the crayfish.

Thus, within a single organism, the basic crustacean appendage is modified to serve no less than six to ten different functions (depending on how one may wish to classify them). If each of these appendages had been originally created for the function which it now serves, then it would be a strange circumstance that all of them are built upon the same pattern as are the legs. For most of these functions are subserved in other groups by organs having nothing to do with appendages. Antennae of molluscs, for example, are constructed on a completely different plan, yet there is no reason to suppose that they function less efficiently than those of the Crustacea. Nor is there any intrinsic reason why mouth parts, copulatory organs, and gills should be based upon a leg-like structure. Yet it is so in the Crustacea. These facts, so puzzling on any other theory, are easily understandable on the basis of the evolutionary concept. Given a primitive crustacean in which all of the appendages are in a simple condition, somewhat like the swimmerets of the crayfish, natural selection should favor diversification of function. There is a general tendency among the higher phyla for centralization of sensory functions in the head region; hence the sensory functions of the anterior-most appendages become intensified at the expense of other functions, and they become antennae and (possibly) eyes. Those appendages nearest the mouth naturally are used for feeding, and so they become specialized for chewing, biting, and food handling. Those appendages nearest the reproductive organs are, in the male, modified for the transfer of sperm to the female. Other appendages continue to serve the original locomotor function in various ways.

If the Crustacea as a whole be considered, then the range of adaptations of these appendages is still greater. In some of the Crustacea, mouth parts are more numerous than in the crayfish described above. In these, there are fewer legs. Why such a relationship should hold is utterly inexplicable except upon the theory that both the mouth parts and the legs have been derived from primitive appendages by adaptive modification. In the barnacles, most of the appendages have been suppressed, and the thoracic appendages have been specialized as plume-like cirri which sweep a current of food-bearing water toward the mouth. In the lobster, the abdominal appendages are flattened out to form broad, oar-like plates, effective as swimming organs. In the crabs, which normally hold the abdomen

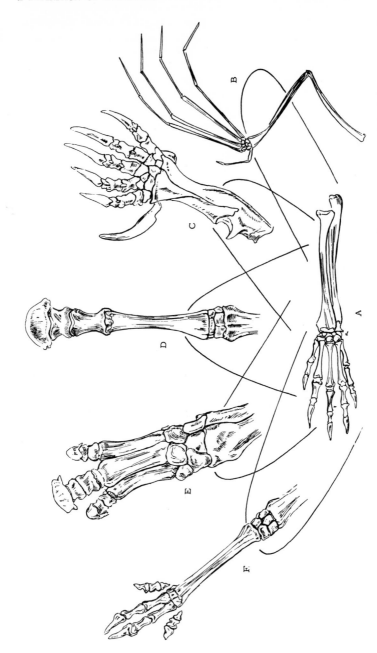

FIGURE 15. ADAPTIVE RADIATION IN THE FORELIMBS OF MAMMALS. *A*, tenrec, a large insectivore with limb structure much like the shrew except for size; *B*, bat wing; *C*, mole; *D*, horse; *E*, rhinoceros; and *F*, deer. Each is modified from the primitive form (approximated by the tenrec) by changes of proportion, fusion of parts, or loss of parts. Not drawn to the same scale. (*C*, *D*, *E*, and *F* redrawn from Flower.)

recurved against the ventral part of the thorax, the abdominal appendages are much reduced or missing entirely, excepting the first two which serve as copulatory organs. But through all of this great range of variation, a single pattern is discernible, a fact which bespeaks an organic relationship of all of the Crustacea.

Adaptive Radiation in the Forelimbs of Mammals. The same principle holds true within each class. The forelimb of the mammals may be taken as an example. In any, there is a single long bone, the humerus, in the upper arm. In the forearm there are two parallel bones, the ulna and the radius. In the wrist there are typically eight carpal bones arranged as two rows of four. Five parallel metacarpals form the skeleton of the palm of the hand, while rows of three phalanges each form the skeleton of the digits, excepting the first digit, which has only two phalanges.

The shrews (order Insectivora, family Soricidae) show a very primitive arm structure (Figure 15). Their close relatives, the moles (family Talpidae) are, however, highly modified for digging. All of the bones of the limbs are short and broad, giving the limb a shovel-like appearance. Thus adaptation is attained by mutual fitting of structure (the shovel-like limb), function (digging), and environment (the subterranean habitat). In the order Chiroptera (bats), the humerus, radius and ulna, and four of the digits are greatly elongated to support the wing membrane. In the ungulates, the humerus is short and heavy. The remaining bones of the forelimb are generally elongated, and the digits are reduced in number. Fusion of bones is quite common in adults, but in the embryos the primitive centers of ossification can be identified. The details naturally differ considerably among the various orders of ungulates.

Examples could be multiplied indefinitely, but the principle remains the same throughout. Within any taxonomic category, all of the members appear to be built upon a common plan, with variation among the various members resulting in the adaptation of each to its mode of life. The higher the category examined, the greater the scope of variation. But the common plan is always discernible. To some of the predecessors of Darwin, this indicated the supernal archetypes. But since Darwin's time, the great majority of biologists have been convinced that close anatomical similarity must be based upon close genetic relationship, while more remote resemblances are based upon more remote blood relationship.

Homology and Analogy Contrasted. Another highly suggestive aspect of comparative anatomy is the comparison of homologous structures which, within a single group, are used for quite different purposes, and analogous structures, which, although quite different morphologically, and developed in different groups, have nonetheless a certain similarity which is based upon adaptation to the same function. If each structure were created for the purpose for which it is now used, analogy should be far more pervasive and important than homology. A few examples will show that this is not the case. A classical example is furnished by the fins of fishes and the flippers of aquatic mammals, such as whales and seals. These structures, characterizing organisms at opposite ends of the vertebrate series, serve the same function. Superficially, they show a close

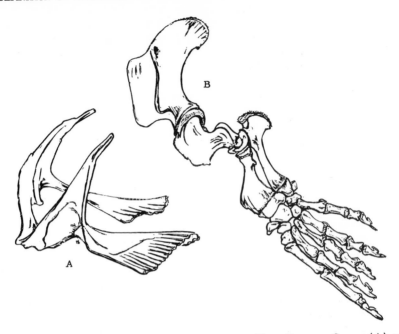

FIGURE 16. ANALOGICAL RESEMBLANCE BETWEEN THE FORE FIN OF A SHARK (A) AND THE FORE FLIPPER OF A SEAL (B). Though superficially similar, the skeletons are very different.

resemblance, for each has the form of a short, broad paddle. But this adaptive form is achieved upon an entirely different basis in the two cases (Figure 16). For the skeleton of the fin has a very simple structure, though perhaps similar to that which has given rise to the limbs of higher vertebrates. But the skeleton of the flipper of any aquatic mammal is, for all its adaptation to a mode of life unusual among the mammals, still very recognizably that of a mammal. The parts are strongly foreshortened, but the typical bones are easily recognized, and they still bear the same spatial relationships to one another as in more typical mammals. Thus even extreme adaptive changes have not been able to destroy the evidence of true homologies.

Again, the vertebrate eye is one of the most complex and efficient visual organs in the animal kingdom. Minor variations in the various vertebrate classes involve changes like those of the limb bones of the seal, that is, identical parts differ in proportions and may function somewhat differently. Thus the muscles of accommodation function in mammals by varying the tension upon the suspensory ligament of the lens, and thus controlling its curvature. In birds, more rapid accommodation is achieved by the same muscles pulling the lens closer to the retina or by relaxation permitting it to move farther away. In all vertebrates, the eye is homologous, constructed of identical materials which are used in similar ways. The most nearly similar eye outside of the Chordata is that of the Cephalopoda (squids, octopi, and their allies). Superficially, the cephalopod eye bears a close resemblance to that of the vertebrates, but detailed examination shows that in every part different materials have been used in different ways. Embryologically, the cephalopod eye develops from the skin, while the vertebrate eye develops from the brain, with the exception of the lens, which is a skin derivative. Yet even the lenses differ fundamentally, for that of the vertebrates is cellular, that of the cephalopods a crystalline secretion of skin cells. Vertebrate and cephalopod eyes are only analogous.

Another classical example is provided by the wings which have been developed independently by insects, reptiles, birds, and bats (Figure 17). All show analogical resemblances because they are adapted to the same function. The insect wing shares nothing but the planing surface with the others, for it is simply a membrane supported by chitinous veins. All of the vertebrate wings are constructed from the typical parts for the fore-limb of a tetrapod (land vertebrate), and to this extent they may be considered homologous. But the several groups of flying vertebrates are not descended from a common ancestor which flew: flight has been developed independently in three different lines of descent. The pterodactyl,

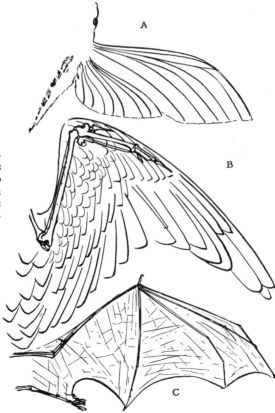

FIGURE 17. ANALOGICAL RE-SEMBLANCES OF THE WINGS OF AN INSECT (A), A BIRD (B), AND A BAT (C). In each, a planing surface is formed from completely different materials.

an extinct flying reptile, had a wing which was formed by a fold of skin stretched between the body, the posterior surface of the arm, and an immensely elongated fifth digit, the other four digits remaining free. In birds, the feathers which form the planing surface are inserted on all of the three major segments of the appendage. The first digit is somewhat independent of the others and can be moved separately by some birds. It bears feathers. The second and third digits are fused together to form the major skeletal basis of the distal part of the wing. The fourth digit is accessory to these, and the fifth is missing altogether. In the bats, a fold of skin again extends from the body to the arm to form the planing surface. But this time only the first digit is free and of typical size, while the other four are elongated to form the major supports of the wing. Therefore, although the wings of all of these flying vertebrates can be regarded as homologous when considered simply as vertebrate append-ages, they are only analogous when considered as flight organs. Different materials have been used for the adaptations to flight.

Vestigial Organs. Another very persuasive aspect of comparative anat-omy concerns *vestigial* or *rudimentary* organs. These are dwarfed and generally useless organs which are found in many plants and animals, relatives of which may have the same organ in a fully developed and functional condition. Perhaps the most widely known example is the ver-miform appendix of man (Figure 18). This small structure, without any known function in man, is notorious as a seat of disease. In other Primates, however, this organ is considerably larger. And in mammals which eat a coarse diet, involving considerable amounts of cellulose, the appendix and cecum form a large sac in which mixtures of food and enzymes can react for long periods of time. The appendix of man is easily understand-able as a degenerating legacy from ancestors with a much coarser diet; but it is inexplicable why a useless and disease-ridden structure should have been created especially to plague him.

Weidersheim has listed nearly 100 such vestigial characters in man, and a few of these may be mentioned. In the inner corner of the eye of all vertebrates there is a transparent membranous fold, the nictitating mem-

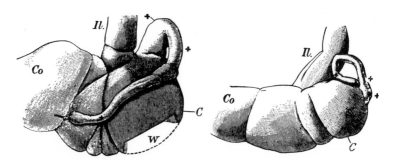

FIGURE 18. THE APPENDIX IN THE ORANG-UTAN AND IN MAN. *il*, ileum; *co*, colon; *C*, caecum; *W*, window cut in caecum; +, appendix. (From Romanes, "Darwin and After Darwin," Open Court Publishing Co., La Salle, Ill., 1902.)

brane. In most vertebrates, this "third eyelid" can be swept clear across the eyeball to cleanse the latter, much like the blinking of a mammal. In birds, it is particularly well developed. Its use can be easily observed if a captive owl is watched by daylight. But in mammals, it forms a mere crescentic fold at the inner corner of each eye, with no known—or probable—function. It is understandable only as a degenerating inheritance from an ancestor to which the nictitating membrane was actually useful, as it now is to the majority of vertebrates. The ear muscles of man present a similar situation. Many mammals move the external ear freely in order to detect sounds efficiently. The complete muscular apparatus for these movements is present in man, but quite vestigial. While school boys sometimes pride themselves on their ability to wiggle their ears, the ability has no real usefulness, and even this limited ability is not shared by everyone. The presence of these muscles, then, suggests descent from an ancestor to which they were really useful.

Similarly, most mammals have a well-developed tail, but this is lacking in all of the higher Primates. It is represented in them by vestigial caudal vertebrae, usually three to five in man. Rarely, a fleshy tail extends a few inches beyond the caudal vertebrae. Whether or not an external tail is present, the muscles which move the tails of other mammals are also present in the Primates.

A final example from man concerns the wisdom teeth. The wisdom teeth, or third molars, are the posterior-most teeth, as well as the last to erupt. In other Primates, these teeth are as sound and as fully developed as the rest of the dentition. But in man, they are far more variable than are the other teeth with respect to size and time of eruption. Frequently, they fail to erupt altogether. And when they are present, they are far more subject to all types of dental defect than are the other teeth. Thus it is probable that these teeth should be regarded as vestigial, and, in view of the frequency with which they fail to erupt, that they will in time be completely lost to man.

Many examples of vestigial characters may be found among lower animals. The external ears of whales are completely of the type found in terrestrial mammals, but they are much reduced in size, and it seems unlikely that they are efficient auditory organs. Also among whales, the hind limbs are completely missing, yet in some species rudiments of the pelvic girdle still remain, but have lost their connection to the vertebral column. In ungulates (horse, deer, and other hoofed animals), the smaller bone of the lower rear leg, the fibula, has been reduced to a mere splint on the larger bone, the tibia. A similar reduction of the fibula has occurred in the birds. Perhaps no feature of the anatomy of snakes is so generally known as their leglessness. No snake shows any vestige of the forelimbs, but some (pythons and boas) have small, ineffective rudiments of the hindlimbs. (See Figure 96, p. 251.) These are capped by claws which show externally, but they are so reduced that they appear at a glance to be scarcely more than raised scales.

Many animals, both vertebrates and invertebrates, have become adapted to life in deep caves, where the light of the sun never reaches. Living thus in perpetual darkness, there is no adverse selection against degenerative changes of the eyes, and in fact blindness is a general characteristic of such cave dwellers. Their eyes show all degrees of degeneration from just short of the typical functional condition to complete absence of the eyes. Examples include the blind, cave-dwelling salamander of central Europe, *Proteus anguineus;* the many species of cave-dwelling fishes of the United States as well as other parts of the world; and the blind crayfishes. The latter have eyestalks which do not, however, bear eyes. While such degenerated eyes are easily understandable on the basis of descent from ancestors with functional eyes, their presence is inexplicable, indeed it is contradictory, on any other theory.

Typical beetles, as every school boy knows, are strong fliers. The island of Madeira, a wind-swept island about 600 miles off of the coast of Portugal and 400 miles to the west of North Africa, has a rich beetle fauna. But a large proportion of these Madeiran species are either wingless or they have much reduced wings which can be of no use for flight. If one makes the assumption that the reduction or loss of these wings has been a result of natural selection, then the selective force may be easily surmised. For beetles in flight would be quite likely to be blown out to sea and lost. Thus a strong selective force should favor any variations toward reduction of wings. Those species which have been on the island and subject to this selection longest should be most numerous in the list of flightless species. That this is probably true is indicated by the facts that almost all of the endemic species are flightless, while most of the flying species are also represented in Europe or Africa, or have close relatives on these continents. Only the evolutionary explanation for these facts can be rational.

When structures undergo a reduction in size together with a loss of their typical function, that is, when they become vestigial, they are quite commonly considered to be degenerate and functionless. But Simpson has recently pointed out that this need not be true at all: the loss of the original function may be accompanied by specialization for a new function. Thus the wing of penguins has become reduced to a point that will not permit flight, but at the same time it has become a highly efficient paddle for swimming. The wings of rheas, ostriches, and other running birds are also much reduced, and have been described as "at the most still used for display of the decorative wing feathers!" But Simpson has observed that the rheas, when running, spread the wings and use them as balancers, especially when turning rapidly. It seems quite probable that this is true of the running birds generally.

A comparable case is afforded by the pineal gland. This small structure grows dorsally from the forebrain, and its histological structure shows a mixture of glandular and nervous characteristics. In the lampreys and in many reptiles, including the very primitive, lizard-like *Sphenodon* of New Zealand, this structure bears a third eye, located on the dorsal surface of the skull. Fossil evidence indicates that this is also true of the Crossopterygii, the group of fishes from which land vertebrates appear to have been derived. Thus the possibility is open that the possession of a pineal eye may have been a primitive character of the vertebrates, and that the pineal gland of most extant vertebrates is in fact a vestigial eyestalk. But if so, it may well have an important new function, for it is commonly regarded as an endocrine gland, although the evidence is incomplete.

EVIDENCE FROM EMBRYOLOGY

Comparative embryology is a specialized branch of anatomy which furnishes evidence for evolution regarded by Darwin as "second to none in importance." Ernst Haeckel brought this field into prominence in the immediate post-Darwinian period with his Biogenetic Law, which states that "Ontogeny recapitulates phylogeny." He believed that embryonic stages corresponded to ancestral adults, and hence provided direct evidence of lines of descent. Recapitulations do occur, but not as Haeckel thought, for resemblances are chiefly between embryos, not embryos and adults, and embryos, too, have adaptive problems.

A striking example occurs among crustaceans. A series of six larval types (Figure 19) strongly resemble adults in a sequence from primitive to advanced. The larvae pass through these stages at successive molts, with larvae of primitive crustaceans stopping early in the series and advanced crustaceans going through most of the stages, as indicated in the chart. Failure of the Cypris larva to appear in the development of higher crusta-

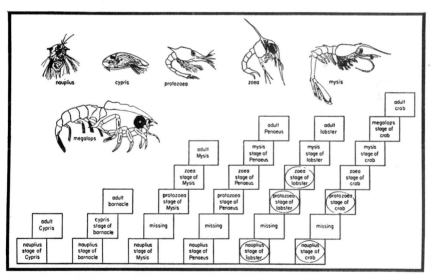

FIGURE 19. DEVELOPMENTAL STAGES OF CRUSTACEANS. At the top are a series of larval types, several of which resemble adults of corresponding groups of crustaceans. The sequences of squares at the bottom indicate the developmental histories of the major crustacean types. Stages in ovals are completed in the egg, but others are free-swimming larvae. (From MacGinitie and MacGinitie, "Natural History of Marine Animals," McGraw-Hill Book Co., Inc., New York, N.Y., 1949.)

45

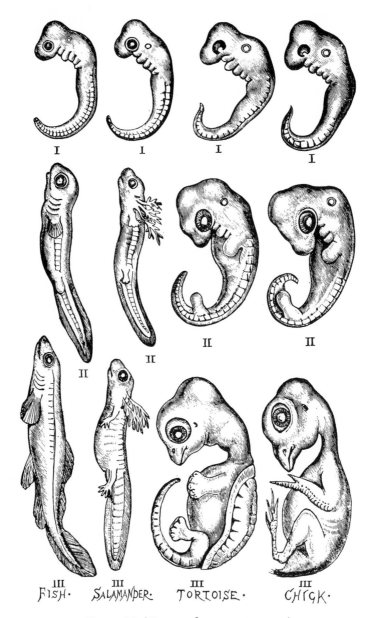

FIGURE 20. (Continued on opposite page.)

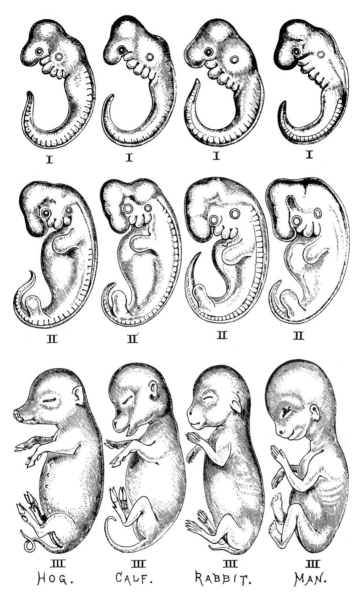

I I I I

II II II II

III III III III
HOG. CALF. RABBIT. MAN.

FIGURE 20. EMBRYOS OF A SERIES OF VERTEBRATES COMPARED AT THREE STAGES IN DEVELOPMENT. (After Haeckel, from Romanes, "Darwin and After Darwin," Open Court Publishing Co., La Salle, Ill., 1902.)

ceans simply means that *Cypris* and barnacles represent a side branch of crustacean evolution.

Examples Among Vertebrates. The ontogeny of man, considered biogenetically, indicates a long, complicated history. The fertilized egg corresponds to a protozoan ancestor, but it soon becomes multicellular, indicating a primitive metazoan grade. Gastrulation forms a coelenterate-like embryo, but this soon becomes triploblastic, like a flat worm. Fundamental chordate characters (dorsal nerve tube, notochord, and pharynx specialized for respiration) are then developed. Fish-like characters, such as gill slits and aortic arches, appear, followed by tetrapod characters, such as the pentadactyl limb and metanephric kidney. Finally mammalian, then primate, and at last specifically human characters appear (Figure 20).

The details of the development of specific systems are sometimes very impressive. The kidneys of vertebrates are all developed from the *nephrotome,* a segmented mass of mesoderm lying on either side of the somites, from which so many of the serially homologous structures of the body are developed. Yet there are three distinct types of kidneys among the vertebrates. All vertebrate embryos first develop a pronephric type of kidney, utilizing only the anterior-most part of the nephrotome. Only in the hagfishes and a few of the bony fishes (and here only in part) does this remain as the functional kidney of the adult. In all other vertebrates (including all of the bony fishes), a mesonephric type of kidney is developed posterior to the pronephros, and the pronephros either degenerates or is partly incorporated into the mesonephros (with the exception noted above). This mesonephros is the functional kidney of adult fishes and amphibians, and of the embryos of reptiles, birds, and mammals. But in all of the last mentioned classes, a third type of kidney, the metanephros, is formed posterior to the mesonephros, and serves as the functional kidney of the adult organism.

Similarly, all vertebrate embryos develop a series (most commonly six) of aortic arches, each of which runs unbroken from the ventral aorta to the dorsal aorta, much as in adult Amphioxus (Figure 21). In the fishes, these arches are modified in several ways, all of which involve the separation of each aortic arch into a ventral afferent branchial artery and a dorsal efferent branchial artery, the two being connected by a capillary network in the gill filaments. In the Choanichthyes, the group of fishes most closely related to the Amphibia, the first arch drops out, and is largely missing in the adult, but its ventral and dorsal roots, together with new growths from them, form the major arteries of the head (the external and internal carotids). The sixth arch has given rise to a *pulmonary* branch which supplies the lungs. This tendency for parts to drop out after having been formed in the embryo, and for the remaining parts to be diverted to completely different functions from the original purely respiratory function, is the principal factor in the embryology of this part of the circulatory system of all tetrapods. Among the urodeles, the main portions of the first and second arches drop out, so that now the carotids arise from the third arch. The third arch is broken by a capillary network

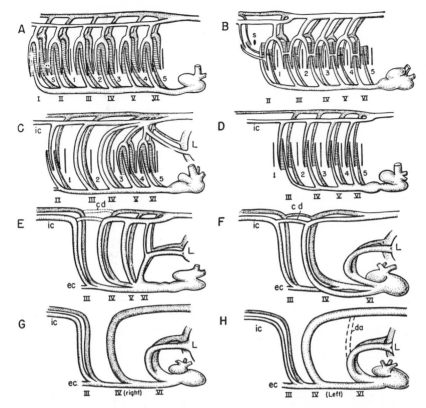

FIGURE 21. DIAGRAMS OF THE AORTIC ARCHES AND THEIR DERIVATIVES IN A SERIES OF VERTEBRATES. *A*, hypothetical ancestor; *B*, shark; *C*, the lungfish *Protopterus*; *D*, teleost; *E*, salamander; *F*, lizard; *G*, bird; *H*, mammal. Roman numerals indicate aortic arches; *s*, spiracle; Arabic numerals, gill slits; *cd*, carotid duct; *da*, ductus arteriosus; *ec*, external carotid artery; *ic*, internal carotid artery; *L*, lung. (From Romer, "The Vertebrate Body," 2nd Ed., W. B. Saunders Co., Philadelphia, Pa., 1955.)

early in development, but it soon becomes continuous again. The dorsal connection between arches three and four disappears, with the result that the ventral connections now appear as common carotid arteries on either side, while the two fourth arches now supply the major circulation to the body. The fifth arch becomes reduced in size and may be lost altogether, while the sixth arch again gives rise to a pulmonary branch. In the Anura and in the reptiles, this process goes a little farther, with the fifth arch being lost completely and with the dorsal part of the sixth arch being lost, so that all of the blood entering the sixth arch must go into the pulmonary artery. The birds have essentially the reptilian system, with, however, the left fourth aortic arch degenerating, thus leaving the right fourth arch to carry the entire systemic circulation. In the mammals, it is the right fourth arch which degenerates and the left one which persists. Thus, of the six original pairs of aortic arches, only three persist in the highest classes of vertebrates. Arch three serves the head region, arch four (but only one of the pair) serves the systemic circulation, and arch six serves the lungs. Yet all six pairs are developed in the embryos of birds and mammals.

A similar story could be told with respect to almost any organ system in any major group. The details differ, but the general facts are the same. When, in the course of embryonic development, a new organ system is formed, its structure is closely similar even in the most widely dissimilar species of the same class, or even phylum in many instances. As differentiation proceeds, detectable differences first become apparent between the embryos of those species of which the adults are most widely different. Thus the very early embryos of fishes and of mammals are quite similar, but they soon become recognizably differentiated. As development proceeds, the similarities of the embryos become progressively restricted to smaller and smaller taxonomic groups until finally the characters which distinguish the adults of closely related species are formed. In many instances, this process is not completed until after birth (or hatching). Thus juvenal robins have the speckled breast which is typical of the adults of most species of the thrush family, to which the robin belongs.

Many examples indicate relationship within as well as between classes. Thus the whalebone whales, which feed by straining minute organisms out of the sea water, have no teeth in the adult stage, yet their embryos have a set of tooth buds which are resorbed without ever erupting. Whale embryos also have a coat of hair which is entirely lost to the adults. The absence of birds' teeth is proverbial, yet their embryos also have ephemeral tooth buds. Such facts are readily understandable on the principle that the basic factors of embryology are determined by heredity, and so are common to related species and groups. Both whales and birds are descended from ancestors which had teeth. The hereditary factors which initiate tooth development are still present and active. But an additional hereditary change (mutation) which acts later in development has been independently acquired in each group since its origin. And this change causes the tooth buds to abort.

Much more radical changes in developmental pattern may also occur, so that only the early embryonic stages can give a clew to the true relationships of the organisms involved. Some extreme examples occur among the cirripedes (barnacles). Typical cirripedes show many typical crustacean characters, so that it is not difficult to understand that they belong in this subphylum. Nonetheless, they are sufficiently aberrant as to have deceived so competent a zoologist as Cuvier, who treated them as a class of the Mollusca. Yet the larvae of all cirripedes are unmistakably crustacean larvae, and it was this fact which finally established the correct taxonomic position of the barnacles. A much more extreme example is presented by *Sacculina*, a very aberrant barnacle which parasitizes crabs. This organism goes through all of the typical developmental stages of a barnacle until it settles down upon the abdomen of its host. At this point, other barnacles would develop the usual adult structures of the group, but *Sacculina* undergoes a degenerative development. Appendages are lost, and the parasite becomes sac-like, with branches invading the tissues of the host like a tumor and absorbing nutriment. The only organ system which is well developed is the reproductive system. This parasite resembles no other organism, and so it defied classification until the larvae were studied. The larvae showed clearly that the parasite could be nothing else but a much degenerated barnacle. The case of the enteroxenid snails which parasitize echinoderms is quite similar. They are worm-like externally. Internally, they have only a much reduced gut and an hermaphroditic reproductive system. Yet they produce typical snail larvae.

Another interesting case which was originally studied by Darwin is that of the loggerhead duck, or steamer duck, *Tachyeres cinereus*. This bird is widely distributed in southern South America, the Straits of Magellan, and the Falkland Islands. The young ducks are fairly good fliers, but the growth of the wings does not keep pace with the growth of the body, and the adults are completely flightless.

Examples Among Plants. The embryology of plants is generally simpler than that of animals, and hence the recapitulation principle is not so abundantly exemplified. Yet there are good examples available in the plant kingdom. The Acacia tree, for example, has highly compound leaves, yet its seedlings have simple leaves like its ancestors. Adult cactus plants have no typical leaves at all, although these may be represented by the needles. But the seedlings have readily recognizable leaves. To give one more example utilizing a leaf character, the live oaks of southern United States, which retain their foliage the year around, are considered to be more primitive than the northern species which are deciduous. However, the saplings of northern oaks commonly retain their leaves during the winter, thus recapitulating what appears to have been an ancestral character.

Jeffrey has found a very interesting and possibly comparable phenomenon in conifers. If conifers are injured and the wound allowed to heal, the new growth may differ histologically from normal tissue. In such instances, the new tissue shows a type of structure which is well known from fossil conifers of the Mesozoic era.

Difficulties of the Biogenetic Law. Because of numerous examples like those described, many biologists of Haeckel's time thought that embryology, when sufficiently known, would be a golden key to problems of phylogeny. Yet there was much unsound biology associated with the Biogenetic Law, and few aspects of evolutionary science have been so heavily attacked in recent years. The reasons are simple enough. The recapitulation theory assumes that embryos need only repeat the past, condensing some stages, eliminating others, without adaptation to the embryonic mode of life. Actually, the embryo must cope with a hostile environment, even as do adults. All pelagic larvae are subject to heavy predation. The first chapter in embryology texts frequently deals with differences in cleavage patterns which are correlated with amounts of yolk in the eggs, an adaptive trait of fundamental biological importance. Fetal membranes of amniotes are an obvious series of embryonic adaptations. Needham has described an interesting series of adaptations of eggs to fresh water, to salt water, or to terrestrial habitats. Each requires different excretory physiology, and the last also requires a cleidoic egg (encased in a shell) and other adaptations for water conservation. Thus mutations can effect embryonic as well as adult stages, and these, too, are subject to natural selection, so that embryonic adaptations become part of the normal pattern of development.

The beautifully simple embryology of the echinoderms played an important role in the establishment of the Biogenetic Law. Yet the recent comprehensive study of echinoderm embryology by Fell reveals extensive differences among various groups of echinoderms, and these differences are referable to embryonic adaptations. Fell even casts doubt on the echinoderm-chordate relationship, for the hemichordate larva does not fit into the scheme of larval relationships which he has worked out. Many biologists, however, feel that this is a result of extensive modifications of larvae by natural selection since the separation of their ancestors.

Another difficulty of the Biogenetic Law was Haeckel's emphasis upon resemblances of embryos to *adult* ancestors. Subsequent study has shown that the resemblance is primarily between embryos of related animals, and only incidentally does this sometimes suggest adults. Thus ontogeny does recapitulate phylogeny in a significant way, but it is ancestral embryonic, not adult, stages which are repeated, and even these may be drastically modified by adaptive mutations which are favored by natural selection.

Von Baer's Principles. In formulating his Biogenetic Law, Haeckel started from von Baer's principles of embryonic differentiation, and these are perhaps much sounder, a better guide to the embryological evidence for evolution. These principles are the following: (1) General characters appear in development before special characters. (2) From the more general, the less general, and finally the special characters appear. (3) An animal during development departs progressively from the form of other animals. (4) Young stages of an animal are like young, or embryonic, stages of lower animals, but not like adults of those animals. The attack upon the Biogenetic Law has never produced evidence that the findings of embryology do not support the facts of evolution: it has merely shown that Haeckel and his followers read into it more than the data will support. A return to the principles of von Baer makes possible a reasonable evolutionary interpretation of the facts of comparative embryology without straining the evidence.

REFERENCES

Davis, D. Dwight, 1949. "Comparative Anatomy and the Evolution of the Vertebrates." In "Genetics, Paleontology, and Evolution," edited by Jepson, Mayr, and Simpson. Princeton University Press, Princeton, N.J. A careful analysis of the role of comparative anatomy in current studies on evolution.

De Beer, G. R., 1958. "Embryos and Ancestors," 3rd Ed., Oxford University Press. A critical review, on a very broad basis.

Fell, H. Barradough, 1948. "Echinoderm Embryology and the Origin of the Chordates," *Biological Reviews of the Cambridge Philosophical Society*, 23, 81–107.

A thorough restudy, leading to the conclusion that the biogenetic interpretation is not valid.

LINNAEUS, CAROLUS, 1758. "Systema Naturae," 10th Ed., Leyden. The starting point of modern taxonomy.

MAYR, E., 1942. "Systematics and the Origin of Species," Columbia University Press, New York, N.Y. A very readable modern treatment of taxonomy and evolution.

ROMER, A. S., 1949. "The Vertebrate Body," W. B. Saunders Co., Philadelphia, Pa. An excellent evolutionary treatment of comparative anatomy.

WILLIER, B. H., P. A. WEISS, and V. HAMBURGER, Eds., 1955. "Analysis of Development," W. B. Saunders Co., Philadelphia, Pa. In Chapter 1, Jane Oppenheimer gives an excellent summary of the history of the controversies which have raged around the relationship of embryology to evolution.

CHAPTER FOUR

Evidences of Evolution III: Comparative Physiology and Biochemistry

FOR MORE THAN TWO CENTURIES before Darwin, the great investigative impetus derived from the Renaissance resulted in the accumulation of a great store of biological knowledge. On the whole, however, it was a disorganized array of data, a burgeoning chaos, for there was no basic biological principle which could integrate the whole field. The publication of the "Origin" caused a revolution in biological thinking because it provided just such a principle, a rational framework upon which the grand scope of biology could be organized. Its success in achieving this is one of the strong arguments for evolution. The biology of Darwin's time was, however, almost exclusively morphological: physiology was in its infancy, and biochemistry did not yet exist. Consequently, these fields were the last to be influenced by the Darwinian revolution in biological thinking. Yet many biologists regard the morphological traits with which we have dealt thus far as simply the more obvious results of specific compounds and processes. In short, they feel that evolution is basically a biochemical phenomenon. It is therefore significant that an array of very cogent evidence for evolution has emerged from these fields.

On the broadest possible level, there is the fact that protoplasm appears to be basically one substance, varied in more or less minor ways from species to species, throughout the living world. It contains very nearly the same elements, compounded into roughly the same proportions of proteins, carbohydrates, fats, water, and supplementary substances. The most basic functions of protoplasm are describable in very similar terms, with few exceptions, throughout the living world. That this should be true is a very impressive fact. It is subject to other interpretations, but it strongly suggests community of origin, with the most fundamental properties of living things remaining rather constant, while variation in less essential respects has produced the immensely varied forms of the living world.

Much the same is the story with respect to the chemistry of the chromosomes, the physical basis of heredity, incompletely though it is known.

Fish sperm comprised the experimental material for the first studies on the chemical constitution of the chromosomes, but subsequent studies on such diverse materials as yeast and mammalian liver cells have all led to similar conclusions. It appears that, throughout the entire living world, the chromosomes consist of basic proteins in combination with nucleic acid. Histone and protamine, which are the simplest types of proteins, predominate, but globulin and some incompletely identified proteins are also present. It may be that these more complex proteins contribute to the diversity of the hereditary material in greater proportion than their quantity would indicate. The nucleic acids are also rather uniform. The basic unit of structure, or nucleotide, consists of phosphoric acid, a pentose (a sugar based upon a five-carbon chain), and a purine or pyrimidine base. These units are highly polymerized (like molecules joined together to form much larger molecules), to form long, double, spiral chains joined to each other by weak bonds between the bases (see Figure 87, p. 228). Nucleic acids differ from one another chiefly in the sequence of base pairs which join together the nucleotide chains, and much evidence now indicates that the specificity of the gene (the unit of inheritance: see Chapter 13) depends upon this trait also. In view of the great diversity of organisms, all of which owe their attributes to the chromosomes which they possess, it is astonishing that the chromosomes themselves should have so uniform a constitution. But this is most readily understood in the same light as is the general unity of protoplasm.

Enzymes and Hormones. Again, closely similar or identical enzymes and hormones are likely to be common to large groups of animals. This is especially true of some of the digestive enzymes. Trypsin, the protein-splitting enzyme, is found in many groups of animals ranging from the Protozoa to the Mammalia. Amylase, the starch-splitting enzyme, is found from sponges to man. The thyroid hormone is found in all vertebrates, and it has been proved to be interchangeable among them. It is well known that beef thyroid is used successfully in the treatment of human thyroid deficiencies. This hormone is also essential for the metamorphosis of frogs. If a frog's thyroid gland is removed surgically, the frog will not metamorphose. Yet feeding of mammalian thyroid tissue to such a frog will correct the deficiency and permit metamorphosis. Even more striking is the case of the melanophore-expanding hormone of amphibians. This pituitary hormone causes the pigmented cells of the skin to expand, thus darkening the color of the animal. It has no known effect in mammals, yet an extract of mammalian pituitary glands is just as effective in amphibians as is their own hormone. Thus the melanophore-expanding hormone of mammals might be regarded as a "vestigial" hormone, the presence of which is understandable only on the basis of descent from an ancestor to which the hormone was useful.

Yet vertebrate hormones are variable. It has long been known that the pituitary growth hormone is not interchangeable among mammals. Beef hormone, for example, is not effective in treating deficiencies of this substance in children. Analysis of growth hormones, isolated and purified from various mammals, has revealed small but functionally significant differences, comparable to the differences of structure with which comparative anatomy deals.

HEMATOLOGICAL CHARACTERISTICS

If a drop of blood from a species having hemoglobin as its respiratory pigment is treated in the proper way, crystals of hematin can be obtained. The details of crystal structure differ from species to species, and they parallel classification in a remarkable way. Crystals obtained from all of the species of a genus share many characteristics, while crystals from members of different classes have characteristics which are mutually exclusive. Thus it is possible, on the basis of hematin crystals alone, to distinguish between the various classes of vertebrates.

Comparative Serology. Much the most impressive physiological evidence is drawn from the field of comparative serology. If a small amount of the blood serum of any animal is injected into a guinea pig (or other test animal), the foreign blood acts as an *antigen*, that is, it causes the production in the serum of the guinea pig of *antibodies* which will precipitate and destroy the antigen if a second inoculation should occur. The

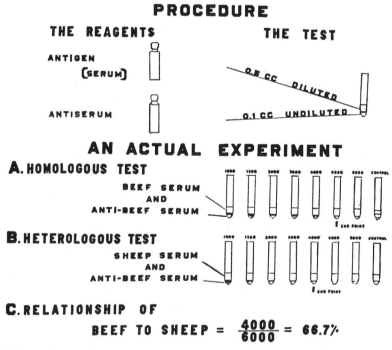

FIGURE 22. PROCEDURE IN MAKING A RING TEST FOR SEROLOGICAL RELATIONSHIP. Undiluted antiserum is layered under diluted antigen in saline solution. A white ring of precipitate indicates a positive reaction. The titer is the highest dilution of the antigen which will give a positive reaction. (From Boyden, *Physiol. Zool.*, V. 15, 1942.)

guinea pig is then said to be *immunized* to the kind of blood which was injected. The precipitation reaction will occur in a test tube as well as in the bloodstream. Thus, if one prepares an antiserum from an immunized animal and adds to it a few drops of antigenic serum, a precipitate will be formed. This can be measured by two principal methods. The first of these is the ring test method (Figure 22). A small quantity of undiluted antiserum is placed in a test tube, and diluted antigenic serum is then carefully layered over it. A ring of precipitate then forms at the interface between the two sera. The greatest dilution of the antigenic serum at which a ring is obtained gives a measure of the strength of the reaction, with a high dilution corresponding to a strong reaction. If, however, the two sera are mixed, the precipitate will make the solution turbid, and the photometric measurement of the absorption of light gives an excellent measurement of the strength of the reaction.

Such antigen-antibody reactions are highly specific. That is, an antibody which precipitates the blood of one species is generally ineffective against the bloods of other species. Yet the specificity is not complete, for serum immunized against the blood of any species A will precipitate the bloods of species related to A, but in ever-decreasing degrees as the relationship grows more distant. For example, a guinea pig may be immunized with

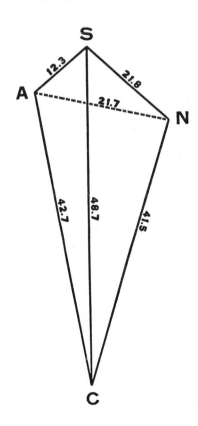

FIGURE 23. RELATIONSHIPS OF *Crypto-branchus* (C), *Amphiuma* (A), *Siren* (S), AND *Necturus* (N) AS DETERMINED BY SEROLOGICAL TESTS. The diagram shows the proportional "distances" between the species. (From Boyden and Noble, *Am. Museum Novitates*, No. 606, 1933.)

blood from the common salamander *Necturus*. If a sample of serum from the immunized guinea pig is then divided among four test tubes, a few drops of serum may be added from *Necturus, Amphiuma, Siren,* and *Cryptobranchus*. The greatest precipitate occurs in the tube to which the *Necturus* serum is added. *Amphiuma* and *Siren* sera give about equal precipitates, but rather less than the *Necturus* serum, while the *Cryptobranchus* serum gives only a slight precipitate (Figure 23). On the basis of other criteria (especially comparative anatomy), it is generally agreed that the first three genera are fairly closely related, while *Cryptobranchus* is a much more primitive salamander.

Similarly, if serum from an animal immunized against human blood were divided among five tubes, and serum added from man, an anthropoid ape, an old-world monkey, a new-world monkey, and a lemur, the amount of precipitate formed would also decrease in that order. Thus, the results of serological tests support the theories of relationship which were originally based upon comparative morphology. They indicate that the serum proteins show varying degrees of homology, just as do gross structures. That this should be just a coincidence is not imaginable, yet it is exactly what would be expected on the basis of Darwin's theory that similar species have been formed by descent with modification from a common ancestor.

The above examples are both taken from among the vertebrates, and indeed this is the group which has been most thoroughly investigated. But extensive studies have also been made on the serological systematics of the Crustacea, Insecta, and Mollusca. Everywhere, the same fundamental result is obtained: animals which had been regarded as closely related on morphological grounds also show close serological affinity. In general, species of a single genus show very close serological similarity; genera of the same family show moderate serological similarity; and families of the same order show slight but detectable similarity. Frequently, different orders of the same class show too little similarity to permit useful comparisons, but there are exceptions, especially among the birds, which have diverged less serologically than have other classes of vertebrates.

In general, then, serology has verified the taxonomy which was worked out on other bases, but, in some cases, serological data have been decisive in resolving difficult problems. Thus the king crab, *Xiphosura*, was long regarded as a crustacean. Serological tests, however, showed that it has little affinity with typical crustaceans but a strong affinity with spiders. Subsequently, it was demonstrated that these aberrant animals share with spiders the fundamental arachnidan morphology, while their resemblance to crustaceans is a result of superficial convergence.

It might reasonably be supposed that only those groups characterized by a blood circulatory system could be investigated by the serological method. But Mez of Königsberg has successfully applied the method to problems in plant taxonomy. The method is to inject proteins from a plant into a rabbit, thus immunizing the rabbit against the proteins of the plant species used. Antibody-containing serum is then prepared and divided among a series of test tubes. To each is added a few drops of a solution of proteins from various plants related to the original test plant. Those which give a precipitate when added in very dilute solution are regarded as being closely related to the test plant, while those which give a precipitate only when added in concentrated solution are regarded as being only distantly related. The results lead to a classification which compares very favorably with that which has been worked out by plant morphologists. Thus the method of comparative serology has unexpectedly wide applications, and it seems probable that the method of Mez could profitably be applied to any group of organisms.

Blood Groups. The well-known A-B-O blood groups of man are based upon antigenic proteins of the red blood cells. Red cells may carry antigen A, or B, or both (AB), or neither (O), and a person's blood group is named accordingly. Any person's serum contains antibodies capable of agglutinating and destroying cells carrying those antigens not present in his own blood. This immune system is inherited upon a simple basis (multiple alleles: Chapter 13). Comparable series of blood groups have been found in many other animals, but only with other Primates, which are morphologically closest to man, does man share the A-B-O groups. Chimpanzees are predominantly group A, but group O also occurs among them. Gorillas and orang-utans are known to possess groups A, B, and AB. Among various species of *Macaca* (the rhesus group) all four blood groups are known. In lower Primates, the same antigenic proteins can be demonstrated in the saliva (as also in man), but not on the blood cells. The inference of relationship is unavoidable.

BIOCHEMISTRY AND RECAPITULATION

Phosphagens. Few chapters in physiology have been so thoroughly investigated as muscle contraction. Energy-rich phosphate compounds play a key role. Very briefly, adenosine triphosphate (ATP) breaks down to yield energy for the contraction. Then a second energy-rich compound, called a phosphagen, breaks down and releases energy for the resynthesis of ATP. These reactions are anaerobic, but the cycle is completed by the oxidation of glucose to provide energy for resynthesis of the phosphagen.

In the muscle of vertebrates, the phosphagen is always a specific compound, creatine phosphate, while in most invertebrates it is arginine phosphate. It is important to determine which characterizes the most primitive chordates, and which characterizes those groups from which the chordates may have arisen. Actually, the Hemichordata, the most primitive group allied to the chordates, has *both* phosphagens, a condition found elsewhere only in certain echinoderms, allies of the sea stars. On embryological grounds, echinoderms were already considered as probably close to the ancestry of the chordates.

This picture is not uncomplicated, for, although most invertebrates have only arginine phosphate, annelids lack this and have instead a substance which is similar to creatine phosphate and may be identical with it. However, serological evidence affirms the relationship of echinoderms and protochordates, while failing to show relationship of the annelids to either of these groups.

Arginine and creatine are very closely related chemically, and the former is actually used by vertebrates in the synthesis of the latter. In the embryos of sharks, arginine is abundant, but its occurrence in adults is more restricted. Thus it is possible that the creatine metabolism of vertebrates is a biochemical recapitulation, comparable to the embryological recapitulations discussed in the preceding chapter.

Visual Pigments. Vision among vertebrates depends upon one or the other of two chemical systems in the rods of the retina. Fresh-water fishes have visual purple, a porphyropsin-vitamin A_2 system, while marine fishes and land vertebrates have visual red, a rhodopsin-vitamin A_1 system. That marine and land vertebrates should both contrast with fresh-water fishes is surprising until one remembers that vertebrates probably arose in fresh water, then migrated from it in the two directions.

However, some anadromous fishes, like the salmon, live principally in the sea, but return to fresh water to breed; while others, like the eel, are catadromous; they live much of their life cycle in fresh water, but return to the sea to breed. And amphibians, of course, may live much of their adult lives on land, but they return to fresh water to breed. Wald and his collaborators have studied the visual pigments in these animals, and the results are most illuminating. Anadromous fishes and amphibians are both hatched in fresh water, and they undergo larval development there. After a metamorphosis involving profound anatomical and physiological changes, they migrate to salt water and to land, respectively, where they live much of their adult lives. Finally, after changes which are partly a reversal of those occurring at metamorphosis, and which Wald has called a second metamorphosis, they return to fresh water to spawn.

Actually tadpoles and young anadromous fishes have porphyropsin in their rods. At metamorphosis, that is, when they are preparing to migrate to land or to the sea, the morphological changes are accompanied by a change to a predominantly rhodopsin visual system. Later, when the mature animals are ready to return to fresh water for breeding purposes, they once again revert to a porphyropsin system. In catadromous fishes, the facts are similar, but the sequence of changes is reversed.

On other grounds, it had already been believed that the vertebrates originated in fresh water, and there the ancestral fishes differentiated, giving rise to fresh-water and marine fishes, and to amphibians. If this is correct, then the sequence of changes in visual pigments is best interpreted as a recapitulation—a condensed repetition of ancestral history. Wald summarizes this story as follows:

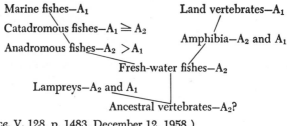

(See *Science,* V. 128, p. 1483, December 12, 1958.)

Similar facts have been worked out in amphibians for some aspects of blood physiology and excretory physiology. Thus it appears that the principle of recapitulation can give meaning to many otherwise inexplicable phenomena of biochemistry and physiology.

Wald has said that, "without the rationalizations of phylogeny, comparative biochemistry is little more than a catalogue." * But evolutionary (phylogenetic) considerations do give meaning to this great and fundamental body of data, and this is cogent evidence for evolution.

* See Wald, George, 1952, in References at end of this chapter.

REFERENCES

Chester, K. Starr, 1937. "A Critique of Plant Serology," *Quart. Rev. Biol.*, 12, 19–46; and 165–190. The principal English language review of Mez's work.

Florkin, Marcel, 1949. "Biochemical Evolution," translated by S. Morgulis. Academic Press, New York, N.Y. A great deal of information in a small volume.

Needham, Joseph, 1950. "Biochemistry and Morphogenesis," Cambridge University Press, Cambridge. This chemical embryology contains a wealth of information applicable to the present chapter.

Smith, H. W., 1953. "From Fish to Philosopher," Little, Brown & Co., Boston, Mass. Evolution from the viewpoint of renal physiology.

Wald, George, 1952. "Biochemical Evolution," in "Modern Trends in Physiology and Biochemistry," Academic Press, New York, N.Y. Much the best treatment of this subject now available.

Wald, George, 1958. "The Significance of Vertebrate Metamorphosis," *Science*, 128, 1481–1490. A review of metamorphosis of visual pigments, with broader speculations.

Weiner, A. S., 1943. "Blood Groups and Transfusions," 3rd Ed., Charles C Thomas, Springfield, Ill. Valuable for material on blood groups in Primates.

Williams, R. J., 1956. "Biochemical Individuality," John Wiley & Sons, Inc., New York, N.Y. While not primarily concerned with evolution, this book gives a wealth of information on variability of biochemical traits.

CHAPTER FIVE

Evidences of Evolution
IV: Paleontology and Genetics

THE VARIOUS LINES OF EVIDENCE discussed thus far all indicate that the species now populating the earth must have been produced by evolution, but generally the only possible direct evidence for a specific line of descent is a series of fossils leading stepwise from an ancestral to a descended species. Hence the science of paleontology, which deals with fossil remains, has unique importance for evolution.

THE GEOLOGICAL TIME SCALE

Before beginning the discussion of paleontological evidence for evolution, however, it is necessary to introduce the problem of geological time. Fortunately, there are methods of determining the ages of rock deposits in the earth's crust. The oldest method is applicable only to sedimentary rocks—those successive layers or strata of rocks which are formed by slow settling out of sediments from the oceans or other large bodies of water. The use of the method for dating purposes is based upon the assumption that those geological processes which are observable in action now are the same ones which have determined the past history of this earth, and that they have in the past acted at rates comparable to those now observable. As applied to the problem at hand, this simply means that the sedimentary rocks of the past were deposited at rates comparable to those which are now being deposited. In a purely comparative way, dating by this means is fairly satisfactory. The deepest strata are the oldest, while the most superficial strata are quite recent. Thick strata represent long-continued deposition, while thin strata represent short periods of deposition. Thus some relative concepts of geological time are readily gained from an inspection of the sedimentary rock deposits. But more exact ideas are difficult to obtain because it is clear that sedimentation is now progressing at very different rates in different parts of the world, and there is no reason to doubt that the variation in the rate of sedimentation has been as great over much of the world's history. It may well have been greater at times. Hence calculations of age based upon the thicknesses of stratified rocks lead to such statements as that the Mesozoic Era began somewhere between 190,000,-000 and 240,000,000 years ago. But the difference between these figures is more than 25 per cent of the smaller one. In addition to this difficulty, the strata have been changed by such geological processes as folding and erosion, so that often the record is fragmentary and confused. There are a few instances in which stratified rocks appear to have been laid down in definite annual layers, the varves, more or less comparable to the annual growth rings of trees. In such cases, the extent of the periods of sedimentation can be determined with great exactness, if the varves actually do represent annual layers. But this is by no means certain. And even if it were certain, the number of known examples is so small that it could have little importance for the general problem of dating geological history.

Although the study of sedimentary rocks has not led to a satisfactory dating system, it has been possible to determine the sequence in which the various strata of the earth's surface have been laid down, from very ancient rocks right up to those of very recent origin. Particular strata are identifiable not only by their position and their physical characteristics, but also by the fossils which they contain. Thus it has been possible to divide geological time into a series of eras, the sequence of which is undoubted. The first two eras, the Archeozoic and the Proterozoic, are not of great interest for the present discussion, because the rocks deposited in these eras contain very few fossils, and those are generally of doubtful character. During the Paleozoic Era, fossils were deposited in great abundance, but only archaic types were present. At first, only invertebrates were represented, but fishes, amphibians, and finally reptiles made their appearance during the Paleozoic Era. The next great era was the Mesozoic, or Age of Reptiles, during which birds and small mammals also arose. Finally, the Cenozoic Era, which is still in progress, has been marked by the rise to dominance of the mammals and man.

The eras are, however, immensely long expanses of time, characterized by progressive differences of flora and fauna, and not infrequently of climate and other physical characteristics. The eras are therefore divided into *periods* of shorter, but still very long, duration. Thus the Paleozoic Era lasted for about 300,000,000 years, but it is subdivided into seven periods, the durations of which vary from 25,000,000 years to 80,000,000 years. Finally, it is sometimes desirable to break up the periods into still smaller divisions, the *epochs*. The Tertiary Period of the Cenozoic Era, for example, lasted for about 74,000,000 years. This is subdivided into five epochs the durations of which varied from 11,000,000 years to 19,000,000 years.

While this system of time measurement is not quantitatively accurate, it is, on the whole, quite workable. For, if a fossil is found in strata from the Cretaceous Period, it is possible to state with complete assurance that it followed Jurassic forebearers and preceded Paleocene descendants, if any descendants were left. The geological time scale, together with some of the characteristics of life at each level, is summarized in Table 1.

The Lead Method. In 1907, Boltwood introduced a method for dating geological strata based upon radioactive elements. The conclusions to

TABLE 1. GEOLOGIC TIMETABLE *

ERA	PERIOD	EPOCH	DURATION IN MILLIONS OF YEARS	TIME FROM BEGINNING OF PERIOD TO PRESENT (MILLIONS OF YEARS)	GEOLOGIC CONDITIONS	PLANT LIFE	ANIMAL LIFE
Cenozoic (Age of Mammals)	Quaternary	Recent	0.025	0.025	End of last ice age; climate warmer	Decline of woody plants; rise of herbaceous ones	Age of man
		Pleistocene	1	1	Repeated glaciation; 4 ice ages	Great extinction of species	Extinction of great mammals; first human social life
	Tertiary	Pliocene	11	12	Continued rise of mountains of western North America; volcanic activity	Decline of forests; spread of grasslands; flowering plants, monocotyledons developed	Man evolving; elephants, horses, camels almost like modern species
		Miocene	16	28	Sierra and Cascade mountains formed; volcanic activity in northwest U.S.; climate cooler		Mammals at height of evolution; first manlike apes
		Oligocene	11	39	Lands lower; climate warmer	Maximum spread of forests; rise of monocotyledons, flowering plants	Archaic mammals extinct; rise of anthropoids; forerunners of most living genera of mammals
		Eocene	19	58	Mountains eroded; no continental seas; climate warmer		Placental mammals diversified and specialized; hoofed mammals and carnivores established
		Paleocene	17	75			Spread of archaic mammal
Rocky Mountain Revolution (Little Destruction of Fossils)							
Mesozoic (Age of Reptiles)	Cretaceous		60	135	Andes, Alps, Himalayas, Rockies formed late; earlier, inland seas and swamps; chalk, shale deposited	First monocotyledons; first oak and maple forests; gymnosperms declined	Dinosaurs reached peak, became extinct; toothed birds became extinct; first modern birds; archaic mammals common
	Jurassic		30	165	Continents fairly high; shallow seas over some of Europe and western U.S.	Increase of dicotyledons; cycads and conifers common	First toothed birds; dinosaurs larger and specialized; insectivorous marsupials

Era	Period			Physical conditions	Plant life	Animal life
	Triassic	40	205	Continents exposed; widespread desert conditions; many land deposits	Gymnosperms dominant; declining toward end; extinction of seed ferns	First dinosaurs, pterosaurs and egg-laying mammals; extinction of primitive amphibians
	Appalachian Revolution (Some Loss of Fossils)					
Paleozoic (Age of Ancient Life)	Permian	25	230	Continents rose; Appalachians formed; increasing glaciation and aridity	Decline of lycopods and horsetails	Many ancient animals died out; mammal-like reptiles, modern insects arose
	Pennsylvanian (Carboniferous)	25	255	Lands at first low; great coal swamps	Great forests of seed ferns and gymnosperms	First reptiles; insects common; spread of ancient amphibians
	Mississippian (Carboniferous)	25	280	Climate warm and humid at first, cooler later as land rose	Lycopods and horsetails dominant; gymnosperms increasingly widespread	Sea lilies at height; spread of ancient sharks
	Devonian	45	325	Smaller inland seas; land higher, more arid; glaciation	First forests; land plants well established; first gymnosperms	First amphibians; lung-fishes, sharks abundant
	Silurian	35	360	Extensive continental seas; lowlands increasingly arid as land rose	First definite evidence of land plants; algae dominant	Marine arachnids dominant; first (wingless) insects; rise of fishes
	Ordovician	65	425	Great submergence of land; warm climates even in Arctic	Land plants probably first appeared; marine algae abundant	First fishes, probably fresh-water; corals, trilobites abundant; diversified molluscs
	Cambrian	80	505	Lands low, climate mild; earliest rocks with abundant fossils	Marine algae	Trilobites, brachiopods dominant; most modern phyla established
	Second Great Revolution (Considerable Loss of Fossils)					
Proterozoic		1500	2000	Great sedimentation; volcanic activity later; extensive erosion, repeated glaciations	Primitive aquatic plants —algae, fungi	Various marine protozoa; towards end, molluscs, worms, other marine invertebrates
	First Great Revolution (Considerable Loss of Fossils)					
Archeozoic		???	???	Great volcanic activity; some sedimentary deposition; extensive erosion	No recognizable fossils; indirect evidence of living things from deposits of organic material in rock	

* Adapted from Villee, "Biology, the Human Approach," 3rd Ed., W. B. Saunders Co., 1957.

which the new method led indicated that the earth was vastly older than had been generally believed, and the method was received with skepticism. But it has since become the standard by which the accuracy of other methods of dating is judged. The method is based upon the fact that uranium 238 will slowly disintegrate to produce lead with an atomic weight of 206 and helium. The rate at which this occurs is calculable. With any definite amount of uranium, one half of the molecules will break down, forming lead and helium, in the course of 4,510,000,000 years. As this figure is independent of the actual quantity of uranium originally present, it is called the "half-life" of the element. Now, if a uranium-bearing rock is found, the ratio of uranium to lead 206 can be determined, and from this, utilizing the half-life, the interval since the formation of the rock can be calculated.

Recently, additional geochemical methods have been developed. Potassium 40 yields calcium 40 and argon 40; rubidium 87 yields strontium 87; thorium 232 yields lead 208; and uranium 235 yields lead 207. Each of these parent elements has its characteristic half-life, ranging from 126,-000,000 years to as much as 60,000,000,000 years.

While the "lead" method is now universally accepted, it has serious limitations. Uranium is not a common element, and it is often found in geological formations which are not readily fitted into the geological time scale. The introduction of additional geochemical methods helps to overcome these limitations, and so there are good grounds for hope that an accurately dated scale of geological time may be available in the near future. For the present, it may be said that the oldest dated rocks are more than 3,000,000,000 years old, and the oldest which might have borne life are about 2,000,000,000 years old. Around three quarters of that 2,000,-000,000 years during which life might have existed passed before the beginning of the Cambrian Period, with which the useful fossil record begins, for a lead measurement of the age of a late Cambrian deposit gave a figure of 440,000,000 years. The next exact determination is in the early Permian Period, at an age of 230,000,000 years. Thus the entire Paleozoic Era probably lasted about 300,000,000 years; the Mesozoic about 130,000,-000 years; and the Cenozoic about 75,000,000 years up to the present. There is an accurately dated deposit from the beginning of the Eocene Epoch which places this at 58,000,000 years ago. The dates are few, but fortunately they are widely scattered in geologic time. But as yet, accurate determinations are not available for the extent of any of the periods or epochs.

The Radio-Carbon Method. Another radioactive method has been developed which shows promise for shorter range determinations of age, up to perhaps 40,000 years. It has been found that living organisms utilize a small, but constant proportion of their organic carbon in the radioactive form. The half-life of radioactive carbon is 5568 ± 30 years. Hence, remnants of bone, wood, or other carbon-containing remains of dead organisms can be assayed for their radio-carbon content. The difference between the average amount in fresh tissue and in the fossil may then be treated as being due to radioactive disintegration, and the age of the fossil calculated from the half-life. This method has proved very useful in the study of late Pleistocene and Recent remains. It has been checked against historical objects of known age, such as wood from Egyptian pyramids, and has been found reliable.

FOSSILIZATION

Any remnant of living forms from the remote past may be regarded as a fossil. While there are many ways in which fossils may be formed, most of them involve the burial of the dead organism. As more and more sediment is laid down above, the depth of the fossil increases, and hence, in a general way, the oldest fossils may be expected to be found in the deepest layers of the earth's surface, while the more superficially located fossils may be regarded as recent in origin. Thus, ideally, one ought to be able to read the story of life in the correct historical sequence by examining the fossil record in sequence from the deepest to the most superficial strata.

But the vast majority of plants and animals are not fossilized after death. Decay and destruction ordinarily await the dying organism. Predators and scavengers may not only eat away the soft parts of the body, but they may also break up skeletal structures beyond all hope of preservation or recognition. It is only the unusual instance in which the organism is rapidly buried or in some other way protected from scavengers and from oxidation that a fossil may be formed. Thus the fossil record, even if completely known, would have to be very fragmentary, because the majority of organisms never take the first step toward fossil formation.

Much the most common method of fossilization is burial in the sediments which are continually deposited on the floor of the oceans and of other large bodies of water. When aquatic organisms die, they may fall into deep sedimentary deposits in which the bodies are protected from scavengers and from oxidation. The soft parts of the body gradually decay and are carried away by the seepage of water. Bones and other hard parts may remain as such, or they may be replaced particle by particle by minerals in the water. As this process continues, the layer of sediment which is being deposited grows ever thicker, and its lower portions gradually harden into rock, the sedimentary, stratified rock which is characteristic of the beds of marine or aquatic deposits everywhere, and which is the hallmark of ancient seas in areas which are now dry land. However, this is not the only means by which organisms may be buried. Dust storms can have the same effect, and will be effective in causing fossilization of terrestrial organisms. Again, volcanic ash may also rapidly bury organisms and thus preserve them as fossils. Pompeii, which was buried by volcanic ash from Mount Vesuvius in 79 A.D., has been extensively investigated in modern times. Whole families, together with their domestic animals, have been preserved as cavities in the ash, from which casts can be made. Desert forms may be dried out by the hot, dry, desert winds, then buried under the shifting sands.

Some special methods of burial are also occasionally effective. If a petroleum spring should occur, evaporation of the more volatile oils will produce first a pool of sticky tar and then one of viscous asphalt. This has happened during the Pleistocene epoch at Rancho La Brea in southern California. Many Pleistocene and Recent mammals and birds have been trapped in this asphalt, and they are among the best preserved of fossils. It appears to have worked in the following ways. Small mammals, herbivores, and birds try to reach the rain pools which occur on the surface of the asphalt. In so doing, they become stuck in the soft asphalt, and predators are then ensnared while attempting to catch the former. Water birds may alight on the water pools and then become entrapped in the asphalt around the edges. Thus Rancho La Brea is one of the richest sources known for well preserved fossils of recent mammals and birds. Because the city of Los Angeles has grown up around it, it no longer entraps the wild fauna of the region, but the Los Angeles fire department is still occasionally called out to rescue a child who has gotten his feet stuck in the asphalt.

Another unusual method of burial is the entrapment of insects in amber (Figure 24). Such fossils are sometimes preserved almost perfectly, so that even histological details are comparable to those of freshly fixed specimens.

FIGURE 24. TERMITES IN MID-CENOZOIC AMBER. (Buchsbaum, R., "Animals Without Backbones," 2nd Ed., University of Chicago Press, 1948. Specimen lent by A. E. Emerson, photograph by P. S. Tice.)

Lastly, organisms may be petrified, that is, their actual tissues may be replaced, particle by particle, by minerals in solution in the waters of the locality. The principal minerals utilized in this type of fossilization are iron pyrites, silica, calcium carbonate, and other carbonates. The most widely known example of petrifaction is afforded by the petrified forests of southwestern United States; however, animal remains may also be petrified. Generally, this preserves only the hard parts of the body, but occasionally soft parts are so well preserved that even fine details of cells can be made out in thin sections. Most of the fossils from sedimentary rocks are of this type: the original material has been replaced by minerals from the surrounding medium.

Types of Fossils. With respect to what is preserved, there are several types of fossils. The whole organism may be preserved, but this is very rare, and is known only for Cenozoic fossils. The best examples are insects in amber, some of the mammals from Rancho La Brea, and the mammoths and other mammals which have been found frozen in the arctic. These are, of course, the ideal fossils. But because of their great scarcity, and because they are all of recent origin, they are less important for paleontology as a whole than are less complete and less satisfactory types of fossils. The vast majority of fossils are petrified or carbonized, and only the hard parts of the organism are preserved. Thus the bones of vertebrates, shells and spicules of invertebrates, and the woody parts of plants are commonly fossilized. Such fossils are, because of the nature of their formation, almost always incomplete. But even beyond this, they are commonly found as broken fragments.

A fossil, however, need not include any part of the original organism. It may be a mere footprint or leafprint which has chanced to escape destruction while the mud or sand in which it lay hardened into stone. Such impressions may subsequently be filled with a hard mineral, thus forming a cast of the original structure much as a sculptor might. Soft-bodied organisms are preserved in this way as readily as are those with hard skeletal structures, and many organisms are known only from the casts and impressions which they have left. Even the feces may be fossilized and studied under the name of coprolites, and these may yield valuable information about the food habits of extinct animals.

Incompleteness of the Fossil Record. One of the important aspects of the fossil record is its incompleteness. One reason for this has already been pointed out: the majority of organisms never take the first step toward fossilization. They are destroyed by predators or scavengers, or else they lie exposed to the elements and deteriorate. But many other factors also contribute to the incompleteness of the fossil record. Of the several methods of fossilization mentioned above, only one, burial in marine or fresh-water sediments, could be expected to occur with great regularity. Because of this, the majority of fossils are found in such sedimentary rocks, and this consequently results in poor representation of terrestrial forms.

The record is also biased by the fact that generally only the hard parts of the body are fossilized. In some instances, such parts are taxonomically useful, while in other instances they are not. Among vertebrates, the skeletons are the most commonly preserved parts, and they are also of immense value taxonomically. Skeletal remains may give quite diversified information about a vertebrate. Obviously, a complete skeleton will indicate the size of the animal; but, in the hands of a competent anatomist, even a single bone, or a mere fragment of a bone, may offer a basis for a reasonable estimate of size. From the scars of muscle attachments on the bones, the sizes and contours of the muscles can be determined. From this it is an easy step to the general appearance of the animal, and its characteristic gait and speed. The skulls give an indication of relative intelligence. The teeth indicate the type of diet eaten. Thus vertebrate skeletons are among the most satisfactory of all fossils. But preservation of hard parts does not always lead to so fortunate a result. Among plants, woody parts are most commonly preserved. But these are of quite secondary importance taxonomically. The flowers, which are of great importance for plant taxonomy, are rarely preserved. And for many groups, fossils are very rare because there are no hard parts at all.

Not all periods in the earth's history have been equally favorable for the formation and preservation of sedimentary rocks. It is well established that the general level of the continents has fluctuated from time to time, with cold ages during which the continental shelves were largely exposed alternating with warm ages, during which the continental shelves and even much of the inland low areas were submerged to form shallow seas. During times of progressive submergence, a newly deposited stratum, together with its included fossils, would be protected by the deposition of additional strata above it. In this way, very thick layers can be formed. But, during periods of elevation, newly formed strata may be quickly raised above the water level, where they become subject to erosion and destruction of fossils by wave action, wind, and rain. The result is that remains from periods of submergence are much richer in fossils than are those from periods of elevation. This is in contrast to the probable relative abundance of species. For elevation of the continents exposes increasing areas and new habitats for colonization. Numbers of species are therefore probably increasing at such times. But during periods of submergence, the habitable world undergoes a shrinkage, with the attendant extinction of many species.

Up to the present, this discussion has assumed that the strata, once formed, remain undisturbed until struck by a geologist's pick. But this is by no means true. The rocks may be profoundly altered in many ways, together with their contained fossils. Mountains may be thrust up; rivers may cut deep gorges through many successive strata; the rocks may be cracked by tensions and one portion slipped over another; glaciation may carry away superficial rocks and lay bare the deeper ones; erosion by wind and water may wear away the rocks; and many other normal geological processes may alter or destroy fossil-bearing strata. All of this is useful in that such processes tend to expose for study deep strata which would be unavailable if left exactly as originally deposited; but all of these processes also result in extensive destruction of fossils.

EVALUATION OF THE FOSSIL RECORD

With the above facts in mind, some evaluation of the fossil record for various groups may be attempted. Characteristics to be treated are the abundance of fossils available for study, their degree of completeness, and the ease with which they can be interpreted. The Protozoa (unicellular, or better, acellular, animals) are generally not fossilized, but those which have calcareous or siliceous shells have been fossilized in immense numbers. These are chiefly the Foraminifera, the Heliozoa, and the Radiolaria, all of them orders of the class Rhizopoda, which is best known by the ameba. Chalk deposits and some limestones are largely formed from the shells of Foraminifera, while siliceous rocks are largely radiolarian in origin. The shells are often recoverable in a good state of completeness, and they are fairly good as far as ease of interpretation is concerned. The Porifera (sponges) are represented in the fossil record by their spicules, which may be either calcareous or siliceous. They are only fair in abundance, but their preservation is often good. However, ease of interpretation is again only fair.

Most of the Coelenterata (hydroids, jellyfishes, corals, and their allies) have left a very scant fossil record of poor quality. But the corals have left numerous fossils, quite complete, and readily identifiable. The Annelida and other worm-like phyla have been fossilized so rarely that such fossils as are available have little value for tracing the history of these groups. Yet a few of the known annelid fossils are surprisingly complete. Because of the importance of several of the worm-like phyla among living animals, this is an especially serious deficiency in the fossil record. Marine Arthropoda are abundantly represented in the fossil record, but insects are rather scantily represented. Many of these arthropod fossils are quite complete, and the ease of interpretation is fair, often good. The Brachiopoda, or lampshells, are a minor phylum, at present consisting of only a few species, mostly confined to tropical seas. In Paleozoic times they were more important. Their fossil record is one of the best in all respects.

The Mollusca, including such animals as sea cradles, octopuses, clams, and snails, have left a very abundant fossil record. The shells are often in a good state of preservation, and in many lines of descent they are readily interpretable. Fossils of the Echinodermata, a group including the starfishes and their allies, are very numerous. In completeness they are excellent, and in interpretability they are fair. Finally, the vertebrates are present in the fossil record in good numbers. Their completeness and interpretability are excellent. The prochordates, however, having no bony skeleton, are not represented in the fossil record at all unless *Jamoytius*, a controversial Amphioxus-like fossil from Silurian rocks of England, is a prochordate.

Yet, fragmentary though the fossil record is, it is a striking thing that it gives clear testimony to the fact of evolution, and considerable detail can be worked out in many lines of descent. The most ancient fossils include only invertebrates. Then primitive fish-like vertebrates appear, and these gradually blend into true fishes, similar to some species now living.

71

Later, amphibians and reptiles appear in the fossil record, and birds and mammals finally appear quite late. Thus the simplest animals appear in the most remote geological eras, while the most complex appear late in geological history. In most major groups (order, class, and phylum), there is marked change from one geological period to the next, but always a particular fauna resembles that of another period near it in time more closely than it does that of any other period remote from it in time. Finally, the fossils of recent organisms blend into our present living flora and fauna, with often the same genera and even the same species being represented.

CATASTROPHISM VERSUS EVOLUTION

Before the time of Darwin, it was customary to explain these facts by assuming that all life had been destroyed by catastrophes from time to time, each catastrophe being followed by a new creation. As more and more paleontological information accumulated, however, it became apparent that the number of catastrophes necessary to account for the known succession of floras and faunas was absurdly large, and further that extinction of different contemporaneous groups was not simultaneous, as would be necessary under the theory of catastrophism. Under Darwin's theory, it was not necessary to assume any catastrophes. Species simply change continuously under the influence of natural selection. The inevitable result is a changed aspect of the flora and fauna from one period to the next, with the difference increasing throughout time. This does not require that the rate of change in different groups, or in different members of the same group, be the same.

EVOLUTION OF THE HORSE

Perhaps the most thoroughly known phylogeny of any vertebrate is that of the horse. The history of the horse, as understood today, covers a period of about 60,000,000 years, beginning in the Eocene Period, and involving twenty genera and a much larger number of species, although most of these are not believed to be in the direct line of descent. The oldest members of the superfamily Equoidea were the eight genera of the family Paleotheriidae, but these are usually not taken into account in discussions of horse phylogeny, partly because of uncertainty regarding their relationship to more advanced equoids, but also because they show as much relationship, or more, to other mammals as to horses. Their "hooves," for example, are quite claw-like.

The usual starting point of horse phylogenies is with *Hyracotherium* (including *Eohippus* of many publications), the oldest known member of the family Equidae. This little animal stood less than a foot high at the shoulder, and browsed in forest underbrush (known from the character of the teeth). The teeth were forty-four in number, and the cheek teeth were only moderately specialized for grinding. The front feet had four toes and a splint, while the hind feet had three toes and two splints (rep-

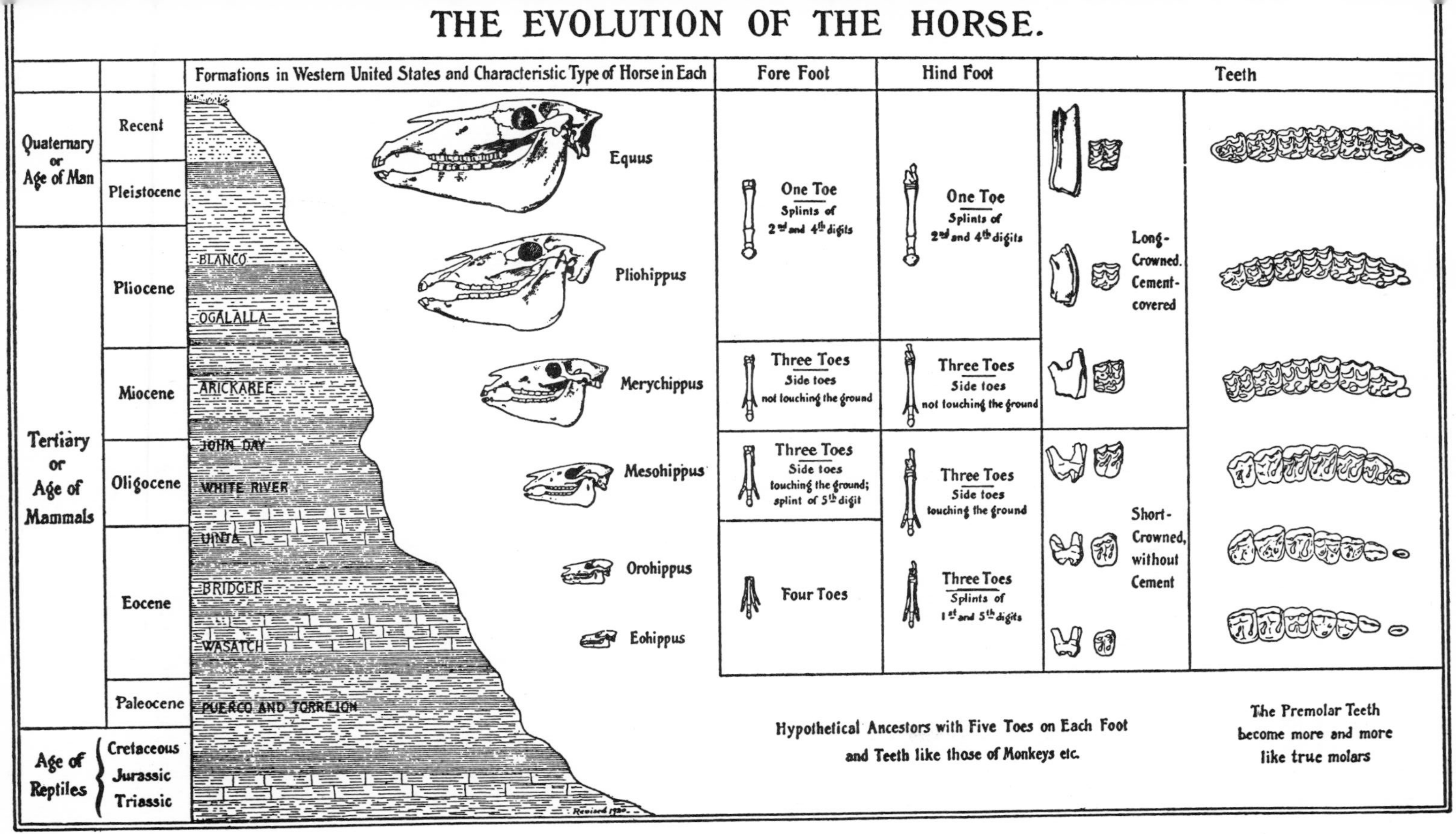

THE EVOLUTION OF THE HORSE.

FIGURE 25. EVOLUTION OF THE HORSE. (From Matthew, W. D., *Quart. Rev. Biol.*, V. 1, 1926.)

resenting the first and fifth toes). Yet already the third toe was more prominent, both front and rear, than the others (Figure 25).

The line of descent appears to pass, with moderate changes at each step, from *Hyracotherium* through *Mesohippus, Miohippus, Parahippus, Merychippus,* and *Pliohippus* to *Equus,* the horses of today and their allies. The changes achieved are great indeed. The size has increased from that of a house cat to that of the Percheron. The cheek teeth, originally low-crowned and rather primitive, have become very high-crowned and highly specialized for grinding coarse, siliceous grasses. These changes have been associated with elongation of the jaws and related parts of the head. The evolution of siliceous range grasses occurred simultaneously, and probably served as a selective force acting upon the horses. The limbs were elongated, and simplified by the loss of some parts (toes) and the fusion of others (metacarpals and metatarsals). The neck became much longer and more mobile.

This phylogeny, emphasizing as it does the six extinct genera believed to be in or near the direct line leading to living horses, should not be interpreted as meaning that horse evolution has gone in a definite direction from the beginning, as though horses had from the first known where they were going and had taken the most direct steps to get there. It must be remembered that the record includes thirteen other genera of the family Equidae, as well as the eight genera of the family Paleotheriidae. These many genera varied in several directions, sharing some characteristics with their cousins in the "main line," but differing from them in many others. Some lines of descent, for example, did not develop high-crowned teeth. At every period in the evolution of the horse there have been widely varying genera and species. Natural selection has simply eliminated the great majority of these.

The validity of this series of horse fossils has been conceded by everyone, but some opponents of evolution have contended that the mere fact that such a series of genera existed in a succession from the most primitive to the most specialized does not require that the latter be descended from the former. They argue that each may have been independently created, and that there is no reason why they should not have been created in an orderly sequence. There is no answer to this, simply because it waives the evidence altogether. One could just as well say that there is no reason why they should have been created in any particular sequence. It seems far more probable that the appearance of descent is there because the specialized horses actually are descended from the primitive ones. But when one considers also the innumerable examples from all major groups of organisms which parallel the above case, the probability that this appearance of descent is misleading becomes very remote.

EVIDENCE FROM GENETICS

The final line of evidence for evolution is drawn from genetics, the science of heredity. It is not intended to review this subject here, for the bearing of genetics on evolution will make up a large part of the succeeding chapters. It may be stated, however, that it has been demonstrated that the genes (hereditary determiners) are quite constant, and are inherited on a statistically predictable basis. They tend, therefore, to keep species constant. However the genes are capable of undergoing a change (mutation), so that the trait determined is different from the original and is just as stable. Hence mutation forms the basis of hereditary variability, which is the raw material of evolution.

REFERENCES

ALDRICH, L. T., 1956. "Measurement of Radioactive Ages of Rocks," *Science,* **123,** 871–875. A brief description of the methods, with some of the recent results.

DARWIN, C. R., 1859. "Origin of Species," Chapter X, "On the Imperfections of the Geological Record." After a century, this is still the finest treatment of the subject.

GREGORY, W. K., 1951. "Evolution Emerging," Vols. I and II, The Macmillan Co., New York, N.Y. An exhaustive presentation of the viewpoint of a paleontologist at the end of a long career.

MOORE, R. C., 1958. "Introduction to Historical Geology," 2nd Ed., McGraw-Hill Book Co., New York, N.Y. Paleontology is well presented here.

ROMER, ALFRED S., 1945. "Vertebrate Paleontology," 2nd Ed., University of Chicago Press. An excellent treatment of the vertebrates.

STIRTON, R. A., 1959. "Time, Life, and Man," John Wiley & Sons, New York, N.Y. A well illustrated, nontechnical introduction to paleontology.

WHITE, E. I., 1946. "*Jamoytius kerwoodi,* a new Chordate from the Silurian of Lanarkshire," *Geological Magazine,* **83,** 89. The original description of this interesting and controversial fossil.

CHAPTER SIX

The History of Evolutionary Thought

IT IS NOT PROPOSED HERE to review the whole history of evolutionary thought. This subject is well covered in works on the history of biology. The present objective is simply to review some of the main trends in evolutionary thought in recent times.

LAMARCK

The only really important pre-Darwinian student of evolution was Lamarck (Figure 26), a French biologist (1744–1829) who began his career as a botanist, but became a zoologist when he was offered an appointment in zoology at the Jardin des Plantes (an institute of general biology, in spite of its name). Lamarck's services to general zoology are manifold, although his name is usually associated with an outmoded theory of evolution. His studies of invertebrates were extensive, and resulted in a greatly improved classification, including the recognition of the invertebrates and the vertebrates as distinct sections of the Animal Kingdom. He came very close to the cell theory thirty-nine years before Schleiden and Schwann formulated it.

Lamarck's systematic studies convinced him that species were not constant, but rather were derived from pre-existing species. To account for this, he devised an elaborate theory which may be summarized in four propositions: (1) Living organisms and their component parts tend continually to increase in size. (2) Production of a new organ results from a new need and from the new movement which this need starts and maintains. (3) If an organ is used constantly, it will tend to become highly developed, whereas disuse results in degeneration. (4) Modifications produced by the above principles during the lifetime of an individual will be inherited by its offspring, with the result that changes are cumulative over a period of time.

Lamarck first published his theory in 1802, and he defended it vigorously until his death. For it, he suffered both social and scientific ostracism, but he had the courage of his convictions. He failed to convince his

FIGURE 26. JEAN BAPTISTE LA-
MARCK. (From Locy, "Biology
and Its Makers," 3rd Ed., Henry
Holt & Co., Inc., 1935.)

contemporary scientists not simply because the temper of the times was
opposed to evolution, for many others were skeptical of the fixity of species, but because of the implausibility of some of his major theses. Thus,
his first principle, the tendency to increase in size, while it is illustrated
by many actual lines of descent, is far from universally true. Many groups
of organisms show no tendency whatever to produce strains leading to
gigantism. And in not a few groups, size reduction has been a prominent
feature of evolution. The second principle, that new organs result from
new needs, is quite manifestly false. In the case of plants, Lamarck believed that the environment acted directly upon the plant, causing the
production of such new characters as might adapt the plant to its environment. In the case of animals, Lamarck believed that the environment
acted through the nervous system; in other words, the desire of the animal
leads to the formation of new structures. In its crudest form, this would
mean that the man who mused "Birds can fly, so why can't I?" should have
sprouted wings and taken to the air.

Lamarck did not present quite such crude examples. He did explain
the long neck and high shoulders of the giraffe on a similar basis, however.
Giraffes browse upon the leaves of trees. He presumed them to have had
proportions much like typical mammals originally, but as they strained to
reach ever higher and higher leaves, their shoulders grew higher and their
necks longer in response to their need. The increase was cumulative from
generation to generation.

His final proposition was the inheritance of characters acquired during
the lifetime of the individual. This is a necessary proposition if environ-

mentally produced modifications are to have any evolutionary significance. However, every serious experimental study designed to test this principle has discredited it, with a single doubtful exception. This exception is presented by a series of experiments conducted by McDougall on learning in rats. Rats were dropped into a tank of water from which there were two exits, one lighted and one dark, but not always the same one. A rat leaving by the lighted exit received an electric shock, while one leaving by the dark exit received no shock. Thus the number of trials required for a particular rat to learn always to select the dark exit constituted a measure of the speed of learning. These rats were then bred, and their descendants were similarly studied. It appeared that the speed of learning increased from generation to generation, and so McDougall concluded that learning, an acquired trait *par excellence,* is inherited. Some serious criticisms have been raised against McDougall's experiments. The genetic constitution of his rats was not properly controlled, so that his initial breeding stock may have been quite mixed as to intelligence levels. Neither the intensity of the light nor of the electric shock was kept constant throughout the experiment. Yet it is quite possible that variations in light intensity influence speed of learning in such an experiment, and McDougall himself demonstrated that speed of learning varies directly with the intensity of the shock. McDougall stated that adequate precautions were taken to select breeding stock for subsequent generations at random. But he did not describe his procedure, and it is entirely possible that the more intelligent rats of each generation were selected to sire the next generation. In this event, inheritance of acquired characteristics would not be necessary in order to explain the improvement in learning ability from one generation to the next. Finally, during the course of the experiment, speed of learning also increased among the control rats, that is among those bred from untrained forebears. Thus it is probable that some unanalyzed changes in the technique of the experiment may have been responsible for all or part of the recorded increase in the speed of learning. But the most serious defect is the fact that repetition of the experiment in other laboratories has failed to produce similar results. In contrast to this paucity of positive evidence for the inheritance of acquired characteristics, countless experiments have led only to negative results. For example, Jewish boys have been circumcised for thousands of years, yet this has not resulted in any tendency whatever toward reduction of the prepuce among Jews. Examples could be multiplied indefinitely, but they all lead to the same conclusion: acquired characters are not inherited.

DARWIN

So when Darwin, jointly with Wallace, brought forward the theory of the origin of species by natural selection, there was no other evolutionary theory to compete with it. The rapidity with which it achieved world-wide acceptance by the majority of competent scientists is generally known, as is also the bitter controversy which it produced among the lay public, as well as among some scientists. It has been said that its rapid acceptance was due to the fact that evolution was "in the air" at the time. Darwin himself has stated in his autobiography that he did not believe this to be true, for he had discussed his ideas with many naturalists over a period of twenty years before the publication of the "Origin of Species," and he had not found any of them seriously inclined to agree with him. It seems more probable that Kingsley was right when he said that "Darwin is conquering everywhere, and is rushing in like a flood by the mere force of truth and fact." Darwin himself attributed his success to the fact that the "Origin of Species" was highly condensed from a mass of data which had been compiled and critically studied over a period of twenty years before publication.

Darwin realized that an understanding of heredity was essential for evolutionary studies, but he apparently never came across Mendel's paper, and he stated in the last edition of the "Origin" that the fundamental principles of heredity were still unknown. To fill the need of a working hypothesis, he devised the theory of pangenesis. According to this theory, all organs produce *pangenes,* minute particles which are carried away by the bloodstream and segregated out into the gametes. Thus every mature gamete contains a pangene from every organ of the animal producing it. In the developing zygote, each pangene tends to cause the formation of a duplicate of the organ from which it originally came. This theory plainly provides for the inheritance of acquired characters. Darwin did not suggest that this theory was correct. He proposed it simply as a working hypothesis which could serve as a starting point for investigation. The theory of pangenesis has been universally discarded.

The history of evolutionary thought subsequent to the publication of the "Origin of Species" may be divided, following Stebbins (personal communication), into three periods: the Romantic Period, extending from 1860 to about 1903; the Agnostic Period, or Period of Reaction, extending from 1903 to about 1935; and the Period of the Modern Synthesis, which began about 1935 and is still in progress. Of course, such dates are purely arbitrary: the characteristics of any period can be demonstrated in some publications of earlier or later date.

THE ROMANTIC PERIOD

The Romantic Period was characterized by extreme enthusiasm for Darwinism, together with an uncritical acceptance of whatever data were claimed to support Darwinism. Negative evidence was given little weight (in contrast to Darwin's own practice), while absurd extremes of interpretation in order to make observed facts fit Darwinian theory were quite common. Leaders of this group in England included T. H. Huxley, Herbert Spencer, and George Romanes, while in the United States David Starr Jordan and Asa Gray were the leaders. As a group, they went to interpretive extremes, reading adaptive significance into every organic structure, even on the most imaginative evidence. This was often based upon excellent anatomical and taxonomic evidence, but experiments to test adaptive values were unusual if not unknown. Yet it must not be

FIGURE 27. DAVID STARR JORDAN. (From the Huntington Library.)

thought that these were second-rate biologists who were blinded by the brilliance of a great man: on the contrary, they were excellent men in their respective fields. Huxley made brilliant contributions to the development of invertebrate zoology, taxonomy, and vertebrate anatomy. Spencer was one of the leading philosophers of his time. Romanes began his career as an invertebrate neurologist, but he soon became exclusively engrossed in evolutionary problems. Jordan (Figure 27) was undoubtedly one of the best ichthyologists who has ever lived. And Gray was a botanist of such stature that his work still has great influence. Nor must it be thought that they never ventured to differ from Darwin, for these men were independent thinkers. Yet the atmosphere of approbation was extraordinary.

It has been said that evolution was born in England, but found its home in Germany. The German evolutionists of the Romantic Period were more strictly Darwinian than their English and American colleagues in the sense that they were generally more thorough and careful collectors of detailed data. The leaders in Germany were Carl Gegenbaur, Ernst Haeckel, and August Weismann. Gegenbaur was a comparative anatomist, and undoubtedly one of the greatest and most influential, for his students held most of the chairs of anatomy in European universities throughout the Romantic Period. He and his collaborators made exhaustive studies, in complete detail, upon all classes of vertebrates, and used the data so obtained in support of Darwinian theory. Much of the phylogeny of the vertebrates, as represented in current textbooks of zoology, is taken from the works of Gegenbaur and his collaborators.

Ernst Haeckel did much less experimental work than did Gegenbaur, yet he did do significant work in anatomy, embryology, and taxonomy.

His studies in comparative embryology led him to broaden the principles of von Baer to make the Biogenetic Law, in support of which he published extensively. He based extensive phylogenies upon embryological evidence, interpreted according to the Biogenetic Law. Bateson has said that this "law" dominated all of the zoology of the last half of the nineteenth century. Haeckel's bona fide scientific work was all done in his youth. In later life, he became primarily a controversialist and popularizer, a fact which is said to have earned him the contempt of Gegenbaur.

August Weismann's first interest was heredity, the aspect of Darwinism which Darwin himself had recognized as weakest. Weismann (Figure 28) also seems not to have heard of Mendel and his work. He was hampered by progressive blindness which became complete before his major works were done. Over much of his career, his graduate students made observations and reported them to him in detail. His data were principally some of the facts of cytology, mainly of mitosis. He reasoned that, since the hereditary mechanism must be orderly, and since only the chromosomes were divided in an exact and orderly fashion in mitosis, the chromosomes must be the physical basis of heredity. The facts of meiosis were not yet known, but he predicted reduction divisions because otherwise the chromosomes would be doubled in number from one generation to the next, an unstable situation. Beyond these propositions, which may have been proven since, his hypothesis of heredity was purely speculative, and

FIGURE 28. AUGUST WEIS-MANN. (From Locy, "Biology and Its Makers," 3rd Ed., Henry Holt & Co., Inc., 1935.)

has been disproven. Unlike Darwin, he was loath to admit of factors other than natural selection in the origin of species.

Also prominent during the Romantic Period were Darwin's cousin, Francis Galton, and Karl Pearson. In many respects, these men were much more akin to the Period of Modern Synthesis, for they laid the basis for the new sciences of statistics and biometry which play so prominent a role in modern evolutionary studies.

THE AGNOSTIC PERIOD

Such uncritical enthusiasm can hardly fail to cause a wave of skepticism and disillusionment in its wake, and so the Agnostic Period set in soon after the turn of the century. Many factors converged to cause this. In part, it resulted from the palpably false extremes of interpretation of evidence which were so common during the Romantic Period. A second factor was the rediscovery of Mendel's laws of heredity. Today Mendelism is the foundation of most studies in evolution, but then the permanence of

FIGURE 29. STATUE OF MENDEL AT BRÜNN, BY CHARLEMONT. (Iltis, "Life of Mendel," W. W. Norton & Co., Inc., 1932.)

the gene seemed to raise formidable obstacles to the origin of new species, so that genetics was regarded as a sort of blind alley at the end of which stood the sign: THE GENE, DEAD END.

A third factor was the work of Johannsen on inheritance of size in beans. He found that, in a seed stock of variable inheritance, selection is highly effective in increasing or decreasing the size of the beans. If a genetically pure line is obtained, however, then selection no longer has any effect. To illustrate, from a pure line with an average weight of 49.2 centigrams, he selected beans weighing 20, 40, and 60 centigrams. The average weights of their offspring were 45.9, 49.5, and 48.2 centigrams respectively. Plainly, selection of the parents had not influenced the average weights of the offspring at all. Johannsen concluded that selection could be effective only in a stock with hereditary variability, but that variations produced by the environment (including nutrition, sunlight, temperature, moisture, etc.) were unimportant for evolution.

An additional factor was the mutation theory of DeVries. In his studies of the evening primrose, *Oenothera*, DeVries had discovered sudden changes of considerable magnitude which behaved like Mendelian genes. He called such sudden hereditary changes "mutations," and he believed that some of his mutants were actually new species, produced at a single step. Thus, *Oenothera lamarckiana* occurred suddenly in a form much larger than normal, and DeVries described it as a new species, *O. gigas*. DeVries, incidentally, was one of the codiscoverers of Mendel's work. Evolution was now conceived as a series of mutations occurring in pure lines. Natural selection found small place or none.

Finally, much of the work of the Romantic Period was taxonomic in character, and now taxonomy had fallen into disrepute. "Taxonomist" became a term of reproach, and such men were regarded as merely biological file clerks. Contributing to this was the fact that many taxonomists were Lamarckian in their viewpoints.

Biologists generally still believed that evolution must be a fact, but they were gravely doubtful either that the main causal factors were known, or that the necessary clews for their discovery were at hand. Their viewpoint is well typified by William Bateson, who began his address before the 1921 convention of the American Association for the Advancement of Science with the statement that "I may seem behind the times in asking you to devote an hour to the old topic of evolution." Later in the same address, he said that "Discussions of evolution came to an end primarily because it was obvious that no progress was being made. . . . When students of other sciences ask us what is now currently believed about the origin of species, we have no clear answer to give. Faith has given place to agnosticism . . . we have absolute certainty that new forms of life, new orders and new species have arisen on earth. That is proven by the paleontological record . . . our faith in evolution stands unshaken."

This, then, was the tenor of evolutionary thinking during the Agnostic Period. Bateson's speech was printed in *Science* in January, 1922. During the year that followed, *Science* published only a single challenge to Bateson's position, and that was written by H. F. Osborn, an elderly zoologist who had reached the peak of his career during the Romantic Period. Scientific interest in evolution was, indeed, at a low ebb.

While evolutionary studies were largely suspended during this time, studies in the many branches of biology which contribute to an understanding of evolution were actively pursued. The result was that the stumbling blocks which caused the reaction were gradually removed, thus paving the way for the Period of the Modern Synthesis. The most important developments occurred in genetics. It became apparent that the large mutations with which DeVries worked were rather rare, while much smaller mutations, comparable to the individual fluctuations of which Darwin wrote, were quite frequent. Further, the large mutations were usually less viable than their normal alleles. Taxonomists, meanwhile, had shown that natural species do not differ from one another by single, striking traits, but rather they differ quantitatively in a large number of traits.

Study of wild species in the laboratory showed that pure lines are rare in nature, being found usually only in self-fertilizing plants; hence the pure line concept could no longer have a serious bearing on evolution. Instead, it appeared that wild species are not only quite variable, but that they commonly take up latent variability (heterozygous recessive genes) "like a sponge" (Tschetverikoff, see Chapter 15). Geneticists and taxonomists both undertook the study of variability in wild species by the statistical methods devised by Galton and extended by his successors.

Finally, a new systematics has developed, in which the lower groups are being studied by the methods of population genetics, ecology, physiology, indeed, by every possible approach in addition to the classical purely morphological approach, and all of this with a view to determining the dynamics of the origin of species. On the higher taxonomic levels, an effort is being made to apply to the problems of phylogeny the knowledge which has been gained at the lower levels.

THE PERIOD OF THE MODERN SYNTHESIS

Thus the bases of the agnostic reaction were gradually destroyed, and the Period of the Modern Synthesis naturally followed. The evolutionary studies of this period have been marked by confidence that the processes of evolution are open to study as well as is the fact of evolution. On the lower taxonomic levels genetic, ecological, geographical, and morphological studies have all been brought to bear upon the problems of the origin of hereditary variation and the origin of species. On the higher levels, paleontologists especially have been applying the new knowledge gained on the lower levels to the problems of phylogeny. While certain types of study have been mentioned above as especially important, this is misleading, for the study of evolution at present is truly the "modern synthesis" of all biological disciplines. It would be very difficult to find any important phase of biology which does not make some important contributions to the study of evolution in the current Period of the Modern Synthesis.

A few men may be mentioned as leaders in the present movement in the study of evolution. Th. Dobzhansky deserves first mention, for the publication of his "Genetics and the Origin of Species" might well be taken as the starting point of the period. Dobzhansky started as an entomologist, but has become a leading Drosophila geneticist, and his book is a cornerstone of the neo-Darwinian theory, to which much attention will be given below. The late R. B. Goldschmidt (1878–1958) was closely associated with the rise of genetics since the rediscovery of Mendel's work. He had wide experience in the study of geographic variation, taxonomy, and physiological genetics. He and his students are the principal proponents of the main alternative to the fashionable neo-Darwinian theory. R. A. Fisher, J. B. S. Haldane, and Sewall Wright have been foremost in the statistical analysis of populations. Ernst Mayr, a systematic ornithologist, has been prominent in the application of taxonomy to problems of evolution. E. B. Babcock's study of the genus *Crepis*, carried out over a period of more than thirty years, is one of the most thorough and comprehensive studies ever made upon a single genus of plants, and it is one of the major supports of the neo-Darwinian theory. Edgar Anderson has been important in the study of hybridization of natural species. G. L. Stebbins has made extensive studies of the range grasses of the western United States, with special reference to polyploidy. Clausen, Keck, and Hiesey have published extensive studies on the behavior of plants when grown in widely different habitats.

Thus the modern synthesis is proceeding on many different fronts, and to this most of the subsequent chapters will be devoted.

REFERENCES

AGAR, W. E., *et al.*, 1954. "Fourth (final) Report on a Test of McDougall's Lamarckian Experiments on Learning in Rats," *J. Exp. Biol.*, **31**, 307–321. Experiments conducted for more than twenty years fail to support the Lamarckian hypothesis.

BATESON, WILLIAM, 1922. "Evolutionary Faith and Modern Doubts," *Science*, **55**, 55–61. A brilliant statement of the agnostic reaction to evolution.

CANNON, H. GRAHAM, 1958. "The Evolution of Living Things," Manchester University Press, Manchester. Neo-Lamarckism presented by one of its proponents.

COLE, FAY-COOPER, 1959. "A Witness at the Scopes Trial," *Scientific American*, **200**, 120–130 (January). This trial was the low point in the history of evolutionary thought in the U.S. This is a most interesting account of the trial.

EISELEY, LOREN C., 1959. "Alfred Russel Wallace," *Scientific American*, **200**, 70–84 (February). An interesting account of the life and travels of the codiscoverer.

GILLESPIE, C. C., 1958. "Lamarck and Darwin in the History of Science," *Am. Scientist*, **46**, 388–409. A searching historico-philosophical analysis.

GOLDSCHMIDT, R. B., 1956. "Portraits from Memory. Recollections of a Zoologist," University of Washington Press, Seattle, Wash. In these pages, some of the men discussed in this chapter live once more.

IRVINE, WILLIAM, 1955. "Apes, Angels, and Victorians," McGraw-Hill Book Co., Inc., New York, N.Y. A fascinating insight into the lives and personalities of Darwin and some of his contemporaries, particularly T. H. Huxley.

LOCY, W. A., 1935. "Biology and its Makers," 3rd Ed., Henry Holt & Co., Inc., New York, N.Y. Short biographies of many of the men cited in this chapter.

NORDENSKIOLD, ERIC, 1928. "The History of Biology," Tudor Publishing Co., New York, N.Y. A scholarly treatment of the history of evolutionary thought is included.

CHAPTER SEVEN

The Two Main Problems of Evolution

EVOLUTION WAS DEFINED ABOVE as "descent with modification." Now *hereditary* modifications must therefore be the basic materials for evolution and the manner of origin of hereditary variations must be the first major problem of evolution. Lamarck attempted its solution with the theses that the action of the environment on an organism tends to produce adaptive modifications, and that these acquired characters are inherited. He failed because both of these theses are easily disprovable.

Darwin side-stepped this problem. He simply accepted without explanation the observed fact that organisms do vary one from another. He did not distinguish between inheritable and non-inheritable variations. Darwin's problem was the action of natural selection in the formation of new species, given a variable progenitor. This is the second major problem of evolution: how the varying arrays of organisms become sorted out into species and higher categories. The role of selection and of other factors in this process must be analyzed. The differences between the several subspecies of a species may be quite as great as those between species of a genus, but the former interbreed freely and blend from one to another in nature while the latter generally do not. The origin and nature of the barriers which account for this difference are therefore of great importance for evolution.

The two main problems of evolution, therefore, are:

(1) The origin of variation, and

(2) The origin of species (and of higher categories).

With these problems the subsequent chapters will be largely concerned. But first, it may be profitable to survey the actual course of evolution as shown by the paleontological record and by other data.

PART TWO

Phylogeny

Evolution above the Specific Level

CHAPTER EIGHT

The Origin of Life and the Differentiation of the Plant and Animal Kingdoms

THE EXPRESSION "from ameba to man" is commonly used as though this encompassed the grand extent of evolution. Yet this cannot be true, for there is a vast realm of life on more primitive levels of organization than the ameba. Within the Protozoa, the Flagellata are now universally recognized as being more primitive than the Rhizopoda (the class to which ameba belongs), and probably ancestral to it. Indeed, many of the Flagellata have chlorophyll (as, for example, the well known *Euglena*) and other typical plant characters, and thus form a link between the Plant and Animal Kingdoms. But the Flagellata are already highly complex organisms, hardly a starting point on the scale of life. The Cyanophyta, or blue-green algae, are still more primitive. In these, there is typically no morphological separation of nucleus and cytoplasm, but, rather, the chromatin is distributed throughout the cell. Nonetheless these organisms are still significantly advanced beyond the bacteria, for the blue-green algae do, by virtue of the catalytic properties of chlorophyll, synthesize sugar from carbon dioxide and water in the presence of sunlight. Yet the bacteria themselves can hardly be called simple. The chemical analysis of their protoplasm leads to a result not too different from that obtained by the analysis of the protoplasm of higher plants and animals. And their morphology and colonial characteristics are sufficiently distinct to serve as guides to identification even where the smallest bacteria are concerned. But there are pathogenic agents so small that they will pass the finest filters, and they are invisible with the best light microscopes. Yet they multiply within the protoplasm of an appropriate host, and the products of their metabolism cause the production of disease symptoms in the host organism. These are the viruses. Viruses have been crystallized, and the crystals are nucleoproteins—very simple when compared to typical protoplasm, but very complex when compared to inorganic, or to most organic, molecules. It is debatable whether the viruses are living. But they are the simplest things with respect to which any such debate might be possible, and so they bring us squarely before the problem of the origin of life.

THEORIES OF THE ORIGIN OF LIFE

Spontaneous Generation. Of the many theories of the origin of life, perhaps the oldest is the theory of spontaneous generation, according to which even complicated forms of life might arise spontaneously from non-living matter. Thus Aristotle believed that mosquitos and fleas arose from putrefying matter. Tadpoles, worms, and many other small organisms have been supposed to arise from mud. Flies were supposed to be formed from putrefying flesh. Before the motor age, every child heard that a horsehair, left in water, would transform into a horsehair worm. And meal worms have often been supposed to arise from flour spontaneously. Even such large and complex animals as rats have been supposed to arise spontaneously from nonliving matter. Redi, an Italian physician of the seventeenth century, attacked the theory of spontaneous generation experimentally, and he left it badly damaged. He exposed meat in containers which were covered over with fine mesh cloth. No maggots appeared on the putrefying meat, but flies laid their eggs on the cloth covers, and maggots developed there. It was evident, then, that the maggots which ordinarily appeared in spoiling meat were not spontaneously produced, but were developed from eggs laid in the meat by adult flies. A century later, similar experiments were performed by the Italian priest Spallanzani, who also showed that, if meat were boiled in a sealed container, no organisms developed in it, even if it had been previously infected. This fact was soon applied to the practical problem of food preservation by means of canning.

After the work of Redi and Spallanzani, the theory of spontaneous generation no longer commanded the respect of biologists. But the discovery of bacteria changed this. Here were organisms simpler than any previously imagined. Their occurrence was practically universal, and it was very difficult to exclude them from any medium suitable for their growth. The possibility that they might be produced spontaneously within every sort of organic medium was most suggestive, and it had many adherents. The famous experiments of Pasteur disproved this completely. Boiled broth was kept in a closed container, with air entering by a capillary tube which was bent to form a trap for solid particles. Thus the broth was freely exposed to oxidation, yet no bacteria appeared in it. Hence it was evident that air-borne bacteria ordinarily infected exposed broth (or other suitable media), and that the bacteria themselves arose only from pre-existing bacteria. This dealt the death blow to the theory of spontaneous generation of complex organisms.

The Cosmozoic Theory. A second theory of the origin of life is the cosmozoic theory, that the original spores of life reached the earth accidentally from some other part of the universe. This theory is completely unsatisfactory for two reasons. First, because of the intense cold, extreme dryness, and the intense radiation of interstellar space, the probability that even the most resistant of living spores could withstand exposure to interstellar space is vanishingly small. Second, the theory does not explain the origin of life at all, but merely changes the scene of origin from the earth to some remote and undefined part of the universe.

Viruses and the Origin of Life. The discovery of the viruses has placed the problem of the origin of life in a new light. The higher forms of life cannot be descended from viruses as the latter are now known, for they are all parasitic, and parasites must always be descended from free-living ancestors. But the viruses show a unique combination of characters of living and nonliving systems. They reproduce and metabolize, as undoubted organisms do. That they reproduce is shown by the fact that inoculation of a susceptible organism with a very small amount of virus-containing solution gives rise to a heavy infection. That they metabolize is shown by the production of disease symptoms in the host by the products of metabolism. The chemical nature of viruses is that of the nucleoproteins, and it has been suggested that they are "escaped" genes. Like the genes, viruses are ordinarily reproduced without change, but they can mutate as genes do, that is, they can undergo an inheritable change which does not interfere with their capacity for self-reproduction. Such a mutation is detectable by a change in the disease symptoms produced by the virus or by a change in the degree of toxicity. Unlike undoubted organisms, however, the viruses do not respire. But the most striking property which viruses share with nonliving systems is the fact that they can be crystallized and stored indefinitely without loss of infective powers. This was first demonstrated in 1935 by the chemist W. M. Stanley, who succeeded in crystallizing the virus which causes tobacco mosaic disease. The crystals turned out to be a nucleoprotein (Figure 30). Chemical purity is indicated not only by crystallization but by a sharp sedimentation boundary when a virus suspension is ultracentrifuged. Finally, viruses can be broken down to a protein and a nucleic acid, both of which are inactive, then they can be recombined to form infective virus again. And so the viruses appear to be homogeneous, or nearly so, in contrast to all undoubted organisms.

Thus viruses are on the border line between the living and the nonliving, even though their parasitism cannot be primitive. The possibility cannot be excluded that there may be free-living viruses, for we know the viruses primarily by their effects (disease production), and no one knows what type of effect should be looked for from a free-living virus. The existence of these bodies which are intermediate between the living and the nonliving and which have fairly simple chemical properties suggests the possibility that something like a free-living virus may have been produced by chemical evolution under the influence of the unique conditions which prevailed when the earth was a young planet just cooling down toward a temperature range which could support life. Such a free-living, self-reproducing unit might be regarded as a single gene. Mutation could then lead to formation of gene-aggregates, with differentiation among the members of each aggregate. Such gene-aggregates could be regarded as independently existing chromosomes, and it has been suggested that some of the smallest bacteria represent such a stage in the evolution of life. Fur-

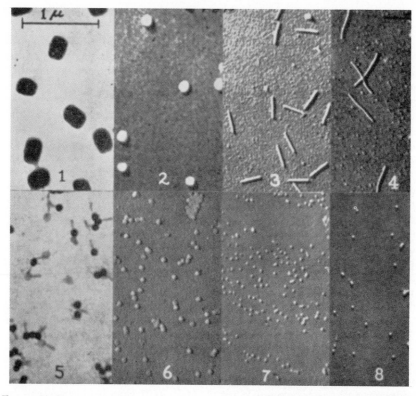

FIGURE 30. ELECTRON MICROGRAPHS OF VIRUSES. 1. Vaccinia virus. 2. Influenza virus. 3. Tobacco mosaic virus. 4. Potato mosaic virus. 5. Bacteriophages. 6. Shope papilloma virus. 7. Southern bean mosaic virus. 8. Tomato bushy stunt virus. (From Villee, "Biology, The Human Approach," 3rd Ed., W. B. Saunders Co., 1957.)

ther mutations might then lead to the accumulation of metabolites around the chromosomes, and the complex so produced could then be regarded as an exposed nucleus. Bacteria in general are sometimes regarded as representing this level of organization. Next, cytoplasm might be acquired but not separated from the nuclear material, as in the blue-green algae, or in the larger bacteria. Finally, mutation could then result in the formation of typical cells with nucleus and cytoplasm separated by a membrane. Within such cells lie all of the potentialities of higher plants and animals. The possible details of such a chemical evolution of the most elementary forms of life have been particularly well worked out by Oparin, and his account will be followed for the most part below. It may be noted at the outset that this is actually a special case of spontaneous generation, though it is a plausible one because it does not involve direct origin of complicated organisms from nonliving matter.

OPARIN AND CHEMICAL EVOLUTION

The earth may have originated as a fragment from the sun. The probable chemical composition of the earth in the earliest geological periods can be surmised from spectrographic analysis of the sun and of stars in early stages of their physical evolution. All of the elements which enter into the composition of protoplasm were probably present as inorganic compounds. Free nitrogen, hydrogen, and oxygen, gases which now form so large a portion of the earth's atmosphere, probably were present at the very beginning, because they are present as free elements in the sun. But they were probably soon lost to outer space, for it is unlikely that the gravitational pull of the earth is great enough to hold such light elements at the high temperatures which prevailed during the earliest ages of the earth's history. So these elements were probably left only in compounds. A large quantity of the hydrogen and oxygen were probably united as water, but this was certainly for long ages present as superheated steam. The hot vapors would rise toward the cold outer layers of the atmosphere, condense and fall as rain, only to be again converted to steam before striking the earth. Gradually the earth cooled sufficiently to permit the rainfall to strike the earth, then to begin the formation of pools and larger bodies of water. Optimum conditions for solubility and reaction thus existed, and the entire earth was a great crucible for random compound formation and re-formation.

Origin of Organic Compounds. In such a situation, where the most important elements of organic compounds—carbon, nitrogen, oxygen, and hydrogen—are reacting at random with each other and with many other elements, forming countless compounds which in turn react in such combinations as may be possible, it is highly probable that, sooner or later, organic compounds or their precursors would appear. It is known that methane (CH_4), the simplest of organic compounds, is present in the atmospheres of some of the cooler stars. And more complex hydrocarbons (compounds of carbon and hydrogen) have been found in meteorites, and so it is certain that they can originate without the intervention of living organisms. These may have been preceded by the compounds of carbon with metals, for such metallic carbides give rise to hydrocarbons when treated with steam. So it is altogether probable that such compounds were formed in abundance while the cooling earth was still much too hot to permit the existence of life. In a similar way, one would expect the formation of ammonia by the reaction of steam and metallic nitrides. Further, cyanogen (CN) and dicyanogen ($NC \cdot CN$) were probably also formed.

Other known pathways for the abiotic origin of organic compounds include ultraviolet irradiation of formic acid, which yields large organic molecules, including amino acids; the action of cosmic rays upon carbon dioxide and water vapor, which yields organic acids; and the action of lightning upon an atmosphere of methane, hydrogen, ammonia, and water vapor, which yields a mixture of organic compounds, including amino acids. All of these reactions have been proven in the laboratory. Using the last-mentioned method, with electrical sparks discharged into an artificial atmosphere in a sealed flask, Miller obtained a mixture of many amino acids and other organic compounds, some in fairly high yields.

Further reactions of the compounds discussed above have great potentialities. The hydrocarbons are not particularly reactive; however, one hydrogen atom in a hydrocarbon molecule could be readily replaced by chlorine or bromine. The new compound would be highly reactive. It might, for example, be hydrolyzed to form an alcohol and an inorganic acid. The alcohol could then be oxidized to form the corresponding aldehyde or ketone. These could, in turn, be further oxidized to form organic acids. The simplest of the aldehydes, formaldehyde, is an intermediate compound in the photosynthetic production of sugar. The reaction to produce glucose ($C_6H_{12}O_6$) from formaldehyde (CH_2O) proceeds extremely slowly in the absence of chlorophyll and sunlight. But it does proceed, however slowly, under the influence of sunlight alone, and, given time on a geologic scale, it is quite conceivable that large quantities of sugar might be accumulated. If a double alcohol, or glycol, were formed, then one alcohol group could be oxidized to form an acid, while the other alcohol group could react with ammonia (NH_3) to give water and an amine. The result would be an amino acid, one of the building blocks of the proteins. The simplest possible amino acid is glycine, $CH_2 \cdot NH_2 \cdot COOH$.

These compounds can enter into reactions which lead to compounds intimately associated with protoplasm. Sugars can be polymerized to form starches, glycogen, and cellulose. The essential precursors of fats are long-chain hydrocarbons in which an end member has been oxidized to form an acid, and glycerine. Glycerine is simply a three-carbon chain in which one hydrogen on each carbon has been replaced by an hydroxyl group ($CH_2OH \cdot CHOH \cdot CH_2OH$). Now if each hydroxyl group reacts with a long-chain organic acid, the resulting compound is a fat molecule. But the most significant possibility is held by the amino acids, for these can react with one another to form aggregates of great molecular weight, the proteins. The acid group of one amino acid reacts with the amino group of another in what amounts to a salt-forming reaction. This is called the *peptide linkage*. As it always leaves an acid radical exposed on one of the reacting amino acids and an amino group on the other, the reaction is subject to indefinite repetition, and thus leads to the great molecular weights of the proteins.

Colloids, Coacervates, and Individuality. By the time the earth had cooled sufficiently to permit the formation of permanent bodies of water, there was probably a large amount of organic material of great variety, and the variety was continually increasing through the natural experiment of reaction between whatever substances could react and happened to be brought together. Many of these large organic compounds would tend to form colloidal solutions in water. Wherever such colloidal particles included electrically active groups, as all proteins do in abundance, water molecules would tend to become oriented around the surface of such a particle. If colloidal droplets of opposite electrical charges should be mixed, then they would be mutually precipitated to form droplets of a complex mixture called a coacervate. Such coacervates would adsorb water on their surfaces to form a sort of membrane, which would thus establish the beginnings of individuality.

Autocatalytic Systems, Genes, and Viruses. These large, complex, colloidal aggregates, the coacervates, continue to undergo random chemical reactions in the course of which still larger aggregates may be formed, or the existing ones may be broken down. It is most probable that some of the systems thus formed would be enzymatic in character, that is, they would tend to increase, to catalyze, the rates of specific reactions. Many enzymes catalyze the reactions of substances unrelated to the enzyme itself. Thus lipase will break down fats to glycerine and long-chain acids, or it will catalyze the synthesis of fats under appropriate conditions. But there are catalysts known which, given a suitable substrate, tend to cause the production of more of the catalyst itself. The duplication of the chromosomes is an example *par excellence*. Such enzymes are said to be autocatalytic. If an autocatalytic substance were to be formed in the primordial seas, it would be expected to increase in quantity at the expense of those organic compounds which lack this property. And thus is the characteristic of self-reproduction introduced. Compounds possessing this property might well be regarded as free genes. We need only add the characteristic of mutability, that is, the capacity for undergoing changes which are reproduced without interfering with the autocatalytic properties of the molecule, and we have the most fundamental characteristics of the hereditary units, the genes. Such a structure would be comparable to a free-living virus, and it is probable that, like the viruses of today, these primitive anlagen of life were nucleoproteins. These nucleoproteins would utilize the complex organic compounds of their environment for the autocatalytic synthesis of more nucleoprotein, that is, for reproduction. Calvin has shown that even in the pre-organismic stages, a sort of natural selection, based upon thermodynamic principles, favors the events described.

Origin of Bacteria. As larger self-reproducing units were formed, mutation might result in the different parts of each aggregate becoming differentiated. Or the same result might be obtained by the coalescence of originally different "genes." At this stage, the structure would resemble some of the smallest known bacteria, and might properly be called an organism. The organisms dealt with up to this point would have to be completely heterotrophic, that is, dependent upon complex food materials present in their environment, and "feeding" would occur only by absorption. The fact that viruses can metabolize and reproduce only within the protoplasm of higher organisms indicates that they are also heterotrophic. They can utilize proteins, starches, fats, and vitamins already present, but they cannot synthesize these from amino acids, sugars, and other simpler organic precursors. On the other hand, all of the undoubted organisms are able to synthesize at least some of their required foods from simpler compounds. Green plants can synthesize their required foods from the elements. Animals cannot do this, but they can synthesize proteins from amino acids, complex carbohydrates from sugars, and some of the vitamins from simpler precursors. These syntheses are possible because the organisms possess enzymes which are specific for the necessary reactions. But these reactions are frequently quite complex, requiring a long series of steps with a different enzyme controlling each step. Thus it has been shown that the synthesis of arginine, an amino acid, by the mold *Neurospora* requires at least seven different genically controlled enzymes. Each of these is useless without all of the others. That such "useless" genes should be preserved by natural selection is incomprehensible. But the probability that all of them should appear simultaneously, and so become useful, is equally incomprehensible.

Horowitz has devised a very clever solution to this dilemma, namely that the enzymes might have been originally acquired in a sequence opposite to that in which they are used by existing organisms. Thus a primitive organism requires a substance *A*, which is abundantly present in its environment, as are also substances *B* and *C*, from which *A* could be synthesized in the presence of an appropriate enzyme. As long as *A* remains abundantly present in the environment, presence or absence of the enzyme can have no bearing on survival. But eventually the food requirements of the growing population of primitive organisms may be expected to outstrip the stock pile of organic compounds which was built up during the abiotic ages. Thus when *A* becomes rare in the environment, those organisms which have the enzyme for the production of *A* from *B* and *C* will have a selective advantage, and will replace the original type. Such *preadaptation* is a common phenomenon in evolution. The enzyme has now become a permanent part of the organism's biological equipment. But *B* might also be synthesizable from *D* and *E* in the presence of the necessary enzyme. Then when *B* becomes scarce, the possession of this enzyme, acquired by mutation, will also have selective advantage. There is no theoretical reason why such a process should not continue until an organism had acquired the ability to synthesize all of its requirements from the elements. This has happened in the case of the green plants. Such an organism is said to be autotrophic.

Diversification of the Bacteria. It seems probable, then, that the primitive free-living viruses and bacteria gradually used up the available supply of proteins and other complex compounds in the environment, and that they simultaneously, step by step, developed the systems of enzymes necessary for the biosynthesis of the same compounds. Or it may be that only the primitive bacteria were sufficiently complex to permit this development, the viruses remaining completely dependent upon preformed compounds. If so, this would account for their obligate parasitism, for only within the protoplasm of undoubted organisms could an environment be found to contain all necessary food substances in the completely elaborated form. The evolution of the bacteria made parasitism possible; the impoverishment of the earth's supply of complex organic compounds made it desirable for such organisms as could not synthesize their food requirements at least in part.

Respiratory Systems and Photosynthesis. The most primitive organisms, living in an environment which included an abundance of the most complex substances which the organisms might require, must have had a metabolism which was mainly catabolic; that is, it consisted of the breaking down of complex compounds to release the energy stored in them. But, as the environmental supply of high-energy compounds was reduced, the development of a new type of metabolism became necessary. This is anabolic metabolism, whereby complex compounds are built up from simpler ones, and energy thereby stored. Expenditure of energy by the organism is required, and so the evolving bacteria explored possible sources of energy, that is, systems of respiration. Several types of respiratory reactions, and correlated nutritive systems, have been exploited by bacteria. These may be classified as heterotrophic and autotrophic systems. The heterotrophic bacteria are generally parasitic, deriving their energy by oxidation of the carbohydrates or other organic compounds of their hosts. Or they may be *saprophytic,* that is, they live by absorption of dissolved organic matter from their environment, much like the primitive ones in the scheme of Oparin. On the other hand, the autotrophic bacteria derive energy from chemical reactions involving simple inorganic compounds. They are thus independent of external sources of high-energy compounds. There are three principal groups of such *chemotrophic* bacteria, the sulfur bacteria, the nitrifying bacteria, and the iron bacteria. Within each group, there are many different species, and several different respiratory reactions. A few examples may be given:

$$2 H_2S + O_2 \longrightarrow 2 H_2O + 2 S + 81,600 \text{ calories per mole}$$
$$S + 2 H_2O + 3 O_2 \longrightarrow 2 H_2SO_4 + 214,000 \text{ calories}$$
$$Na_2S_2O_3 + 2 O_2 \longrightarrow Na_2SO_4 + S + 75,000 \text{ calories}$$
$$2 NO_2 + O_2 \longrightarrow 2 NO_3 + 46,000 \text{ calories}$$
$$2 NH_3 + 3 O_2 \longrightarrow 2 NO_2 + 2 H_2O + 156,000 \text{ calories}$$

Iron bacteria oxidize ferrous compounds to ferric compounds with a much lower energy yield than is indicated for the above reactions. But the best of these is vastly inferior to photosynthesis, the autotrophic mechanism of all green plants, and perhaps the real basis for the evolution of the higher plants and animals, for this is the mechanism which has made it possible for the world of life to tap the great reservoir of radiant energy from the sun. The photosynthetic reaction may be summarized as follows:

$$6 CO_2 + 6 H_2O + 677,000 \text{ calories} \longrightarrow C_6H_{12}O_6 + 6 O_2$$

This large energy reserve is subsequently released by the sugar metabolism of the organism. Photosynthesis, then, for which the chlorophyll of green plants is the catalyst, not only vastly increases the energy potentially available to organisms, but it also releases oxygen from its compounds, thus making possible the oxygen respiratory systems of animals and of some bacteria. Photosynthesis is not generally known among the bacteria, but some of the sulfur bacteria contain a green pigment which is capable of absorbing sunlight, and this makes some contribution to the metabolism of these bacteria.

Oparin was of the opinion that this chemical evolution of life could have occurred only once, because it would require a sterile environment.

That is, it could not occur in a world already inhabited by organisms ready to seize upon any beginnings of organic compounds to utilize them as food. But, as Plunkett pointed out ten years earlier, this is an assumption for the proof of which no data are available, and it is entirely possible that life is in process of origin on earth at the present time. On the other hand, Plunkett also acknowledged that there is no evidence that this is true.

EVOLUTION OF THE BACTERIAL CELL

The problems of the development of the nucleus, mitotic division (cell reproduction), and the evolution of sex are by no means clear. It has already been mentioned that the most primitive bacteria may be more or less equivalent to a single chromosome. Further evolution certainly involved the accumulation of a quantity of accessory material. Some authorities have described the bacteria as cells in which cytoplasmic and nuclear materials are intermingled, while others regard the bacterial cell as an isolated nucleus without cytoplasm. Perhaps the observable facts can be adequately described in either way until much more critical data are obtained. In either case, the evolution of the bacteria must have involved an increase in the quantity of the genic material, and increasing differentiation within it. Study of the mechanics of cell division among the bacteria is hampered by the small size of most bacterial cells. It has long been generally believed that simple fission, without qualitative division of the genes, is the rule among bacteria. This requires a genetic system in which rather few equivalent genes are scattered throughout the cell, so that division need not result in genetically unlike offspring. Thus occasional mutation and the immediate action of natural selection on the mutants would be the only means of evolution.

Reports of indications of mitotic divisions, that is, qualitatively equal divisions of chromosomes, have been published for many bacteria. Some of the published photographs are quite convincing, and the possibility is open that mitotic division, so fundamental for all of the higher plants and animals, may have been first evolved among the bacteria. Genetic evidence also indicates serial arrangement of genes on bacterial chromosomes. Nothing even suggestive of sexual reproduction, the formation and fusion of gametes, has ever been observed among the bacteria, but Lederburg has obtained genetic evidence that it does occur, because recombinations of characters occur when different mutant strains of the same species are "crossed." And so it may be that even sexual reproduction was first developed among the bacteria. If so, this is certainly among the most important events in the whole story of evolution, for sexual reproduction makes possible the accumulation of variability in the heterozygous state (that is, the two genes of a pair are different, with the possibility that one may mask the effect of the other). Also, sexual reproduction makes possible the reshuffling of the genes already present and the testing of different combinations by natural selection. The result is a much greater probability of the achievement of good adaptation, particularly under changing conditions, than could otherwise be obtained.

Bacteria and Phylogeny. The phylogenetic relationships of the bacteria are by no means clear, primarily because of the great difficulties of taxonomic studies on the bacteria. Figure 31 illustrates typical bacterial morphology. Two major alternative theories of the phylogenetic sequence within the bacteria have been presented. The first of these maintains that the autotrophic bacteria are the more primitive ones, and that these have given rise, through a series of intermediates, to the heterotrophic bacteria. The iron bacteria have sometimes been regarded as the most primitive of all, perhaps because of their low energy output. Adherents of this theory believe that the most primitive bacteria must have been autotrophic because there could have been no source of organic compounds for a heterotrophic organism. This theory was generally held before the publication of Oparin's work. But, as this demonstrated the plausibility of the thesis that organic material might have been accumulated by a slow process of chemical evolution long before the advent of life, the theory that the heterotrophic bacteria may be more primitive, with the autotrophs derived from them, gained prestige. According to this idea, the several chemosynthetic mechanisms represent a natural trial-and-error experiment, resulting in a superior new energy source which culminated in the development of photosynthesis.

The knowledge of the relationship of the bacteria to higher organisms is also unsatisfactory, but some indications of possible relationships are at hand. As has been mentioned above, some of the sulfur bacteria contain a green pigment, *bacteriochlorophyll* or *bacteriochlorin*. This pigment is

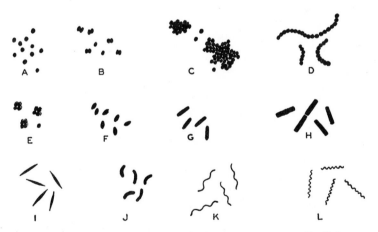

FIGURE 31. MORPHOLOGIC TYPES OF BACTERIA. A, micrococcus; B, diplococcus; C, staphylococcus; D, streptococcus; E, sarcina; F, coccobacilli; G and H, commonly occurring rod forms; I, fusiform bacilli; J, curved rods or vibrios; K and L, spiral forms. (From Jordan and Burrows, "Textbook of Bacteriology," 16th Ed., W. B. Saunders Co., 1954.)

chemically quite similar to chlorophyll, and it has been shown by van Niel to have photosynthetic activity. Hence there is a reasonable probability that the blue-green algae, the most primitive of the green plants, have been derived from the sulfur bacteria. Some other possibilities have also been suggested. The spirochaetes have a type of motility which suggests that of some of the Protozoa, but they do not have a well-defined nucleus. They have been variously classified with the bacteria and with the Protozoa. Although their general biology appears to tie them in more closely with the bacteria, it is possible that they are in fact a transitional group. Finally, the slime bacteria show some similarities to the slime molds, and it has been suggested that this is based upon phylogenetic relationship. However, the cells of slime bacteria live in a secreted nonliving slime, while slime molds are multinucleate masses of protoplasm. Any or none of these hypotheses may be correct. There is no direct evidence available, and perhaps there never will be, for these events must have preceded by long ages the formation of the oldest rocks which bear fossils of any value for study. Further, the lowest plants, the algae and fungi, include many groups of quite diverse nature, their primitiveness being their principal common character. While some of these may have had a common origin from the bacteria, it is possible that each of these phyla of primitive plants may have arisen independently from a different bacterial ancestor.

SUBKINGDOM THALLOPHYTA

The plant phyla which formerly were grouped together as a single phylum Thallophyta are those which are most closely related to the bacteria. The Thallophyta are now treated as a subkingdom, including all plants which reproduce without the formation of an embryo within the ovary of the maternal plant. This subkingdom includes ten phyla ranging in complexity from the bacteria to algae nearly as complex as the simpler vascular plants. Of these, three phyla, the Schizomycophyta or bacteria; the Myxomycophyta or slime molds; and the Eumycophyta or true molds and yeasts comprise the old group Fungi. The relationships of these groups are not at all clear, but the latter two appear to be terminal groups, that is, they have not given rise to others. As was already indicated, the slime molds may have been derived from the slime bacteria. But the slime molds move by ameboid movement. They reproduce by means of flagellate, ameboid swarm cells. As these swarm spores fuse in pairs, a type of sexual reproduction is represented. Many mycologists feel that the Myxomycophyta are more closely similar to the Protozoa than to any other group, and they have even been classified as Protozoa under the name Mycetozoa. Those who support this viewpoint regard them as at least transitional between the two kingdoms. The Eumycophyta are similarly obscure as to their origin. In structure and function, they show many parallels to the green algae, and so they have been thought to be derived from this group. But their zoospores resemble those of the slime molds and show ameboid movement. Hence it is possible that the Myxomycophyta have given rise to the Eumycophyta. These flagellated zoospores closely resemble some

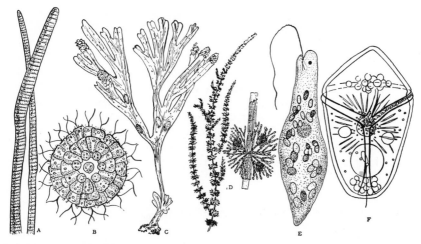

FIGURE 32. REPRESENTATIVE ALGAE OF SEVERAL PHYLA. *A, Oscillatoria,* a blue-green alga; *B, Synura,* a golden-brown alga; *C, Fucus,* a brown alga; *D, Batrachospermum,* a red alga, showing growth habit and details of one whorl; *E, Euglena; F, Amphidinium,* a dinoflagellate. (*A, C,* and *D* redrawn from Weatherwax; *B* and *D* redrawn from Fuller and Tippo; and *F* redrawn from Kofoid and Swezy.)

Protozoa, and the viewpoint that the Eumycophyta may be of Protozoan origin is currently gaining much support.

The remainder of the old phylum Thallophyta comprises seven phyla, the various types of algae (Figure 32), ranging in complexity from the extremely simple, unicellular blue-green algae to the green algae of large size and complexity only a little less than that of the vascular plants. Most of these phyla are of uncertain origin and have given rise to no further groups, hence they need not be discussed here in detail, even though some of them have attained a considerable degree of specialization. These are the phylum Chrysophyta, including yellow-green algae, golden-brown algae, and diatoms; the phylum Pyrrophyta, including cryptomonads and dinoflagellates; the phylum Phaeophyta or brown algae which Dillon and Hutner believe to be of especial importance for the origin of animals; and the phylum Rhodophyta or red algae. Three more phyla of algae (this term means seaweeds, and is based upon the larger members of this diverse group) are of more especial interest for the present discussion: these are the Cyanophyta or blue-green algae; the Euglenophyta, including *Euglena* and its allies; and the Chlorophyta or green algae.

The Cyanophyta. The Cyanophyta or blue-green algae are undoubtedly the most primitive of all green plants. The individual plant is always unicellular, but these may form colonies of moderate size. There is, however, no differentiation within such colonies. Cytoplasm is much more abundant than in the bacterial cell, but there is still no morphological separation of nucleus and cytoplasm. No evidence of sexual processes has ever been observed in the blue-green algae, reproduction apparently occurring exclusively by simple fission. In all other green plants, the chlorophyll is confined to cell organelles, the plastids, but that of the blue-green algae is diffusely spread throughout the outer part of the cell. Their bluish color is caused by a blue pigment, phycocyanin, and a red pigment chemically similar to that of the red algae may also be present. Unlike most other algae, they lack flagella, and are in fact generally non-motile. The oldest known fossils, from rocks on the order of a billion years old, are remains of blue-green algae, and their descendants of today are regarded as little-changed survivors from that remote time. Whether other groups of algae have been derived from blue-green ancestors is unknown, but the fact that the blue-green algae share with more advanced groups so fundamental a trait as the possession of chlorophyll at least indicates that this is a possibility.

THE EUGLENOPHYTA AND THE SEPARATION OF THE KINGDOMS

The Euglenophyta, typified by the common *Euglena* of elementary biology laboratories, show many advances over the blue-green algae. They have a definitely organized nucleus which is separated from the cytoplasm by a membrane. The chlorophyll is no longer free in the cytoplasm, but rather it is concentrated in numerous ovoid bodies, the chloroplasts. Further, the color is grass-green rather than blue-green. Unlike the green algae, the Euglenophyta are not provided with a cellulose cell wall. The cells are provided with one or two flagella, and they are active swimmers. There is a gullet at the anterior end, yet it appears that *Euglena* is autotrophic. Near the gullet there is a red pigmented eyespot, which seems to be sensitive to light. Reproduction is always by simple mitotic division, although sexual reproduction has been reported for one genus (*Scytomonas*).

The Euglenophyta are also described in zoological works as the order Euglenoidina of the class Flagellata and the phylum Protozoa. This class includes representatives of most of the unicellular algal phyla as well as some indisputably animal flagellates. The group as a whole shows a curious mixture of plant and animal characteristics.

But before taking up this subject, it may be well to consider what are the typical differences between plants and animals. Generally speaking, the mode of life of animals is aggressive, while that of plants is passive. Animals are heterotrophic, eating other organisms to obtain the complex organic compounds which they require as foods. Plants, on the other hand, are generally autotrophic, being able to synthesize all of their food requirements from the elements. But there are exceptions in both kingdoms. Many of the higher plants, such as the sundew (*Drosera*), have developed mechanisms for the capture and digestion of insects, and many animals are saprozoic, that is they absorb decaying organic matter from their environment. In plants, differentiation of organs is predominantly external, while in animals it is predominantly internal. In plants, growing tissue, the meristem, is present at all stages of the life cycle, whereas most animals have a definitely limited growth. Finally, plants are generally sedentary, and the individual cells are surrounded by a rigid cellulose wall, while such rigid walls are generally lacking on animal cells, and the animal moves about freely in its environment.

As measured against such a group of criteria, *Euglena* and its allies are difficult to place. *Euglena* itself is supplied with an abundance of chlorophyll, yet it can be raised on completely inorganic media only with difficulty. Traces of amino acids or peptones facilitate culture. Nonetheless, there is no evidence that *Euglena* ever ingests other organisms; it seems more probable that its normal nutrition is predominantly holophytic (by photosynthesis) with a supplement being obtained saprozoically. Some other flagellates, however, are completely holophytic. Others lack chlorophyll, and these of course cannot be autotrophic. Some of them are entirely saprozoic, but some ingest other organisms in typical protozoan fashion and so may be said to be holozoic. Thus the whole range of nutritional possibilities occurs within a single group, and opposite extremes may occur within a single genus. Stored food in plants is ordinarily starch, while in animals it is either glycogen (similar to starch) or fat. In the euglenoids, it is paramylum, a carbohydrate different from both starch and glycogen. The embryological criteria of external or internal organ formation and presence or absence of a continuously growing meristem obviously have no applicability to unicellular organisms. As might be expected of an animal, *Euglena* moves freely in its environment, but its near relatives include sedentary forms. Finally, while Euglenoids lack a cellulose cell wall, they do have a pellicle which in some species is quite rigid.

One result of this mixture of plant and animal characteristics among the unicellular, flagellate organisms is a much confused area in taxonomy. Some biologists have treated the whole array as plants, a procedure which makes it necessary to treat as plants such organisms as the trypanosomes, blood parasites of vertebrates which do not show any plant-like characteristics. Others have tried to designate some forms as plants and others as animals on the basis of the above criteria or similar ones. This involves one in the absurd situation of assigning different members of the same genus to different kingdoms in some instances. Often the whole series of algae and Protozoa have been lumped together as a single kingdom Protista, with only the Metazoa left in the Animal Kingdom and only the vascular plants and the bryophytes left in the Plant Kingdom. This is again unsatisfactory because the higher algae are obviously much more closely related to the vascular plants than to many of the Protozoa, and conversely the animal nature of many of the Protozoa, such as the ciliates, is not open to doubt.

Copeland has made a very radical suggestion, but with much good reason. He suggests that the living world be divided into four kingdoms. Kingdom I, the Mychota, would include the bacteria, and the blue-green algae, that is, all organisms in which the nuclear-cytoplasmic differentiation is not complete. Kingdom II, the Protoctista, would include most of the algal phyla, the fungi, and Protozoa. Thus the kingdom Protoctista includes all of those primitive organisms from which higher plants and animals may have arisen, as well as their relatives which have given rise to no further groups. The kingdom Plantae is thus restricted to green algae, vascular plants, and bryophytes, while the kingdom Animalia is similarly restricted to the Metazoa. While there is much to justify this idea, it presents some serious difficulties. First, it is subject to the same criticism as is the Protista, that the extreme members of the Protoctista are just as clear-cut plants and animals as are the vascular plants and Metazoa, respectively. This system also substitutes three areas of confusion for one in the conventional system. For it would, under Copeland's system, be necessary to decide whether an organism were a mychotan or a protoctistan; a protoctistan or a plant; or a protoctistan or an animal.

It seems altogether probable that this confusion exists because the organisms concerned are closely related to the actual common ancestors of the higher plants and animals, ancestors which need not be thought of as either plant or animal, though they are usually assigned to the Plant Kingdom. The living algae and Protozoa then represent various stages in the evolution of the characteristic differences between the kingdoms, together with many specific adaptations to the unicellular (or better, acellular) grade of construction. There is, theoretically, no reason why there should be a sharp line of separation between those developing along plant-like lines and those developing along animal-like lines. Indeed, this would be in conflict with the idea of origin by evolution. This mixture of plant and animal characters among microorganisms has led Dillon to recommend that the entire world of life be included in one kingdom, Plantae. Perhaps the most important feature of *Euglena* and its allies is this intermediacy between the kingdoms, suggesting as it does the probability that the euglenoids may be primitive organisms, fairly close to the stem group from which both plants and animals have come.

THE CHLOROPHYTA

The Chlorophyta or green algae are an extraordinarily varied group, the simplest members of which are unicellular, but there is a definite separation of nucleus and cytoplasm, and the chlorophyll is contained in a single plastid. They may show considerable specialization of different cells in the multicellular species, and the higher green algae may attain large size. While the more primitive species reproduce by simple fission, sexual reproduction and alternation of generations are well developed in the phylum. The green algae appear to be on or near the main line of evolution leading to the higher plants, and hence the great interest which attaches to this phylum.

Chlamydomonas and the Origin of Sex. Sex probably originated in an unknown green alga, which may have resembled the living *Chlamydomonas* (Figure 33). Each plant consists of a single cell. It has a well-defined nucleus and a single large chloroplast. It swims by means of two flagella which are located at the anterior end of the cell. The cell is protected by a heavy cellulose wall. The plant may reproduce by simply dividing to form two, four, or eight *zoospores* (so called because they are

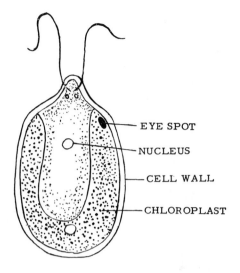

FIGURE 33. *Chlamydomonas.* (Redrawn from Weatherwax, "Plant Biology," 2nd Ed., W. B. Saunders Co., 1947.)

EYE SPOT

NUCLEUS

CELL WALL

CHLOROPLAST

actively swimming cells, like an animal). These zoospores are then released by the dissolution of the cell wall, and each swims away, an independent plant, like the parent in every respect except size. This difference is soon bridged by growth. However, sexual reproduction may also occur, for the parent plant may divide to form eight, sixteen, or thirty-two gametes, cells which resemble the zoospores and the adults, except that they are much smaller. Like the zoospores, these gametes are released into the water, where those from different parent cells unite in pairs to form zygotes. The zygote forms a thick wall about itself and remains quiescent for a time. It is in this highly resistant encysted condition that the plant survives unfavorable conditions such as the drying of ponds. In time, the zygote again becomes active. It then undergoes two divisions, the meiotic or reduction divisions, with the production of four zoospores, which are released to form adult algae.

Chlamydomonas appears to be very close to the origin of sex, and may afford some insight into that problem. It may be noted that most of the life cycle is passed with only the haploid number of chromosomes, for the reduction divisions proceed immediately when the zygote becomes active. Haploidy, that is single representation of each type of chromosome or genetic factor, was undoubtedly the normal situation for all organisms before the origin of sex, and it is still the normal thing for organisms which do not reproduce sexually. Diploidy is a necessary consequence of sexual reproduction, for the union of two gametes can have no other result. Gametes could not be reduced below the haploid condition without qualitative loss of genetic material. It appears that in these organisms, so close to exclusively haploid ancestors, meiosis serves primarily to restore the ordinary, physiological, haploid chromosome number; while diploidy is introduced as a temporary concomitant of a "new" method of reproduction. Why sexual reproduction should ever have been developed under such circumstances is not clear. A "hunger theory of sex" has been proposed, according to which the gametes are simply undersized spores, individually lacking the food and energy necessary to complete development. They therefore pool their resources by means of two-by-two fusions, and thus each zygote obtains a sufficient supply of the materials necessary for development. This theory would be more satisfying were it not for the fact that the four zoospores produced by the zygote are also very small cells, yet they complete their development satisfactorily without further fusions. Whatever the original stimulus to sexual reproduction may have been, the great selective advantages which have made it so nearly a universal property of all higher plants and animals are clear. Sexual reproduction causes a relatively rapid reshuffling of the various possible recombinations of characters within a species, with the possibility that the most favorable combinations may be formed and tested by natural selection. Further, diploidy, which results from sexual reproduction, makes it possible to accumulate a store of genetic variability in the heterozygous state. Diploidy may also be physiologically advantageous because of the increased production of nuclear enzymes, or in other less obvious ways.

The degree of sexual differentiation varies greatly in different species of *Chlamydomonas*. The adults are morphologically identical, and cannot be sexed by inspection. In some isogamous species, the gametes are also morphologically identical. Yet they are physiologically differentiated, for fertilization is possible between some pairs of clones (asexually produced descendants of a single cell) but not between others. The male-female alternative of higher plants and animals is not present here, for as many as eight mating types (sexes?) have been identified in some species of *Chlamydomonas*. Moevus believes that all of these are based upon varying concentrations of the isomeres of a single substance, namely the *cis*- and *trans*- forms of dimethyl crocetin (a fairly simple hydrocarbon, which is, curiously, based upon isoprene, the unit of structure of synthetic rubber). In other species, both macrogametes (egg-like) and microgametes (sperm-like) are produced, both of these being flagellate and actively motile. In some of these species, all possible types of fertilization (large—large, large—small, and small—small) occur, while in others, one macrogamete and one microgamete always form a pair. In such cases, one speaks of heterogamy. Finally, there are species in which the macrogamete is nonflagellate and must be sought by the flagellate microgametes, thus approximating the condition of egg and sperm in the higher animals, or ovule and pollen in the higher plants (oogamy).

Closely related to *Chlamydomonas* is the family Volvocidae, known to all students of elementary biology as the standard example of an evolutionary trend toward increasing complexity of colonies. The simplest members of this family (*Gonium*) consist of four to sixteen or thirty-two identical cells, any of which can form gametes. But these are advanced beyond *Chlamydomonas* in that the gametes are always morphologically different from the vegetative cells. The most specialized member of the family, *Volvox*, comprises as many as 40,000 cells arranged as a hollow sphere (Figure 34). Most of these cells are purely vegetative and sub-

FIGURE 34. *Volvox.*

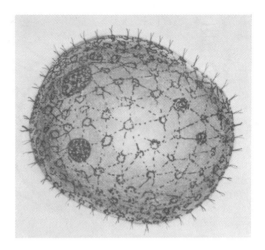

stantially identical. The sex cells are localized in antheridia (sperm) and oogonia (eggs). The eggs of *Volvox* are always large, nonmotile cells, and the sperm are always actively swimming flagellate cells which seek out the egg. As both of these are produced by a single colony, the colony may be regarded as hermaphroditic, a condition which is very much more common in plants and in lower organisms generally than it is among higher animals. It may be, therefore, that hermaphroditism is primitive.

Multicellular Individuals. In another line of descent, which may be typified by *Ulothrix, Draparnaldia,* and *Ulva,* the green algae have developed not merely colonies of substantially independent organisms, but multicellular individuals, the various cells of which are interdependent (Figure 35). These are the filamentous algae, the typical "seaweeds" of laymen, and it is from this group of algae that the bryophytes and vascular plants appear to have developed. *Ulothrix* has the form of a simple, unbranched, multicellular filament. The basal cell is specialized as a holdfast to anchor the plant to a rock or other substrate. Cell divisions occur, but, whereas this results in asexual reproduction in the unicellular algae, in *Ulothrix* and its allies mitosis results in growth without reproduction. Reproduction may occur either sexually or asexually. Asexual reproduction occurs by fragmentation of the plants with subsequent regeneration by each fragment, or it occurs by the formation of zoospores from the vegetative cells of the plant. These zoospores are not unlike those of *Chlamydomonas,* except that they have four flagella instead of two. Each zoospore develops into a plant like the parent plant. But any vegetative cell of the plant may also give rise to gametes. These are smaller than the zoospores, and they have only two flagella. All of the gametes are identical morphologically, but physiologically there must be a sexual differentiation, for zygotes are always formed by the union of two gametes from different plants. After a quiescent period, the zygote undergoes the two meiotic divisions, forming four zoospores which then develop into the

107

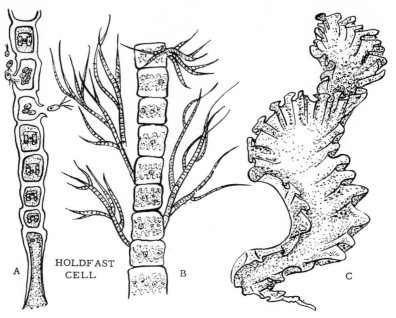

HOLDFAST
CELL

A

B

C

FIGURE 35. *Ulothrix* (A), *Draparnaldia* (B), and *Ulva* (C). (Redrawn from Fuller and Tippo, "College Botany," 2nd Ed., Henry Holt & Co., 1954.)

adult plants. Thus most of the life cycle of *Ulothrix* is passed with only the haploid chromosome complement, while the zygote alone is diploid.

Draparnaldia is advanced over *Ulothrix* principally by the more complex development of the vegetative body. While the latter is composed only of simple, unbranched filaments, the former has a major, basal filament from which many branches arise, and from these there are secondary branches. Reproduction again is accomplished either by zoospores or by isogametes.

Ulva, the common sea lettuce, is advanced over the previously discussed algae in several ways. The vegetative body forms large, leaf-like sheets which, for the first time, are more than one cell thick. Reproduction is again either by zoospores or by isogametes. But there is a significant development in the life cycle. In all of the algae discussed above, the reduction divisions occur in the zygote, with the result that in the whole life cycle only the zygote is diploid. In *Ulva*, however, the divisions of the zygote are ordinary mitotic divisions, with the result that a diploid plant is formed. Some of the cells of the adult plant then undergo the meiotic divisions, which result in the formation of haploid zoospores. This diploid plant which reproduces by the formation of haploid zoospores is called a *sporophyte*. The spores then develop into haploid plants much as do the spores of other algae. These haploid plants then reproduce by isogamy. This haploid, gamete-forming plant is called a *gametophyte*. The sporophytic and gametophytic generations of *Ulva* are morphologically indistinguishable. This alternation of a diploid, sporophytic generation with a haploid, gametophytic generation is one of the most fundamental features of plant biology. Alternation of generations is first introduced with the origin of sex, and it is present in all sexually reproducing algae; but in the lower algae, the sporophytic generation is represented only by the zygote. In *Ulva*, and in many other green algae, however, the sporophyte is as highly developed as is the gametophyte.

Thus within the algae at large, and more especially within the green algae, advances have been made which approach the condition of the simpler vascular plants. The nucleus has been delimited from the cytoplasm by a membrane, and the mitotic mechanism has been perfected. The grass-green chlorophyll is no longer masked by the blue phycocyanin, and it is contained in chloroplasts like that of higher plants, not dissolved in the cytoplasm as it is in the blue-green algae. The most primitive algae are unicellular, but colonies of increasing complexity have been formed, leading finally to true multicellular individuals. The simplest of these are unbranched filaments, but these have given rise to branched and re-branched plants, and finally to large, fleshy plants with considerable differentiation of tissues. These may include root-like, stem-like, and leaf-like structures. Cellulose walls are present in some of the algae. Also, a great range of reproductive mechanisms has been developed among the algae. The most primitive algae reproduce only asexually, either by simple fission, or by the formation of clusters of zoospores. Isogamy, the most primitive form of sexual reproduction, probably began with the pairwise fusion of undersized zoospores. Later, these gametes became differentiated into small microgametes and large macrogametes, both of which were motile (heterogamy). Finally, the macrogamete became a very large, nonmotile cell which was sought by the microgamete. Meanwhile, the alternation of generations was developed, with the diploid generation, at first a minor incident in the life cycle, becoming increasingly prominent.

The most advanced characters of the algae are all carried over into the vascular plants. There is no single alga in which all of these characters are present, yet it is clear that the trend of development in the green algae especially is toward the type of organization characterizing the vascular plants, and it is highly probable that the most primitive vascular plants were derived from green algae. But the evolution of the vascular plants and the bryophytes will be taken up in the next chapter.

The events discussed in this chapter are undoubtedly among the most important in the whole history of evolution. In point of time, they must have occupied much the greater portion of the entire history of life. Yet all of these events must have occurred long before the earliest known useful fossils were formed. Thus it is altogether probable that decisive fossil evidence on the problems discussed in this chapter will never be obtained, and that these subjects must always remain speculative, even though some inferences may be made with a fair degree of probability on the basis of primitive or archaic organisms now living.

REFERENCES

(NOTE: Names in parentheses indicate men whose work is more extensively covered in the references than in this book.)

BALDWIN, E. J., 1957. "Dynamic Aspects of Biochemistry," 3rd Ed., Cambridge University Press. A clear and penetrating analysis of some of the problems discussed in this chapter.

BLUM, H. F., 1951. "Time's Arrow and Evolution," Princeton University Press. Thermodynamic considerations relevant to this chapter. (Horowitz, van Niel.)

CALVIN, MELVIN, 1956. "Chemical Evolution and the Origin of Life," Am. Scientist, 44, 248–263. In a very readable essay, Calvin develops the thesis that, on chemical and thermodynamic grounds, chemical evolution of primitive organisms was inevitable.

COPELAND, H. F., 1947. "Progress Report on Basic Classification," Am. Naturalist, 81, 340–379. An interesting viewpoint on the kingdoms of organisms.

FRAENKEL-CONRAT, H., and R. C. WILLIAMS, 1955. "Reconstitution of Active Tobacco Mosaic Virus from its Inactive Protein and Nucleic Acid Components," Proc. Nat. Acad. Sci. (Washington), 41, 690–698. The technical report on a most important experiment.

FULLER, H. J., and O. TIPPO, 1954. "College Botany," 2nd Ed., Henry Holt & Co., Inc., New York, N.Y. An excellent text, and the source of the classification of plants here used. (Moevus.)

MILLER, S. L., and H. C. UREY, 1959. "Organic Compound Synthesis on the Primitive Earth," Science, 130, 245–251. Two of the leading students of the origin of life here summarize their views.

OPARIN, A. I., 1957. "The Origin of Life," 3rd Ed., translated by Ann Singe. Academic Press, New York, N.Y. The main source of the theory presented in this chapter.

STANLEY, W. M., 1948. "Achievement and Promise in Virus Research," Am. Scientist, 36, 59–68.

WHITTAKER, R. H., 1959. "On the Broad Classification of Organisms," Quart. Rev. Biol., 34, 210–226. This interesting reassessment of the kingdoms was published too late for consideration in the text.

CHAPTER NINE

The Main Lines of Evolution among Land Plants

IT IS ALMOST A POINT of definition for the algae that they are aquatic plants, although some of them have invaded moist habitats on land. The useful fossil record begins with the Cambrian period of the Paleozoic Era, over 500,000,000 years ago, and the Cambrian record, so far as plants are concerned, consists entirely of a wide variety of algae and bacteria. In fact, the early Paleozoic is often referred to as the Age of Algae and Invertebrates. But fossils from the Silurian period, beginning about 360,-000,000 years ago, include primitive land plants, and it is probable that their first appearance was in the preceding period, the Ordovician. When these first colonists left the waters to invade the more difficult but more varied habitats on land, the algae remained the dominant members of the earth's flora. But soon (geologically speaking) the land dwellers surpassed their aquatic progenitors. One of the crucial problems which had to be solved before plants could invade the land was the protection of the zygote against drying. In all land plants, this is accomplished, with important differences in the details, by the retention of the zygote and the developing embryo within the sex organs of the maternal plant. For this reason, the land plants are known collectively as the subkingdom Embryophyta. This subkingdom includes only two phyla, the Bryophyta and the Tracheophyta, but the latter is greatly varied.

THE BRYOPHYTES

The phylum Bryophyta is a relatively small group comprising the mosses, the liverworts, and the hornworts. These are the amphibians of the plant world, for they have met only minimum requirements of adaptation to the terrestrial environment. They are restricted to wet habitats, and all of them require water for reproduction, at least as a film over the surface of the plant in which sperm can swim. The bryophytes share with other land plants certain adaptations which permit them to utilize terrestrial habitats. As already mentioned, the embryos, which are always multicellular, are retained within the female sex organs, thus protecting them from drying. The plants are always oogamous, that is, the egg is a large, nonmotile cell which must be sought by the sperm. The sex organs of both sexes include a jacket layer of sterile, protective cells. All of the aerial parts of the plant are covered by a waxy cuticle which protects the plant against drying. Finally, alternation of generations is well developed, with the sporophyte being substantially a parasite upon the gametophyte in the bryophytes, for the former is always attached to the latter, and it contains inconsequential amounts of chlorophyll, if any.

Most of these characters, other than retention of the embryo in the maternal sex organs, are already known in the algae, although no alga has all of them in combination. Other bryophyte characters tie these plants in with the algae more closely. Like the algae, the bryophyte body is a *thallus,* a rather simple cell mass which is not differentiated into roots, stems, and leaves. Structures which resemble these parts are present, but the same may be said of many of the algae. The gametophyte is larger than the sporophyte, as in most algae, and photosynthesis is largely restricted to the gametophyte.

Life Cycle of a Moss. A typical moss may serve as an example (Figure 36). If a spore (the agent of dispersal) falls on suitable ground, it germinates to form a network of threads not unlike a filamentous alga and is called the protonema. The protonema sends fine, root-like rhizoids into the substrate for attachment, and also for procurement of water and salts. Buds on the protonema develop into shoots bearing leaf-like structures. At the tips of these shoots, the sex organs, antheridia in the male or arche-

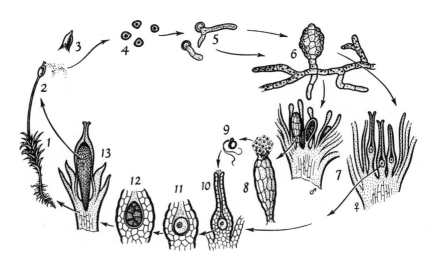

FIGURE 36. THE LIFE CYCLE OF A MOSS. 1. Mature gametophyte with a sporophyte (2) growing upon it and discharging its spores. 3. The cap of the spore capsule. 4. Ripe spores. 5. Germinating spores. 6. Young gametophyte with protonema. 7. Portions of male and female gametophytes showing antheridia and archegonia. 8. Antheridium discharging sperm (9). 10. Archegonium with egg. 11, 12, and 13. Developmental stages of the sporophyte. (From Pauli, "The World of Life," Houghton Mifflin Co., 1949.)

gonia in the female, develop. In some species, the sexes are separate (dioecious), while in other species both male and female shoots may occur on the same plant (monoecious). In either case, the sperm are released and swim to the archegonia, where fertilization occurs. The embryo now develops within the archegonium, and remains attached there by a foot when it grows to maturity. In addition to the foot, the mature sporophyte consists of a stalk and a terminal capsule which contains the spore mother cells. Here the meiotic divisions occur, and the spores are released to start the cycle over again. The gametophyte, consisting of the haploid protonema, rhizoids, shoots, and leaf-like structures is much larger than the sporophyte, and is independent because of the possession of chlorophyll. The sporophyte, consisting of foot, stalk, and capsule, is smaller and dependent on the gametophyte because it is lacking in chlorophyll, or nearly so. In neither the gametophyte nor the sporophyte are there any specialized vascular (conducting) tissues, although a limited amount of conduction of foods and water does occur. Nor are there any fibrous tissues. Lack of these elements (vascular and fibrous tissues) is probably the most important single factor which has prevented the development of large plants in this phylum.

Phylogeny of the Bryophytes. In view of the simplicity of the bryophytes, and in view of the many characters which they share with the green algae, it is highly probable that the bryophytes were derived from an ancestor among the green algae, most probably a tide-flat dweller for which terrestrial adaptations would already be advantageous. One group of bryophytes, the hornworts, suggests relationship to higher plants. While the gametophyte is a small, simple thallus, the sporophyte is larger and quite differentiated, much the dominant phase of the hornwort. It is well supplied with chlorophyll, but still depends upon the gametophyte for absorption of water and minerals from the substrate. The capsule has a central axis of elongate cells which suggests vascular tissue. The epidermis has stomata like those on the leaves of vascular plants. With such an array of traits suggesting those of vascular plants, it is difficult to avoid the inference that modern hornworts may be but little changed from an ancestral stock by which bryophytes gave rise to vascular plants. Nonetheless, this inference appears to be misleading, for vascular plants appear in the fossil record in the Silurian, and the bryophytes appear more than 100,000,000 years later. Obviously, if these data are correct, bryophytes cannot be ancestral to tracheophytes. And so the weight of opinion among botanists now favors direct origin of the vascular plants, the phylum Tracheophyta, from a chlorophytan ancestor.

ORIGIN OF THE VASCULAR PLANTS

The phylum Tracheophyta is an ancient and much varied group, which includes the dominant plants of today. It is subdivided into four subphyla, the subphylum Psilopsida, including the most primitive of vascular plants, of which all but two genera are extinct; the subphylum Lycopsida, or club mosses; the subphylum Sphenopsida, or horsetails; and the subphylum Pteropsida, including ferns, conifers and their allies, and the flowering plants. How the vascular plants arose from their algal ancestors is not known, and the fossil record throws very little light upon this important question. Nearly sixty years ago, the French botanist Lignier published a highly speculative theory upon this subject, and in the meantime such fossils as have been discovered have been consistent with his theory, even if it cannot be said that they proved it. According to Lignier's theory, the ancestor of the vascular plants must have been a green alga characterized by branching filaments, and this plant must have been a tide-flat dweller. As the land was elevated, tide pools became isolated and then dried up. Much of the flora became extinct. But if one or more of the branches of such an alga should penetrate the ground, it might become transformed into a root system capable of supplying the plant with water and minerals. Some of the branchings might then straighten out, leaving a main stem or trunk with branches. Because the entire plant is no longer immersed in water, a conducting system is now necessary, and only those plants which develop one can survive. Thus the stems become thickened, and the ends flatten out to form specialized organs for photosynthesis, the leaves. Now the conducting system must operate in both directions, carrying water and salts up from the roots and carrying organic compounds down from the leaves. Finally, only those plants which developed a cuticle over the aerial parts could escape extinction by drying. While critical proof of this theory is still lacking, evidence accumulated in the past sixty years is consistent with it, and no more probable theory has yet been proposed.

Subphylum Psilopsida. The subphylum Psilopsida includes two orders, and perhaps the sharpest distinction between them is simply one of time: the Psilophytales are known only from rocks of Silurian and Devonian age, while the Psilotales are represented by two living genera, *Psilotum* and *Tmesipterus*. A series of fossils connecting these two orders is lacking, and it has been suggested by some botanists that they are actually unrelated. But the similarities of the plants of these two orders are so close that it seems very improbable that they could be coincidental. The first discovered member of the group, *Psilophyton*, was described in 1858 by Sir William Dawson on the basis of fossils from the Devonian of Canada. The plant looked much like a small, rootless, leafless shrub (Figure 37). It did not fit into the botanical classifications then in use, and Dawson's discovery was ignored or disparaged. But in 1917 Kidston and Lang discovered three genera of similar plants in a silicified bog of Devonian age at Rhynie, Scotland. The preservation of the Rhynie fossils was so good that the cell walls can be seen in thin sections. The fossils are complete and abundant. In the meantime, at least three more genera of psilophytes have been discovered, so that there is now a wide variety of material available for study of this very primitive group which may be the connecting link between the green algae and the vascular plants.

Psilophyton was a small plant growing to a height of about three feet. It consisted almost entirely of a green, branched stem. True leaves were absent, but were suggested by numerous small spine-like projections of

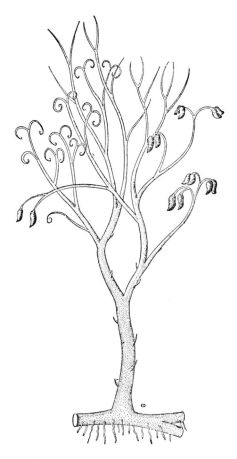

FIGURE 37. *Psilophyton.* (After Dawson, from Fuller and Tippo, "College Botany," 2nd Ed., Henry Holt & Co., Inc., 1954.)

the superficial layers of the plant. The green stem was the main photosynthetic organ. Sporangia were located at the tips of some of the branches. Such shoots arose not from true roots, but from rhizomes, which were nothing but subterranean stems bearing rhizoids for absorption of water and salts. There were vascular bundles in the stems. Thus *Psilophyton* turns out to be remarkably similar to the hypothetical primitive land colonist of Lignier.

Current opinion among botanists favors the Psilophytales as ancestral to the remaining three subphyla of vascular plants, not only because the Psilophytales are obviously extremely ancient and primitive vascular plants, but also because different genera show tendencies toward specialization in the direction of each of these subphyla. Thus some psilophytes, such as *Psilophyton* and *Asteroxylon,* had small scale-like leaves which were developed as projections of the superficial layers of the stem. The

115

Lycopsida (club mosses) are characterized by similar scale-like leaves. Another genus, *Hyenia,* had whorled branches, and may have been ancestral to the Sphenopsida (horsetails). Finally, some genera, such as *Pseudosporochnus,* had the branch tips flattened out, and it is generally believed that this led to the formation of broad leaves, as in many of the Pteropsida.

Thus the psilophytes appear to occupy a key position in the evolution of the higher plants. They are the most primitive vascular plants known— one might almost say they are the most primitive vascular plants imaginable—and it is highly probable that they arose from the green algae. The psilophytes appear to have given rise to the three higher subphyla independently, that is, each of the higher subphyla has arisen from a different psilophytan ancestor, so that these represent parallel developments rather than a sequence from primitive to advanced. It has also been suggested that the bryophytes may have been derived from the psilophytes, but this does not seem very probable in view of the fact that it would have required a reversal of the tendency toward dominance of the sporophyte, as well as a suppression of the vascular bundles. Current opinion favors the independent origin of both the bryophytes and the psilophytes from the green algae. On the other hand, spores resembling those of land plants

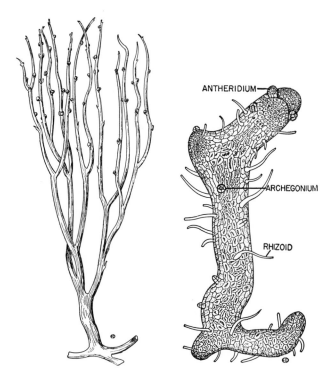

FIGURE 38. *Psilotum* SPOROPHYTE (LEFT) AND GAMETOPHYTE (RIGHT) (After Lawson, from Fuller and Tippo, "College Botany," 2nd Ed., Henry Holt & Co., Inc., 1954.)

are known from the early Cambrian, and it may well be that even the Devonian psilophytes are late remnants of much more ancient stages in the origin of the higher plants.

The living Psilotales (Figure 38), *Psilotum* and *Tmesipterus*, are quite similar to *Psilophyton*. In nature, they are confined to tropical and subtropical regions, but they can be grown in greenhouses anywhere. The major morphological difference between the living and extinct species is that the sporangia are located in the axils of the "leaves" of the living plants, whereas they were located at the tips of branches in the extinct species. Another point of considerable interest relates to the alternation of generations. Only the sporophytes of the psilophytes have been preserved in the fossil record. But the gametophytes of the modern genera are known. The gametophyte looks very much like a fragment of rhizome, bearing numerous archegonia and antheridia which are rather similar to those of the bryophytes. These may give an indication of the character of the gametophyte of the extinct psilophytes, but there is no assurance that they do. In any event, the Psilotales are regarded as the little-changed descendants of the ancient psilophytes.

Subphylum Lycopsida. The Lycopsida, comprising the club mosses, appear to have been derived from psilophytes in the Devonian period. All of the living lycopsids (there are only four genera) are small plants, generally less than a foot high, but this has not always been true, for during the Carboniferous period (Mississippian and Pennsylvanian combined) there were giant club mosses as tall as 135 feet. These were the dominant plants of the time, and their fossil remains are a major part of the coal beds. But these giant club mosses became extinct in the Permian period, perhaps as a result of inability to adapt to the severe Permian climate, for this was a time of extensive glaciation.

The club mosses (Figure 39) are advanced over the psilophytes in many respects. The differentiation into root, stem, and leaves is complete. Their leaves, which are small and spirally arranged, are supplied with vascular tissue. The sporangia are enclosed in specialized leaves called sporophylls, clusters of which occur at the tips of branches. Such clusters are called cones or strobili. The life cycle of *Lycopodium*, one of the living genera, is quite simple, and perhaps not very different from that of the psilophytes. The spores are all alike, and are scattered by the wind. Those which fall on favorable ground germinate to form small, subterranean, thallus-like gametophytes. These are monoecious. Sperm swim to the archegonia to fertilize the eggs. The zygote then produces a young sporophyte which is at first a parasite upon the gametophyte. But a root is soon formed and the gametophyte decays. In another living genus, *Selaginella*, the life cycle is modified in a fashion quite suggestive of the seed plants. There are two types of spores, megaspores which are contained in sporophylls in the lower part of the strobilus, and microspores which are contained in the upper sporophylls. While still contained within the sporangia, these megaspores and microspores germinate to produce female and male gametophytes respectively. The gametophytes are thus parasites upon the sporophyte, although the female gametophyte may contain some

117

FIGURE 39. A CLUB MOSS, *Lycopodium*, INCLUDING AN ENLARGED VIEW OF A SINGLE STEM. (From Weatherwax, "Plant Biology," 2nd Ed., W. B. Saunders Co., 1947.)

chlorophyll. When the microspores are released, they fall down onto the lower sporophylls. When wet by rain or dew the microspore wall splits and sperm are released. These swim to the archegonia of the female and fertilize the eggs while the female gametophyte is still contained in the spore wall, and hence in the parental sporophyte. The embryo is thus produced in the female gametophyte while the latter is still contained in the sporophyte, a condition very similar to that of the seed plants.

One might expect that a group with so many progressive characters would have produced further, more progressive descendants, but this has not been the case. The Lycopsida are a terminal group. Soon after its origin, the group produced the dominant plants of the coal swamps, but these became extinct before the end of the Paleozoic Era, and only four genera of small, insignificant lycopods have survived to the present. These are often referred to as living fossils because their closest relatives are long-extinct plants.

Subphylum Sphenopsida. The Sphenopsida, the horsetails, have played a smaller role in the history of the plant kingdom than the Lycopsida. Like

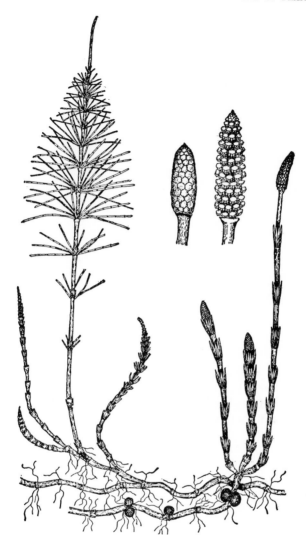

FIGURE 40. A HORSETAIL, *Equisetum*. (From Weather-wax, "Plant Biology," 2nd Ed., W. B. Saunders Co., 1947.)

the latter, they probably arose in the Devonian from psilophyte ancestors by the differentiation of the body into definite root, stem, and leaves. The stem of the horsetails has a pith region which is commonly hollow. The stem grows by means of concentrations of meristem at definite nodes, which can be easily disjointed, to the delight of many a child (Figure 40). The leaves are arranged as whorls, and they are relatively small. Nonetheless, the weight of opinion favors their origin by flattening of

119

FIGURE 41. CARBONIFEROUS FLORA. (Courtesy of the Chicago Natural History Museum.)

the tips of branches rather than as emergences from the stem. The leaves are not adequate for the photosynthetic needs of the plant, and the entire stem has retained this function.

The origin of the horsetails from psilophytes is strongly indicated not only by the similarity of shoots and leaves but by the fact that *Hyenia* is an excellent intermediate between the two groups and might with equal justification be assigned to either. From the Hyenia stock there were developed herbaceous species, comparable to the horsetails of today, and gigantic, tree-like species as much as forty feet tall. These flourished during the Carboniferous, and they form an important part of the coal beds (Figure 41). They became extinct during the Triassic, the first period of the Mesozoic Era. The modern horsetails comprise a single genus, and it is uncertain whether they are descended from the herbaceous species of Paleozoic times, or whether they have been derived from the horsetail trees by reduction.

SUBPHYLUM PTEROPSIDA

Ferns. The last of the subphyla of the phylum Tracheophyta is the Pteropsida, comprising ferns, conifers, and flowering plants. Like club mosses and horsetails, the ferns—class Filicineae—appear to have developed from psilophyte ancestors of Devonian age. Some of the psilophytes, for example *Pseudosporochnus* and *Protopteridium,* show a tendency toward leaf formation by flattening of whole branches, and it is probable that the ferns were derived from such psilophytes. Like all tracheophytes except the psilophytes, the plant is differentiated into true roots, stem, and leaves. The stem is generally subterranean, and so is a rhizome, and small roots arise all along its length. The leaves are large fronds, and these are the only aerial parts of the plant. The sporangia occur in clusters called sori on the undersurfaces of the leaves. The spores, when released, germinate on moist ground to form monoecious gametophytes, which are always small, thallus-like plants. The sperm of one plant swim to the archegonia of another, and there fertilization takes place. A new sporophyte then develops from the zygote.

Although the ferns most familiar in temperate latitudes are all of moderate size, in the tropics there are tree ferns in which the stem forms an erect trunk as high as seventy-five feet. The leaves form a palm-like cluster at the top. The most primitive order of ferns, the Coenopteridales, developed great forests of fern trees during the Carboniferous period. These, together with the giant club mosses and horsetails (erroneously called fern allies) formed the dominant vegetation of the times, and their remains have come down to the present as coal beds. Like the psilophytes, these primitive ferns bore the sporangia at the tips of the branches. Like so many other ancient groups, these giant ferns became extinct during the Permian, but not before giving rise to the three modern orders of ferns.

Gymnosperms. The second class of the subphylum Pteropsida is the class Gymnospermae, the dominant members of which are the conifers, but their allies include the extinct seed ferns, the cycads or sago palms, the maidenhair trees, and the very aberrant order Gnetales. The class may have arisen in the Pennsylvanian period from some group of ferns via the seed ferns. The seed ferns were long classified with the typical ferns, but accumulating data have shown that their affinities are closer to the gymnosperms. The basic datum is the fact that they did reproduce by seeds, an innovation of the greatest importance. The seed ferns formed an important part of the Carboniferous forests, but they later declined in importance and finally became extinct during the Jurassic period, but not without leaving more advanced gymnosperms as descendants. The most successful of these is the order Coniferales, including all of the well-known evergreens. These reached their greatest development in the Mesozoic Era, and have been declining since. Three other orders have survived to the present time, however. These are the Cycadales, or sago palms, with nine living genera; the Ginkgoales, represented only by a single species, *Ginkgo biloba*, which has survived only because it has for centuries been cultivated in Chinese temple gardens; and the Gnetales, including three genera of very aberrant plants, of which the most important is perhaps *Ephedra* from which the drug ephedrine is obtained.

Unlike the lower tracheophytes, gymnosperms are generally large, woody trees, although some are shrubs. Most are evergreens. But their most important characters are concerned with the reproductive cycle. At the outset, the gymnosperms produce two types of spores, megaspores which develop into female gametophytes, and microspores or pollen which develop into male gametophytes. In spite of their names, the two types may be of equal size, or the microspores may actually be the larger. The cones in which the spores are formed consist of specialized sporophylls spirally arranged about a central axis. They are thus modified leaf clusters. The microspores divide while still within the spore wall to form the male gametophyte. The pollen is shed in great numbers while in the four-cell stage, and is carried by the wind, sometimes for great distances. Some of the pollen will reach female cones and become stuck in a sticky fluid exuded by the ovules, complex structures which include the female gametophyte. The pollen produces a tube-like growth which enters the ovule. Down this pollen tube pass the two sperm nuclei, one of which fertilizes the egg. The developing embryo (sporophyte) is pushed into the mass of the female gametophyte, which is now called the endosperm and which serves as nutritive material for the embryo. This endosperm is, in turn, surrounded by a seed coat, which is actually a part of the parent sporophyte. The seed is shed, and if it falls on favorable ground the seedling may develop into a mature sporophyte.

Some significant aspects of this reproductive cycle should be indicated. For the first time in the phylogenetic series, reproduction is independent of water. The pollen is carried to the female cones by wind, and the sperm —mere nuclei rather than flagellate cells—are carried to the egg by the protoplasmic pollen tube. Both the male and female gametophytes are reduced to minute structures consisting of only a few cells, and thus the relative predominance of the sporophyte in the life cycle has become great. The gametophytes lack chlorophyll and are completely dependent upon the sporophyte. Finally, the seed, a new structure in the phylogenetic series, consists of an embryo (sporophyte) contained within the endosperm (gametophyte) which is in turn contained within a seed coat which is sporophytic tissue of the parental generation. Thus the appearance is very much as though the embryo sporophyte were produced directly by the parent sporophyte, with the gametophytes being simply organs of the parent sporophyte. Only comparison to lower plants reveals the true situation.

The Flowering Plants. The third and final class of the subphylum Pteropsida is the class Angiospermae, the flowering plants. This is much the dominant class of the world's flora today, including on the order of 10,000 genera and 195,000 species. The variety of the angiosperms is unlimited, ranging from great trees to grass; generally they are land plants, but they have become adapted to almost every available habitat, including marine; and while they are typically free-living green plants, not a few have become parasitic and some are saprophytic.

The angiosperms share the major characters of the gymnosperms, being differentiated into true roots, stems, and leaves, and reproducing by means of true seeds. They have a highly developed vascular system. Fertilization is by means of pollen which are independent of water. The sporophyte is much the dominant generation, the gametophyte being minute and completely dependent. In addition, the sporangia of the angiosperms are within flowers, which are modified cones surrounded by modified and

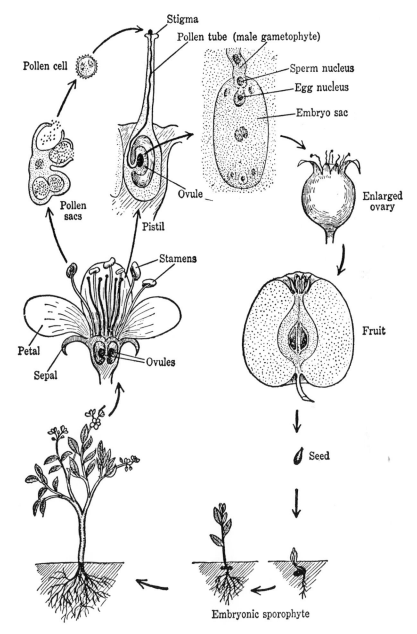

Stigma

Pollen tube (male gametophyte)

Pollen cell

Sperm nucleus

Egg nucleus

Embryo sac

Pollen sacs

Ovule

Enlarged ovary

Pistil

Stamens

Fruit

Petal

Ovules

Sepal

Seed

Embryonic sporophyte

FIGURE 42. THE LIFE CYCLE OF A FLOWERING PLANT. (From Young, Hylander, and Stebbins, "The Human Organism and the World of Life," Harper Brothers, 1938.)

often highly ornamental leaves. The microspores develop into pollen (male gametophytes) with only three nuclei, one tube nucleus and two sperm nuclei. The macrospores develop into female gametophytes with only eight nuclei, including the egg nucleus and two polar nuclei which fuse to form the fusion nucleus. The macrospores are completely enveloped within modified sporophylls, the carpels, which become the fruit when mature. Pollination may be by wind as in the gymnosperms, but it is commonly effected by insects, occasionally by birds, and rarely by water (in the case of aquatic plants). A unique characteristic of the angiosperms is the phenomenon of double fertilization. As usual, a single sperm nucleus unites with the egg nucleus to form a diploid zygote from which the embryo develops. But in addition to this, the other sperm nucleus unites with the fusion nucleus to form a triploid (3 n) cell from which the bulk of the endosperm is formed. The life cycle of a typical angiosperm is illustrated in Figure 42.

About a century ago, Darwin called the origin of the flowering plants an "abominable mystery," and there the problem still rests. While botanists can agree that the angiosperms must have been derived from some primitive gymnosperm stock, there is nothing but speculation as to which stock may have been the correct one. The fossil record is of little help. The angiosperms suddenly appear in considerable numbers in the Cretaceous period (late Mesozoic) with no intermediate plants in older rocks. Whatever their origin, the angiosperms quickly became the dominant plants of the world, and they appear to be still on the increase. The class Angiospermae includes two subclasses, the Dicotyledoneae and the Monocotyledoneae, which are separated mainly on the basis of embryological characters. The monocots include the grasses, lilies, palms, and orchids, while the dicots include the great majority of flowering plants. Anatomically the dicots appear to be the more primitive, and so it is probable that the monocots were derived by reduction of various parts from some primitive dicot.

THE MAIN LINES OF PLANT EVOLUTION

The main lines of plant evolution may now be summarized with the aid of Figure 43. Whatever the origin of the viruses and the bacteria may have been, the seven phyla of algae must have arisen from bacterial ancestors, probably from among the autotrophic bacteria. Whether the different algal phyla arose independently or from a common stock is scarcely indicated by any available evidence. The Cyanophyta or blue-green algae are the earliest plants found in the fossil record; they are extraordinarily primitive in their lack of a distinct nucleus or plastids; and they contain chlorophyll. In view of these facts, it is plausible that the Cyanophyta may be ancestral to some other algal phyla, including the Chlorophyta, yet there is very little evidence that this is actually the case. The origin of the two phyla of molds, Myxomycophyta and Eumycophyta, is also much in doubt. They may have come from bacterial ancestors, or from algae by loss of chlorophyll, or even from Protozoa. However much vexed the

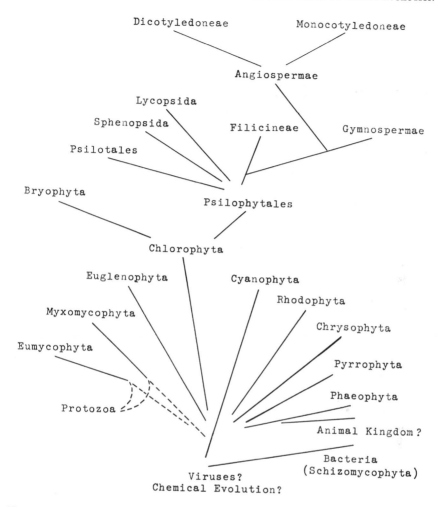

FIGURE 43. A SUMMARY OF PROBABLE LINES OF PLANT EVOLUTION. The dotted lines present highly problematical alternatives. (Redrawn from Fuller and Tippo.)

problem of the origins of these lower groups may be, it appears to be quite clear that the land plants arose from the green algae, with the two phyla, Bryophyta and Tracheophyta arising independently. The bryophytes have become differentiated into three minor groups, the mosses, liverworts, and hornworts, but have not produced any more progressive types of plants. The primitive tracheophytes (psilophytes), on the other hand, having first succeeded in colonizing the land, quickly gave rise to three more subphyla, the Lycopsida, the Sphenopsida, and the Pteropsida. The first two were dominant groups for a time, but were reduced to minor groups about the end of the Paleozoic Era. The pteropsid line was at first represented only by the ferns, but these gave rise to the gymnosperms which in turn appear to have given rise to the angiosperms. The more primitive subclass of angiosperms, the dicots, has given rise to the more specialized monocots.

Some general trends of evolution in the Plant Kingdom may be noted. Among the most primitive plants, increasing complexity of the nuclear apparatus, culminating with the typical cell structure as found in all plants above the blue-green algae, is a major trend. Among the algae, development of the colonial habit and multicellularity is one major trend. A second is the evolution of sexual reproduction, together with the alternation of generations. Finally, a transition from aquatic to terrestrial habitats occurred, but little specific information is available regarding the steps in this process. This required the development of vascular and strengthening tissues, the development of which is one of the most characteristic features of the land plants. These tissues are, however, found only in the sporophyte, and so the sporophytic generation has increased in size and functional importance relative to the gametophytic generation until finally, in the angiosperms, the gametophytic generation is recognizable as distinct from the sporophyte only by comparison to the lower plants. This trend is illustrated in Figure 44. In a general way, there has been a tendency to increase in size. This is obviously so among the algae, where the most primitive species are all unicellular and the more advanced ones may be (but need not be) multicellular. While the most primitive land plants were smaller than large algae, all of the classes of tracheophytes have produced species much larger than the largest of the algae. Nonetheless, it is far from axiomatic that evolution is necessarily accompanied by pro-

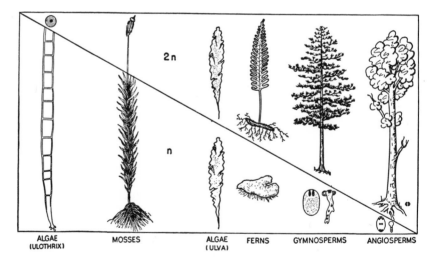

FIGURE 44. DIAGRAM SHOWING THE GRADUAL INCREASE IN SIZE AND IMPORTANCE OF THE SPOROPHYTE, AND THE CORRESPONDING DECREASE IN SIZE AND IMPORTANCE OF THE GAMETOPHYTE. (From Fuller and Tippo, "College Botany," 2nd Ed., Henry Holt & Co., Inc., 1954.)

gressive size increase. The most successful members of many groups are characterized by small size. The club mosses and horsetails, for example, were once represented by great trees, but only the small members of the group have survived to the present. Among the angiosperms, the trees seem to have been primitive, and from these the shrubs and grasses have been derived. This has been accompanied by the development of the annual habit, presumably as an adaptation to prevent extinction by winter-killing, for a dormant seed may easily survive severe weather which would kill a mature plant. Thus development of small size, rapid growth and maturity, and the annual habit comprise an adaptation to arctic and subarctic conditions, and such plants appear to be on the increase in temperate lands as well.

REFERENCES

ANDREWS, H. N., 1947. "Ancient Plants and the World They Lived In," Comstock Publishing Co., New York, N.Y. A readable summary of paleobotany.

AXELROD, D. I., 1959. "Evolution of the Psilophyte Paleoflora," *Evolution*, 13, 264–275. Evidence is adduced that the higher plants arose much earlier than generally supposed, and from algal ancestors.

BOLD, H. C., 1957. "Morphology of Plants," Harper Brothers, New York, N.Y. A penetrating treatment, going far beyond the minimum presented in this chapter.

FULLER, H. J., and O. TIPPO, 1954. "College Botany," 2nd Ed., Henry Holt & Co., Inc., New York, N. Y. An excellent text, and the source of the classification used in this chapter (Lignier, Kidston, and Lang).

THOMAS, H. H., 1936. "Paleobotany and the Origin of the Angiosperms," *Botan. Rev.*, 2, 397–418. Documentation for the statement that the origin of the angiosperms is still "an abominable mystery."

CHAPTER TEN

The Main Lines
of Animal Evolution

As POINTED OUT in Chapter 8, the Protozoa, the most primitive phylum of animals, appears to have been derived from primitive flagellate algae, and within single genera among the euglenoids there may be some species which show predominantly plant characters and others which show predominantly animal characters. Thus there is a high probability that the flagellates are close to the point of separation of the two kingdoms, if, indeed, the separation is complete here, for there are flagellates like *Trypanosoma* which are undoubtedly animals; there are others like *Chlamydomonas* which are undoubtedly plants; and there is the great intermediate group, typified by *Euglena,* which defies any indisputable assignment to the kingdoms.

DIVERSIFICATION OF THE PROTOZOA

Once the Protozoa were established, they became greatly diversified, and their relationships are by no means certain. The class Flagellata itself includes a wide range of structural and ecological types even when only indisputably animal types are considered. Feeding may be by engulfing food with pseudopodia, or by a simple mouth. The parasitic forms are generally saprozoic. Several orders of flagellates deserve especial mention. The order Protomonadina includes a wide variety of small, colorless flagellates, typically with two flagella, one of which trails along the side of the animal. Reproduction is always asexual. In the free-living members of the order, this is by simple fission, but among the parasitic species there may be multiple divisions which are difficult to distinguish from spore formation of the Sporozoa. It seems highly probable that the Sporozoa, all of which are parasitic, were evolved from this group. But sexual reproduction was also developed among the ancestors of the Sporozoa. The best known protomonads are the members of the genus *Trypanosoma,* all of which are blood parasites of vertebrates. Quite as important for evolution, however, are the choanoflagellates. These are protomonads which have a protoplasmic collar encircling the base of the flagellum. Its theoretical importance lies in the fact that sponges have similar cells, called choanocytes, and that colonial choanoflagellates, such as *Protero-spongia*, resemble simple sponges.

The most highly specialized flagellates are probably the orders Polymastigina and Hypermastigina. The polymastigotes are generally parasitic on the digestive tract of arthropods or vertebrates. The simpler members generally have four flagella, but the more complex members may have large numbers. There may be only a single nucleus, or there may be many. Complicated cytoplasmic organelles may be present. Reproduction is generally reported to be by simple fission, but Cleveland has reported sexual reproduction in some species, including the unique feature of reduction of the chromosomes at a single division. The hypermastigotes are morphologically the most complex flagellates. There are numerous flagella arranged in definite ways. Cytoplasmic organelles are very complex (Figure 45). All hypermastigotes are parasites or commensals in the digestive tract of termites or cockroaches. They are essential to the nutrition of the hosts, for which they digest cellulose, for it has been shown that the host starves if the protozoans are removed experimentally.

The order Rhizomastigina is of especial interest because the members of the order appear to be intermediate between the Flagellata and the Sarcodina, ameba and its allies. Many flagellates are capable of ameboid movement, but the members of this order, although they do possess a flagellum, are also permanently ameboid, and so they make a nice connecting link between the Flagellata and the Sarcodina. *Mastigamoeba* is a good example.

Ameba and Its Allies. The class Sarcodina is also highly diversified, and it is by no means certain that it is really a unified group. That is, different orders of the Sarcodina may have arisen from different flagellate ancestors. By far the best known member of this class, or perhaps of any protozoan class, is *Amoeba proteus*, the familiar study material of every elementary biology laboratory, and the traditional example of a primitive animal. *A. proteus* is quite typical of its order, the Lobosa. But, in addition to such simple, free-living, ameboid organisms, the Lobosa includes parasites like *Entamoeba histolytica*, which parasitizes the digestive tract of man; and there are free-living but shelled organisms such as *Difflugia*. The nutrition of all of these, including the parasites, is holozoic. Reproduction by binary fission is the rule, but sexual reproduction has been reported for one species of *Amoeba*, and it is common among the shelled species. The order Lobosa is generally regarded as a terminal group in evolution, yet the idea that the Myxomycophyta or the Eumycophyta or both may have been derived from this group is gaining prestige among mycologists (see Chapter 8). And some zoologists treat the Myxomycophyta as protozoans under the name of Mycetozoa.

The remaining orders of the Sarcodina are all terminal groups which are characterized by elaborate and often beautiful calcareous or siliceous shells, and by many slender, semipermanent pseudopodia, the characteristics of which are peculiar to each order. These are the Foraminifera and the Radiolaria, both of which are abundantly represented in the fossil

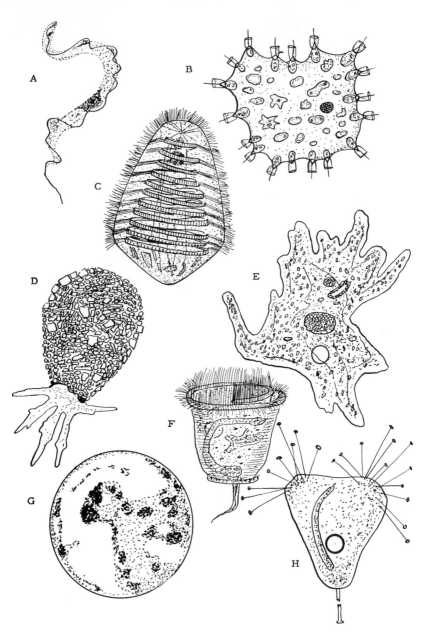

FIGURE 45. REPRESENTATIVE PROTOZOA. A, *Trypanosoma; B, Proterospongia; C, Macrospironympha; D, Difflugia; E, Amoeba; F, Vorticella; G, Plasmodium* infested red blood cell; *H, Acineta*. *A–C* are flagellates; *D* and *E* are sarcodinans; *F* is a ciliate; *G* is a sporozoan; and *H* is a suctorian.

record even into pre-Cambrian times, and the Heliozoa, which are less adequately known as fossils.

Ciliata and Suctoria. In contrast to the classes Flagellata, Sporozoa, and Sarcodina, all of which are fairly clearly interrelated, the two remaining classes of Protozoa show no evidence of relationship to the above named classes. These are the classes Ciliata and Suctoria, which some protozoologists would prefer to assign to a separate phylum. The ciliates have a definite shape, maintained by a pellicle, as do some flagellates. There is a definite anteroposterior axis, and the animal is symmetrical, sometimes radially, sometimes bilaterally, and sometimes an irregular deviant from one of these types of symmetry. Nutrition is holozoic, with other minute organisms being ingested by a mouth. Cilia are commonly arranged in specialized tracts which sweep a food-bearing current toward the mouth. It is often said that the Protozoa are the simplest of animals. While this may be debated, there is little doubt that ciliates are the most complicated of cells. The cilia themselves may be arranged as coordinated tracts, or they may be fused in sheets to form undulating membranes, or they may be fused in tufts to form appendage-like cirri. Whatever the arrangement of the cilia, their movements are coordinated by a complicated network of fibrils, the neuromotor system, which exceeds in complexity the simplest nervous systems of Metazoa. The trichocysts, small bodies which can be discharged to produce filaments, underlie the pellicle. Their function is uncertain, but defense and attachment while feeding have been the principal alternatives suggested. The food vacuoles, formed at the mouth, pass through the body by a regular route and leave at a definite point. In some, it would scarcely be an exaggeration to say that there is a digestive tube and an anus. There are two nuclei, a micronucleus which is concerned with heredity and a macronucleus which is concerned with metabolic functions. This led Kofoid to state that it would be as logical to call a whale unicellular as to call a *Paramecium* unicellular. Finally, the ciliates have developed a unique type of sexual reproduction called conjugation. The details are quite complicated, but in substance it amounts to this, that the maturation divisions result in a stationary nucleus and a wandering nucleus in each of a pair of conjugants, and the wandering nucleus of each fertilizes the stationary nucleus of the other.

The most primitive order of ciliates, the Holotricha, is completely clothed in cilia, and the animals are strong swimmers. But feeding by means of currents seems to lead to a sessile mode of life, for specialization in the ciliates has generally resulted in attachment to the substrate, restriction of the cilia to limited areas usually related to feeding, and specialization of the ciliary tracts to form membranelles, undulating membranes, or similar structures. The well known *Vorticella* is a good exemplar of such tendencies. In a single order of ciliates, the Hypotricha, specialization has been in the direction of increased efficiency of locomotion. In this order, a band of typical cilia still creates a current for feeding purposes, but locomotion is based upon the use of fused tufts of cilia, the cirri, which function in a leg-like fashion.

Finally, the class Suctoria is a small group which is undoubtedly related to the ciliates. Adult suctorians have no cilia, nor have they any other locomotor organelles. Nor do they have a mouth, for they use protoplasmic tentacles for the capture and ingestion of food. They are generally permanently attached to their substrate. Up to this point, they show no affinity with the ciliates. However, they do have the two types of nuclei and sexual reproduction by means of conjugation, phenomena which are unknown elsewhere in the world of life except for these two classes. Further, the suctorian zygote becomes a free-swimming, ciliated organism which only later settles down on the substrate and adopts the typical suctorian mode of life. This is interpreted as embryological recapitulation of ancestral history.

ORIGIN OF THE METAZOA

Hyman has said that there is no direct proof of the origin of the Metazoa from the Protozoa, yet the discussion of the origin of the Metazoa (multicellular animals) usually revolves around the question of which protozoan stock would seem to be the most probable progenitor of the Metazoa. Two broad possibilities exist by which the Metazoa could have been formed from the Protozoa. The first is that repeated nuclear division without cytoplasmic division might have led to formation of a plasmodium, like some of the Heliozoa. Formation of cell membranes would then result in multicellularity, and differentiation might then lead to the true multicellular individual. The second method is the differentiation of cells within a colony of Protozoa, comparable to *Volvox*, leading to interdependence and individuality.

A very different, third possibility has been suggested. In protozoan colonies each cell ingests food, but even in simple metazoans, a new method of feeding is used, with a digestive tract feeding for the whole organism. This transition might be difficult. Hardy suggests that simple plants, like *Volvox*, living in an environment deficient in nitrates and phosphates, may have satisfied the deficiency by capturing smaller organisms. Increasing utilization of this nutritive pathway, together with loss of photosynthesis, would then lead to a simple metazoan. Insectivorous plants show the feasibility of such a nutritive mechanism, and the fact that unicellular plants seem to have given rise to protozoans more than once lends plausibility to the suggestion that multicellular plants may have achieved animalization at least once.

Paleontology is of no help in this problem, for the Metazoa were already well established at the beginning of the Cambrian. It is therefore probable that the origin of the Metazoa will always be speculative. But most zoologists favor the flagellates as the most probable progenitors of the Metazoa. The reasons for this are many. The flagellates are a highly variable group which appear to have given rise to many other groups of plants and to several, perhaps all, other groups of Protozoa. Further, some groups of flagellates show a tendency to form colonies of ever-increasing size and complexity. The evolution of sex has occurred here, and the colonies are definitely divided into somatic and germinal "tissues." Oogamous reproduction is the rule for such colonial flagellates. The sperm are at once similar to some simpler, noncolonial flagellates and to the typical sperm of Metazoa. Such colonies may also show anteroposterior differentiation. These large, highly specialized colonies occur principally among the plant flagellates, yet colonies of a highly suggestive character also occur among the animal flagellates. None of these reasons is conclusive, yet collectively they carry considerable weight.

However much the origin of other Metazoa may be disputed, it seems almost certain that the Porifera (sponges) were derived from choanoflagellates. It is a short step from the structure of the colonial choanoflagellate *Proterospongia* to that of the simplest sponges. *Proterospongia* (Figure 45) consists of a small mass of gelatinous material in the surface of which are imbedded choanocytes (collared, flagellate cells), and in the interior of which are ameboid cells. The choanocytes can withdraw the collar and flagellum and move into the interior to become ameboid cells. In order to change this to the structure of a simple sponge it would be necessary only to develop a system of channels through the gelatinous mass, to let these channels be lined with choanocytes, and to let the outer surface be covered by a simple epithelium. No organ systems are present, the functions of the sponge being carried on by the component cells individually.

Phylum Porifera. Cooperative activity is at a minimum, and it has often been debated whether a sponge is a true organism or a colony of unicellular organisms. There are enough evidences of cooperative activity to swing the balance of zoological opinion in favor of the former alternative. Thus, while some sponges have no definite shape, many have rather complex shapes. Most sponges produce skeletal elements in the form of spicules which may be quite complex, and so exactly formed in every case that they are among the best taxonomic characters. Further, the beat of the flagella in the canals is not obviously coordinated, yet the water ·current is unidirectional, and so there must be at least a limited control of the flagellar beat.

Sponges can reproduce asexually by formation of groups of cells, the gemmules, but they also reproduce sexually by means of typical eggs and sperm. The zygote develops into a free-swimming larva which is flagellated. After swimming for a brief time, it settles down to become a sponge. Because this larva has typical (not collared) flagella, some zoologists believe that the sponges were derived from some typical flagellates, with choanocytes being a secondary development. Yet this is not unexpected in view of the recapitulation principle, for the choanoflagellates themselves must have been derived from typical flagellates.

Evolution within the Porifera has taken the forms of elaboration of the canal system and of the supporting spicules or fibers. As relationships within the phylum are not at all clear, it may be more profitable to go directly to the problem of the relationships of the Porifera to other animals. The Porifera are a terminal group. Further, they are so different from all other Metazoa in the absence of unified tissues, in the physiological independence of the individual cells, and in their embryology and adult anatomy, that it is generally believed that their origin from the

Protozoa must have been independent of that of the rest of the Metazoa. The Porifera are therefore assigned to a separate branch of the subkingdom Metazoa, the branch Parazoa, in contrast to the Eumetazoa.

The Gastrea Theory. Speculation on the origin of the Metazoa has been dominated by the Gastrea theory of Haeckel, a theory based on the literal application of the Biogenetic Law. In its original form, Haeckel interpreted the egg as corresponding to an ameboid ancestor, possibly to *Amoeba* itself. As evidence in favor of this, he pointed to the ameboid eggs of sponges and of some coelenterates. Other types of eggs he assumed to be secondary specializations. The egg, of course, undergoes the cleavage divisions which result first in a solid morula and then in a hollow ball of cells, the blastula. The morula was interpreted as corresponding to a simple hypothetical ameboid colony, the *Synamoeba*, while the blastula was supposed to correspond to a colonial ancestor, the *Blastea*, more or less comparable to *Volvox*, but ameboid rather than flagellate. The modern exponents of this theory assume flagellate rather than ameboid ancestors for the reasons stated above. There being only a single layer of cells in the *Blastea*, all cell functions were at first shared by all cells, but then a division of labor occurred, with the posterior cells assuming the nutritive functions. These cells then invaginated, with the result that the organism became a two-layered gastrula, having an outer layer of flagellated cells (ectoderm) and an inner layer of digestive cells (entoderm). Haeckel called this hypothetical organism the *Gastrea*, and he believed it to be ancestral to all Eumetazoa. Some of the coelenterates he regarded as living gastreads. Next, the *Gastrea* developed a third cell layer, the mesoderm, between the first two, and he regarded all of the structures of the higher phyla as derived from these three layers. The development of the bottom-feeding habit led to elongation of the body and the formation of primitive worms, similar to the living Turbellaria. From these the higher phyla were developed.

The Gastrea theory is a beautiful simplification and synthesis of a vast amount of embryological, morphological, and taxonomic data, and it is almost without a serious competitor. As a result, it was, until recently, presented in nearly every elementary textbook of zoology. Unfortunately, however, as Hyman [*] has pointed out, "it is probably one of those simplifications that are too beautiful to be true." As explained in Chapter 3, embryology is not a safe basis for construction of pedigrees, especially if comparison is made between embryos of advanced species and adults of their supposed ancestors, as Haeckel did in this case. At most, embryology should be treated only as one of several corroborative lines of evidence. But there is the further difficulty that even the embryological evidence does not give unequivocal support to the Gastrea theory. For in the coelenterates, the group which is closest to the hypothetical *Gastrea*, gastrulation ordinarily occurs not by simple invagination of the posterior cells, but by the inwandering of many cells from all parts of the blastula. And

---

[*] By permission from Hyman, L. H., "The Invertebrates," Vol. I, McGraw-Hill Book Co., Inc., 1940.

this does not result in the immediate formation of a typical gastrula, but rather it consists of a ball of ectodermal cells filled by a solid core of entodermal cells. This type of larva is called a *planula*. Only later does this entodermal core hollow out and a mouth (blastopore) break through to form a typical gastrula. While the type of gastrulation with which Haeckel dealt is known, for example in the starfishes, it is not widespread in the Animal Kingdom, and it appears to be a secondary modification rather than a primitive character.

It is plausible, then, that the ancestor of the Eumetazoa may have been a blastula-like colonial flagellate, in which there occurred a differentiation between somatic and reproductive cells, as in living *Volvox*, and then a differentiation between digestive cells and locomotor cells, with the former type moving into the interior of the organism to form either a gastrula or a planula. Yet it is probable that decisive evidence on this basic question will never be obtained. That this primitive metazoan was not identical with any living type is almost certain.

The three most primitive living metazoan phyla are the Mesozoa, the Coelenterata (= Cnidaria), and the Ctenophora. These phyla are generally radially symmetrical (having one differentiated axis), or else biradially symmetrical (having two differentiated axes). Their general grade of organization is more advanced than that of the Porifera, for, while there are no organ systems, there are two well-defined tissues, the ectoderm and entoderm (or epidermis and gastrodermis). In most coelenterates and in the ctenophores, there is between these layers a jelly-like mass, the mesoglea, which also includes some cells. Thus it is not strictly true, as is often stated, that these phyla have only two cellular layers.

Phylum Mesozoa. The correct phylogenetic position of the Mesozoa is a much vexed question. From a structural viewpoint, they are the simplest of metazoans, consisting of an outer, generally ciliated, layer of cells enclosing a core of internal, reproductive cells (Figure 46). They thus resemble a planula, but the internal cells˙ are not digestive cells. Van Beneden named the group in 1877 with the intention of indicating his judgment that the group was extremely primitive, intermediate between the Protozoa and the rest of the Metazoa. On the other hand all Mesozoa are parasitic, and their life cycles are somewhat similar to those of the digenetic trematodes. Many zoologists treat the group as a degenerate offshoot from the flatworms. If it could be shown decisively that the char-

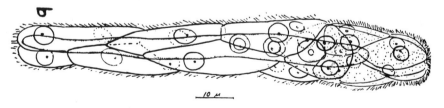

FIGURE 46. *Conocyema deca*, A TYPICAL MESOZOAN. (From MacConnaughey, *J. Parasitol.*, V. 43, 1957.)

acters of the Mesozoa are primitive rather than degenerate, the group would assume great phylogenetic importance, for it could then be reasonably argued that the group must be but little changed from the remote, pre-Cambrian ancestor of the Metazoa. It would prove that the Metazoa were derived from a planula- rather than from a gastrea-type ancestor, and it would leave the Gastrea theory very badly damaged. But the available evidence does not furnish a basis for a final decision on the taxonomic position of the Mesozoa. W. K. Brooks has said that "suspended judgment is the greatest triumph of intellectual discipline," and this appears to be an appropriate place for that achievement. Hyman sets them off as an independent branch of the Metazoa, the branch Mesozoa.

Phylum Coelenterata. The Coelenterata have traditionally been treated as the most primitive of the Eumetazoa. Haeckel regarded them as the source of the flatworms, and hence of all higher phyla. The principal reason for his viewpoint is the obvious resemblance of a hydroid polyp to a gastrula. For the hydroid consists of two simple cell layers, with no organ systems and with only a trace of noncellular mesoglea. The anatomical structure of the polyp could be derived from that of the gastrula simply by the elongation of the body and the drawing out of a circlet of tentacles around the mouth. Food is taken in and waste residues expelled by the mouth, which is simply the blastopore of the blastula. Food is still digested by the protozoan method, that is, the cells of the gastrodermis engulf food particles, and digestion is carried on intracellularly. But enzymes are also secreted into the gastrovascular cavity, and much digestion occurs there. Muscle tails may be formed in connection with either the epidermis or the gastrodermis. A nerve net is formed from epidermal elements. Thus there is a high degree of tissue differentiation, but no organ systems. Outstanding specializations of the coelenterates include the development of nematocysts, organelles for food procurement and defense; and the alternation of a free-swimming, sexually reproducing medusa generation with a sessile, asexually reproducing polyp generation. This is unrelated to the alternation of generation of plants, for both generations are diploid. Superficially so different, the polyp and the medusa are structurally very similar, for one can derive the medusa from the polyp by simply inverting the latter, greatly increasing the amount of mesoglea and its cellular contents, and drawing the circlet of tentacles away from the mouth, as illustrated in Figure 47.

Haeckel assumed that the ancestral coelenterate was a polyp, because of the ease with which a polyp can be derived (in theory) from the Gastrea. But study of what appears to be the most primitive order of hydroids, the Trachylina, has led to the conclusion that the medusa phase is primary and the polyp derived. Once formed, the Coelenterata diverged along three major lines, each comprising a class of the phylum. In the most primitive class, the Hydrozoa, both generations are generally well developed. The class includes hydroids such as the *Hydra* of elementary laboratories and the much more typical marine colonial forms, such as *Obelia*. There are also some in which the polyp generation is reduced, as in the order Trachylina. In the class Scyphozoa, including the jelly fishes,

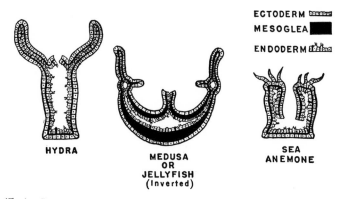

ECTODERM
MESOGLEA
ENDODERM

HYDRA

MEDUSA
OR
JELLYFISH
(Inverted)

SEA
ANEMONE

FIGURE 47. AN INVERTED JELLYFISH COMPARED TO *Hydra* AND TO A SEA ANEMONE, showing their structural similarity. (From Hunter and Hunter, "College Zoology," W. B. Saunders Co., 1949.)

the medusa is much the more prominent generation, with the polyp being reduced or absent altogether. The final class, the Anthozoa, includes only the polyp phase, the medusa being suppressed entirely. The class comprises the sea anemones, the corals, and their allies. Many of these, including all of the corals, secrete a calcareous exoskeleton, because of which they have left an excellent fossil record going back to the Ordovician. All three classes are ancient, and it is probable that they diverged in pre-Cambrian times from a primitive hydrozoan type not dissimilar to the Trachylina.

Phylum Ctenophora. The Ctenophora are a small phylum, only about eighty species being known. All are small, marine animals. They are commonly known as "sea walnuts" or "comb jellies" because of the presence of comb-like plates of cilia. They share some important characteristics with the coelenterates. Thus they are radially or biradially symmetrical in contrast to all higher phyla, including the Echinodermata, in which bilateral symmetry is primary, and radial symmetry derived. They are at the tissue grade of construction, with an abundant mesoglea separating the epidermis from the gastrodermis. In both phyla, the gastrovascular cavity is present, and its branches distribute food through the body. But the ctenophores lack nematocysts, and there is no alternation of generations. Unlike the coelenterates, they are hermaphroditic. There seems to be little doubt that the coelenterates gave rise to the Ctenophora, yet all attempts to relate them to any of the classes of living coelenterates have failed. It seems most probable that the Ctenophora were derived from the same ancient, pre-Cambrian, trachyline stock that gave rise to the three classes of the Coelenterata, and at about the same time.

PRIMITIVE BILATERAL PHYLA

Some of the ctenophores have become elongate and flattened, and it has been suggested that they are related to the ancestors of the flatworms, the phylum Platyhelminthes. But the majority of zoologists believe that the appearance of homology is misleading. Haeckel believed that a primitive hydrozoan was the ancestor of the bilateral phyla, basing his opinion as usual upon the Biogenetic Law. The evidence is insufficient, and it seems at least as probable that flatworms were first derived from the same planula-like stock which gave rise to the coelenterates. But in the flatworms, the middle cell layer (mesoderm) became more highly developed, with organized muscle layers, reproductive system, and an excretory system. Yet no coelom, or body cavity, developed within the mesoderm, as it does in most higher groups. The nervous system is formed from the ectoderm, although it is imbedded in the mesoderm. It is not a diffuse nerve net as in the coelenterates, but rather it is somewhat centralized, being organized about cerebral ganglia at the head end and longitudinal cords. There are organized sense organs, including eyes. Food is still distributed to the various parts of the body by the branches of the digestive tract, which is now called an intestine rather than a gastrovascular cavity. Thus the flatworms are much advanced beyond the tissue grade of construction which characterizes the radiate phyla, for definite organ systems are present, principally in the mesoderm.

Hadzi has urged a radically different phylogeny. He believes that multinuclear ciliates became acoelous flatworms by formation of cell membranes. In evidence, he points out that both are ciliated; hermaphroditism of flatworms he finds homologous with conjugation of ciliates; trichocysts of ciliates he believes are represented by sagittocysts, rod-like inclusions in some of the epidermal cells of acoelous flatworms. He believes that the Anthozoa were derived from flatworms by adoption of the sedentary life, and that the other coelenterates were then derived from anthozoans. The higher Metazoa were also derived from flatworms, and are grouped into only four phyla, a procedure which unites some highly diverse groups.

At this point it may be well to recall the admonition of Hyman [*] that "the exact steps in the evolution of the various grades of invertebrate structure are not and presumably never can be known. Statements about them are inferred from anatomical and embryological evidence and in no case should be regarded as established facts." Although the phylum Platyhelminthes, and especially the most primitive class of this phylum, the Turbellaria, is commonly treated as the stem group from which the higher phyla arose, this is by no means established. Closely related to the Platyhelminthes, and on the same general level of organization, is the phylum Nemertinea, a small group of marine flatworms. This group is less well known than the former because it is predominantly marine, and because it is a difficult group to study. But the Nemertinea show some characters which qualify them for special consideration as potential forerunners of the higher invertebrates. For the first time, there is an anus present, so that the digestive system is said to be complete. Also, there is a simple blood circulatory system, and the blood contains hemoglobin. Particularly

---

[*] By permission from Hyman, L. H., "The Invertebrates," Vol. I, McGraw-Hill Book Co., Inc., 1940.

important according to Kofoid is the fact that the nervous system is based upon cerebral ganglia and eight longitudinal nerve cords, two dorsal, two ventral, and two on each side. He has pointed out that this lends itself to the formation of the principal invertebrate nervous systems by the development of the ventral cords and the suppression of the others; and to the formation of the chordate nervous system by the development of the dorsal cords and the suppression of the others. But the fact that the nemerteans capture their food by means of an extensible proboscis, a mechanism found in no other group, suggests that they are a terminal group.

Yet another possibility is that a third group, an unknown phylum of very primitive flatworms derived from the primitive planula by the development of the mesoderm may have been ancestral to both the Platyhelminthes and the Nemertinea. Whatever their origin, once formed, the platyhelminths diverged along three main lines of descent. The first of these is represented by the class Turbellaria, comprising the free-living flatworms, of which *Planaria* is the best known example. The other two classes have become greatly modified for parasitism, and most of them have developed complicated life cycles to permit their transfer from one host to another. These are the class Trematoda, the flukes, which have retained all of the organ systems of the Turbellaria, and are generally internal parasites of vertebrates; and the class Cestoda, the tapeworms, which have undergone an extreme degenerative evolution. They are all intestinal parasites of vertebrates.

On the same general level of organization as the above discussed acoelomate groups, there are a number of minor phyla, of very uncertain relationships, which are characterized by the possession of a pseudocoel. A coelom is, by definition, a body cavity formed within mesoderm. A pseudocoel, on the other hand, is a remnant of the cavity of the blastula, and it may be partially filled by large, vacuolated cells. The taxonomy and relationships of these pseudocoelomate organisms are very uncertain. It seems best here simply to acknowledge that they exist, and that they are among the most primitive of bilaterally symmetrical animals, without attempting to unravel their relationships. These are the Rotifera, a phylum of microscopic animals which are known to every student of elementary biology as fascinating contaminants of most protozoan cultures; the phylum Gastrotricha, which are small, worm-like animals of wide distribution; the phylum Kinorhyncha, comprising a group of minute, marine, worm-like animals; the phylum Nematoda, a very important group, the so-called round worms, which includes many little-known free-living worms as well as some of the best known parasites of both plants and animals; the phylum Nematomorpha, or horsehair worms, consists of long, slender worms which are parasitic during larval life, but are generally free-living when mature; and the phylum Priapuloidea, comprising a few species of small, externally segmented marine worms. These six phyla are sometimes grouped together as classes of a single phylum, Aschelminthes. The phylum Acanthocephala, or spiny-headed worms, are generally small worms the young of which are parasitic on invertebrates, while the adults are parasitic on vertebrates, including man. A final pseudocoelomate phylum is altogether different from these worm-like phyla. This is the phylum Entoprocta, a group of colonial animals, often forming encrusting colonies. Formerly regarded as a class of the phylum Bryozoa, it is now clear that they differ fundamentally from the Bryozoa. The intestine is U-shaped, and both the mouth and the anus are included within a ring of ciliated tentacles, the lophophore, which sets up a current for plankton feeding. This gives them a rather hydroid-like appearance superficially.

THE PROTOSTOMOUS PHYLA

The remaining major phyla of the Animal Kingdom can be arranged in two diverging lines of descent, largely on the basis of embryological criteria. One line culminates in the Annelida, Arthropoda, and Mollusca, while the other culminates in the Echinodermata and the Chordata. Certain minor phyla can be associated with one or the other of these main lines with varying degrees of satisfaction. Haeckel believed that the echinoderm-chordate line was derived from the Turbellaria, while Kofoid believed that both lines were derived from the Nemertinea. It is at least as likely that both were derived from the unknown, primitive, acoelous flatworm from which both the Platyhelminthes and the Nemertinea were probably derived.

The cleavage divisions of the annelid-arthropod-molluscan line are both spiral and determinate. In spiral cleavage, spindles are at right angles to those of the preceding division, so that the cells of each layer alternate with the next layer like bricks in a building. This is not true of the large, yolky eggs of arthropods, yet their position in this series is secure, for their origin from annelids is clear (see below). Determinate cleavage proceeds according to a set pattern, with the part of the body to be formed from each blastomere fixed from the start. Destruction of a blastomere results in a deficient larva. One can designate which blastomere will form ventral surface, which the gut, and so on. The mesoderm is formed from stem cells which multiply to form a pair of ventral bands, growing forward from the posterior end of the larva. The coelom is formed by splitting of these bands, hence these phyla are said to be *schizocoelous*. The blastopore becomes the mouth of the adult, and hence this whole series of phyla is called Protostomia.

Generally, though by no means always, development leads to a trochophore larva. This is a more or less spherical larva with an *apical tuft* of cilia dorsally, and a prototroch, or girdle of cilia around the equator, by which it can swim weakly. There is a digestive system consisting of a mouth, a short foregut, an enlarged stomach, a short hindgut, and an anus, a complete digestive system. Some mesoderm is present, and some mesodermal organs, such as a kidney, are formed. This type of larva is characteristic of the annelids and molluscs, but unique larval stages have been developed by the arthropods.

Other characters are also held in common by the members of these phyla, but those mentioned are sufficient to indicate the probability of relationship among them. The actual relationships are, however, largely a matter of conjecture. Origin from some primitive, acoelous flatworm is probable, yet there is little basis for a decision as to which of the possible groups of flatworms was the ancestor, beyond the fact that spiral cleavage is also found in the Nemertinea and in the polyclad Turbellaria. Because of the widespread occurrence of the trochophore larva, it is generally supposed that an independent, trochophore-type animal was an ancestor intermediate between the flatworms and the present-day phyla of the annelid-arthropod-molluscan series. The only evidence for this is the interpretation of the embryological evidence according to the Biogenetic Law, and it has already been pointed out how insecure this is. Yet, although it may be true, direct evidence is not likely to be obtained, for such an organism would be very poor material for fossilization, and it would have to be sought in pre-Cambrian rocks, for the major invertebrate phyla were all present in the Cambrian.

Phylum Mollusca. Whatever the source, the phylum Mollusca was early established as a group divergent from the others. This is a great phylum, the second largest in the Animal Kingdom, having around 80,000 species encompassing a great variety of forms, and inhabiting almost every type of marine, freshwater, and terrestrial habitat. All of the extant classes of Mollusca were present already in the Cambrian, and literally hundreds of species are known. The classes were just as distinct then as now, and so paleontology is of no help in deciding what the relationships within the phylum may be.

In addition to the general characters of all members of the protostome line, the phylum Mollusca is distinguished by many structural characters. In all, the body is divided into four regions: a muscular foot, a head, a visceral hump, and a mantle which generally secretes a calcareous shell. Basically, they are bilaterally symmetrical, but this is obscured in the adults of some classes. The coelom is very much reduced. Mollusca are traditionally described as unsegmented, yet in 1957 *Neopilina*, a new species from the deep waters of the Gulf of Mexico, was described. This animal has several segmentally arranged organ systems, and its affinities are with a class, Monoplacophora, which had been thought to be extinct since the Devonian. It indicates that the primitive molluscs may have been segmented. On the basis of anatomical evidence, the class Amphineura is regarded as very primitive, but not necessarily as the progenitor of the more specialized classes. This class comprises the chitons or sea cradles, animals which are familiar parts of the fauna of rocky sea coasts, but there are no freshwater species. The chitons have a broad, flat, elongated foot, over which lies the visceral hump and at the anterior end of which is the head. The mantle covers the visceral hump and typically secretes a series of eight calcareous plates, the valves, which may be beautifully ornamented.

The class Gasteropoda, which includes the snails, slugs, limpets, and their less well-known allies, has developed a much enlarged visceral hump. In most gasteropods, this hump has grown asymmetrically, with the familiar coiling as a result. Correlated with this, there has been a torsion of the visceral hump through 180°, so that structures which were originally posterior have moved to an anterior position, and vice versa. The class Scaphopoda, or tooth shells, is a small group which is very much specialized for digging. All species are marine, and there are only a few of them. The class Lamellibranchiata includes the bivalves, the familiar clams, mussels, and their allies. They are compressed bilaterally, and enclosed within a two-valved shell, hinged dorsally, and secreted by the two lobes of the mantle. The gills are immensely enlarged to form broad ciliated tracts which produce a current for feeding upon plankton and detritus. The head is very much reduced. The foot is generally wedge-shaped, unlike the broad, flat foot of the more primitive molluscs, but it can be protruded between the valves to serve a locomotor function. In some species, it is specialized for burrowing.

The class Cephalopoda includes the octopi, the squids, the chambered *Nautilus,* and its extinct allies. They are the most complex of mollusca, comparing favorably with the most advanced insects and vertebrates in degree of complexity. The visceral hump, like that of the gasteropods, is much enlarged, but it is still symmetrical. The mouth is surrounded by a ring of tentacles for seizing food. The head and foot are fused so completely that there is no agreement among specialists as to which these tentacles are derived from. There is a siphon by which water may be forcibly ejected from the mantle cavity. The coelom is better developed than in most molluscs. The nervous system is highly centralized, and efficient eyes, superficially quite similar to those of vertebrates, are present. The more primitive members of the class have a chambered shell of which the animal always lives in the most recently secreted chamber. Each chamber corresponds to a stage in the growth of the animal, like the successive moults of an arthropod. The shelled cephalopods are today represented only by a single genus, *Nautilus,* but this is the last remnant of a once dominant group in the oceans of the world. The nautiloids appear in the fossil record in the late Cambrian, and rapidly assumed a dominant position in the marine fauna. They reached their peak of development in the Silurian, and then gradually declined. During the Devonian, they gave rise to another suborder, the Ammonoidea, with which they competed unsuccessfully for a long period of the earth's history. The ammonites were the dominant marine molluscs throughout most of the Mesozoic Era, yet they dwindled and became extinct near the end of the Cretaceous. Meanwhile the nautiloids survived in small numbers, and are known from a single genus at the beginning of the Cenozoic Era. No longer being in competition with the ammonites, the nautiloids underwent a rapid evolution at the beginning of the Cenozoic, for seven new genera appear during the Paleocene epoch. But only one of these, *Nautilus,* has survived to the present time. Meantime, forms with the shell much reduced and internal, the octopi and squids, have become the principal cephalopods.

Phylum Annelida. The second main branch of the protostome line is the annelid-arthropod branch. The phylum Annelida is best known by the common earthworms, but it includes a wide variety of worms arranged in four classes. All annelids are segmented, that is, the functional and structural units of the body are repeated serially along the length of the body, and the segments may be marked externally. A well-developed coelom separates the digestive tract from the muscular body wall. There is an advanced nervous system based upon a pair of cerebral ganglia and a pair of ventral cords on which ganglia are located in each segment of the body. Whenever there is a larval stage, it is a trochophore. Most annelids have a thin cuticle, and the typical species have segmentally arranged chitinous bristles called chaetae.

The phylum Annelida is best typified by its most primitive class, the Polychaeta. These take their name from their numerous chaetae, which arise from limb-like lobes, the parapodia, on each body segment. There is typically a distinct head which may have appendages. The sexes are separate, and fertilization is external. There is a trochophore larva. The polychaetes are adapted to a wide range of habitats, including pelagic and bottom-dwelling, surface-crawling and burrowing, and permanent tube-dwelling, and they show very striking morphological adaptations. In numbers of species, they undoubtedly exceed the remaining classes combined. Yet they are less well known because they are almost exclusively marine.

Better known, but far less typical of the phylum, is the class Oligochaeta, including the earthworms and many small freshwater annelids. The class appears to have been formed from the polychaetes by a process of reduction and simplification. The head region is much reduced and never includes appendages. The chaetae are reduced both in size and in numbers, and they are no longer set on parapodia. Oligochaetes are all hermaphroditic. Development is direct, there being no larval stages. The embryos develop in a cocoon, an adaptation to terrestrial life. The class is much more uniform than is the Polychaeta.

The class Archiannelida is a small group of extremely simplified marine annelids. These are all small worms which generally lack parapodia and chaetae. They are ciliated, like the young of polychaetes. Their nerve cords retain the primitive connection with the epidermis. The coelom is only slightly developed. The Archiannelida were named with the intention of inferring that this was the most primitive class of the phylum and the probable progenitor of the other classes. But embryological and morphological studies have led to the conclusion that this class is not truly primitive, but rather has been derived from the Polychaeta by extreme simplification.

The class Hirudinea, or leeches, appear to have been derived from oligochaete ancestors, for they share many characters with that class. But all leeches are ectoparasites, and they are extensively modified for that mode of life.

Phylum Arthropoda. The protostome series is climaxed by the great phylum Arthropoda, the most successful of all groups of animals if numbers of species be an indication of success, for about three fourths of all species of animals are arthropods. This is also the most varied of phyla, for there are arthropods adapted to every imaginable habitat from the abyssal depths of the ocean (crabs and pycnogonids) to aerial heights (insects). Perhaps the key to arthropod evolution is their early develop-

143

ment of a thick, chitinous cuticle. This made necessary the joints—thin places in the cuticle—from which the phylum takes its name. In order to move the hard pieces of the cuticle, the continuous muscular wall, inherited from polychaete ancestors, became broken up into specialized muscles, attached to ingrowths of the cuticle. The limbs, being jointed, are far more adaptable than the parapodia of the polychaetes, and they have become specialized for many functions, including sensation, feeding, and locomotion. This thick cuticle undoubtedly minimizes water loss by evaporation, and it may be that this is a major factor in their successful invasion of the land, for no other invertebrate phylum has so large a portion of terrestrial species. The concentration of the nervous system and the sensory organs in the head region (cephalization), begun in the polychaetes, is carried much further in the arthropods. The coelom is much reduced, and has become largely replaced by a hemocoele. Segmentation is prominent externally, but less so internally.

As might be expected with so large a phylum as the Arthropoda, there is no general agreement on the taxonomic rank to be accorded to the major divisions of the phylum. For present purposes, five major subphyla will be recognized, and no attempt will be made to treat their component classes because of the vastness of the groups. These are the Trilobita, the Crustacea, the Myriapoda, the Insecta, and the Arachnida. These well-defined groups, with the exception of the Trilobita, are universally familiar, and so need not be defined here.

The trilobites are of especial interest because they are the most ancient arthropods known, and because their structure was very generalized and could conceivably have given rise to the other major groups of arthropods. The trilobites were dominant from the Cambrian through the Silurian. They then declined until the Permian, from which period only a single species is known. With this, the group became extinct. The trilobite body was enclosed in a chitinous skeleton, and was divided into three longitudinal lobes by two longitudinal furrows, hence the name of the group. The body consisted of a head and a segmented trunk. The segments were generally movable, but a variable number of posterior segments were joined together to form a rigid unit, the pygidium. There was a single pair of antennae. The remaining appendages were all simple, undifferentiated, biramous appendages. None were specialized as mouth parts, but all had gnathobases, basal processes which could be used for biting. All species were marine. A typical trilobite is illustrated in Figure 48. Nothing is known of the internal structures.

Fossil remains of transitional organisms from trilobites to the other major arthropod types are entirely lacking and it may be that the several groups arose independently. Indeed, this would avoid some perplexing inconsistencies. Comparative morphology indicates that the crustaceans, myriapods, and insects form one line of descent, while the arachnids must have diverged very early. In the Crustacea, this transformation has involved the division of the body into a highly fused cephalothorax and an abdomen in which the original segmentation is retained. The appendages have become greatly diversified, while still retaining the biramous plan

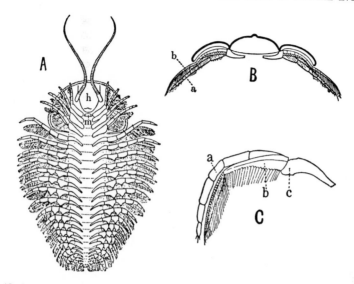

FIGURE 48. A TRILOBITE, *Triarthrus becki*. *A,* ventral view. *B,* section through a thoracic segment. *C,* a thoracic leg. *a,* endopodite; *b,* exopodite; *c,* protopodite with gnathobase. (After Beecher, from Borradaile and Potts, "The Invertebrata.")

(Chapter 3). Almost all crustaceans are aquatic, but there are some terrestrial species, such as the sowbugs. Like the trilobites, the crustaceans are represented in early Cambrian rocks, and it is not improbable that they arose in pre-Cambrian times.

The Myriapoda, including the millipedes, centipedes, and their allies, may have been derived from a crustacean progenitor by the reduction of the exoskeleton and the loss of the gills. The latter were replaced by a system of tracheae, small tubes which carry air to the tissues of the body, thus permitting direct respiration. While not well represented in the fossil record, yet they are known from rocks as old as the Devonian. Insects probably also arose in Devonian times or earlier, but good fossils are first seen in rocks of Pennsylvanian age. They may have been derived from myriapods, or they may have come directly from crustacean progenitors. Whichever ancestry they may have had, the tracheal system is always present and is undoubtedly one of the important adaptations which has permitted successful invasion of the land. The body became divided into three well-defined regions, head, thorax, and abdomen. The appendages of the head segments are all specialized either for sensation or for eating. The thorax bears three pairs of walking legs. Dorsally, it bears two pairs of wings, but one or both pairs may be lacking. The abdominal appendages are all suppressed. The immense variety of insects which has been produced is indicated by the number of known species, approximately 660,000, or more than the combined numbers of species of all other living groups. And the numbers of individuals of many of these species are truly immense.

The subphylum Arachnida includes a highly varied array of organisms which have developed along lines quite divergent from the series just discussed. It comprises the horseshoe crabs, scorpions, spiders, harvest men, ticks, mites, and their allies. It is difficult to derive them from the trilobites, but it may be possible. There is a tendency toward suppression of segmentation, with only two body regions, the anterior prosoma and the posterior opisthosoma being marked off. There are six pairs of appendages, of which four pairs are walking legs. Unlike all other arthropods except the trilobites, none of the appendages are specialized as jaws. Gnathobases of the anterior limbs serve this function.

The class Pycnogonida, the sea spiders, is a small and little-known group which is usually placed with the Arachnida, simply because of their superficial resemblance to spiders. Yet their morphology is utterly different. Hedgpeth has studied the group thoroughly, and has concluded that they are undoubtedly arthropods, but that they are so widely divergent that they cannot reasonably be grouped with any of the other arthropod types. Even less justification can be found for the common practice of making the Tardigrada a class of arachnids. This is a small and little-known group of minute, freshwater organisms which shows some relationship to the Arthropoda, and has usually been treated as a class of the Arachnida. But specialists in the field feel that it should be regarded as an independent phylum of uncertain relationship to the Arthropoda.

The Onychophora—A Unique Evolutionary Link. It was mentioned above that the derivation of the Arthropoda from the Annelida is more certain than is the derivation of any other phylum. This depends upon the Onychophora, a group of about eighty species all of which are assigned to a single genus, *Peripatus* (Figure 49). This group of animals shows a peculiar mixture of annelid and arthropod characters. Among the annelid characters may be mentioned the general appearance of the organisms, for they look much like polychaetes in which the parapodia do not bear chaetae. The cuticle is thin like that of annelids, and the muscles of the body wall are continuous. The excretory organs of both annelids and onychophorans are mesodermal tubules segmentally arranged (coelomoducts), while those of the Arthropoda are usually entodermal or ectodermal in origin. The reproductive ducts of the Onychophora are ciliated, but cilia are unknown among the Arthropoda. The eyes of the annelids and onychophorans are simple, whereas those of the arthropods are compound. On the other hand, in contrast to the annelids, the Onychophora and the Arthropoda have jaws derived from appendages. The coelom in each is much reduced and largely replaced by a hemocoel, while the coelom of the annelids is highly developed. The circulatory system of the Onychophora also resembles that of the Arthropoda rather than that of the Annelida. Finally, the respiratory system of the Onychophora consists of a set of tracheae, a characteristic known nowhere else but in the Arthropoda.

Because of this strange mixture of characters, the taxonomic position of the Onychophora has always been a much vexed question. They were originally treated as a class of the Annelida. But because of the presence

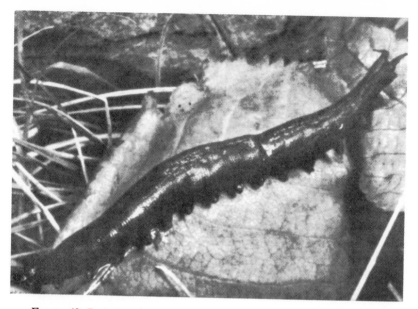

Figure 49. *Peripatus.* (Courtesy Ward's Natural Science Establishment.)

of a hemocoel and especially because of the tracheal system, they are now generally treated as a class of the Arthropoda. But these are not the only possibilities. They are sometimes treated as an independent phylum intermediate between the Annelida and the Arthropoda. And Light has urged that, as they are not properly separable from either of the major phyla, the entire annelid-onychophoran-arthropod series ought to be recognized as one great phylum under the name Articulata. With no other series of phyla are such considerations possible and hence the statement with which this discussion began, that the origin of the Arthropoda from the Annelida is more certain than the origin of any other phylum. It is an unfortunate fact that the Onychophora are scantily known in the fossil record.

Minor Protostomes. A few minor phyla also show the general characters of the protostome line. Two of these, the Sipunculoidea and the Echiuroidea, are worm-like burrowers of the tide flats. They are generally treated as minor annelids, an arrangement which is more convenient than accurate. Yet they probably are more closely related to the Annelida than to any other phylum. The three remaining phyla, the Bryozoa, the Phoronida, and the Brachiopoda, are more difficult to place. Like the Entoprocta, they feed by means of a lophophore, but unlike that phylum they have a coelom. The Bryozoa are small, colonial animals which superficially resemble the Entoprocta, but actually differ from them fundamentally. The Phoronida includes only two genera. The animals are elongate, worm-like creatures which dwell in tubes on tide flats. When the tide is in, the lophophore projects into the water to feed on plankton and detritus. The

147

phylum Brachiopoda, or lampshells, bears a superficial resemblance to the molluscs because of their bivalve shells, but these are dorsal and ventral rather than right and left. Internally, they do not suggest the molluscs at all. They have a very prominent lophophore. These marine animals are present in the earliest Cambrian deposits, and perhaps the greatest interest in the group derives from the fact that a single genus, *Lingula*, has persisted from the Ordovician to the present time, a span of 400,000,000 years. It may be the oldest genus in existence. These three phyla were formerly grouped together as a single phylum, the Molluscoidea. Yet they have little in common with one another except the lophophore, and there is little indication that any of them have a close relationship to the Mollusca. Hence it seems best to treat them as independent phyla of protostomes, of uncertain relationships to the larger phyla.

THE DEUTEROSTOMOUS PHYLA

The other major branch of coelomate animals is the Deuterostomia, and it comprises only five phyla: Chaetognatha, Pogonophora, Echinodermata, Hemichordata, and Chordata. This series contrasts with the Protostomia in that group of embryological characters which was used above to characterize the latter. The cleavage divisions are not spiral, and neither are they determinate. The mesoderm is not formed from stem cells, but rather from outpocketings of the entoderm of the gut. This simultaneously establishes the coelom, which is said to be *enterocoelous* (meaning simply that the cavity is established from the gut). The original blastopore becomes the anus, and a new mouth is formed in this series of phyla, hence the name Deuterostomia.

Development does not lead to a uniform larval type. The Chaetognatha have a unique larval type. The Echinodermata have several types of larvae, but all of them first pass through a *Dipleurula* stage to which especial theoretical significance is attached. A typical gastrula is formed, then the entoderm and ectoderm first fuse at one end, then break through to form a mouth. The blastopore is now the anus. The digestive tube now buds off an anterior vesicle which first divides into two lateral compartments, and finally into three segments on each side. These are the coelomic pouches. The cilia which covered the blastula and gastrula evenly now become concentrated in a series of bands arranged around the concave ventral surface of the animal. This is the Dipleurula larva. Because of its universality among the echinoderms, it is generally believed that the echinoderms must be descended from a Dipleurula-like ancestor which was bilaterally symmetrical and free-swimming.

The Hemochordata have a type of larva, the Tornaria, which is quite similar to the Dipleurula, and even more similar to the Bipinnaria larva of starfishes. When first discovered, the Tornaria was described as a larval starfish, and only much later was the error discovered. This resemblance of the larvae is one of the major arguments for a relationship of the Hemichordata to the Echinodermata. Larvae are not of general occurrence among the Chordata, but the tunicates and amphibians have tadpole larvae while the cyclostomes have a unique larva, the Ammocoetes.

Phylum Chaetognatha. The Chaetognatha are a small and uniform group of marine worms, the arrow worms, which show little evidence of relationship to any larger group of animals. They are included here because of conformity to the general deuterostome characters, yet they show no more specific affinity to any of the remaining deuterostome phyla, and it seems probable that the phylum branched off very soon after the formation of the deuterostome line. They show a superficial resemblance to Amphioxus, but this is undoubtedly misleading. Aside from a few doubtful specimens, the phylum is not represented in the fossil record.

Phylum Pogonophora. The Pogonophora is a phylum of deep-sea worms, only recently discovered and still but little known. The body consists of small protosoma and mesosoma, and a very elongate metasoma. The protosoma may bear one or more tentacles. It has an unpaired coelom, drained by a pair of nephridial ducts. In the other body segments, the coelom is paired. A nervous mass and ring in the protosoma give rise to a paired dorsal nerve cord. The circulatory system consists of two longitudinal vessels. Musculature is made up of subepidermal longitudinal fibers. There is no digestive system whatever. Larvae are unknown. What little is known of these animals seems to ally them with the Deuterostomia.

Phylum Echinodermata. The Echinodermata are best known by the starfishes, and these perhaps typify the phylum well. All echinoderms have secondarily established radial symmetry in the adult, after beginning life as a free swimming, bilaterally symmetrical larva. The radial symmetry of the adults is generally based upon a pentamerous (five radial segments) plan, or upon a plan derived from such. The development of radial symmetry may have been related to the change from a pelagic to a sessile mode of life by primitive echinoderms, for radial symmetry is a general characteristic of sessile organisms. The fossil record of the echinoderms is one of the best, going clear back to early Cambrian times, and excellent phylogenies can be constructed within each of the five extant and two extinct classes. But the record does not throw light upon the origin of the phylum, nor upon its possible relations to other phyla. The questions depend, at present, entirely upon the embryological evidence, with all of its limitations.

Phylum Hemichordata and the Origin of the Chordata. The Hemichordata are a small phylum of worm-like marine animals which have been extensively studied because of their supposed relationship to the Chordata. In Haeckel's phylogeny of the vertebrates, the hemichordates were given as the next stage after the primitive flatworm. They were originally classed as a subphylum of the Chordata, because they show the three basic diagnostic characters of the Chordata: a dorsal nerve tube, a pharynx modified for respiration, and a notochord. Yet each of these is equivocal. The dorsal nerve tube is confined to the collar region, and the main nervous system is a ventral nerve cord like that of many invertebrates. The pharynx is pierced by numerous gill slits, yet it appears that these function primarily as exit pores for the feeding current rather than

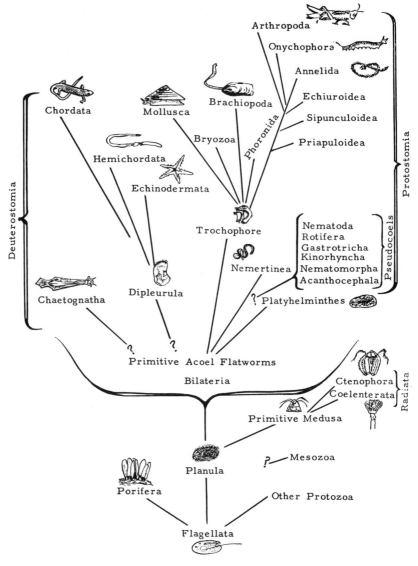

FIGURE 50. PROBABLE MAIN LINES OF ANIMAL PHYLOGENY. (Redrawn from Hyman.)

as respiratory organs. Yet some respiration probably does occur in the gills. But feeding may have been the primary function in the gills of true chordates also. Finally, the notochord is a small structure developed as an outgrowth from the digestive tract into the proboscis. It has also been homologized with the pituitary gland, and so its taxonomic value is doubtful. It seems probable that these worms are related to the chordates, but how is not clear. Their relationship to the echinoderms, as indicated by a comparison of the Tornaria and Bipinnaria larvae, has already been discussed.

As the phylum Chordata, including as it does the vertebrates and man, will be discussed in the next chapter, it need not be taken up here beyond the obvious statement that this phylum forms the climax not only of the obscure (from the viewpoint of relationships) Deuterostomia, but also, perhaps, of the entire Animal Kingdom. Figure 50 summarizes one viewpoint on animal phylogeny.

SOME EVOLUTIONARY GENERALIZATIONS

Now that the greater part of the course of evolution has been sketched, and before taking up that part of the story which is most closely related to those who study it, it may be well to give some consideration to broad tendencies and principles of evolution. At the outset, evolution is not always "upward" or "progressive." Many examples are known of the evolution of simpler or more degenerate types from originally complex types. Thus the fungi may have evolved from algae by the loss of chlorophyll. Grasses have evolved from lily-like ancestors by simplification of parts, especially of flowers. Mistletoe, an angiosperm parasitic upon trees, has undoubtedly evolved from free-living ancestors. Similarly in the Animal Kingdom, many examples of retrogressive or degenerative evolution are known. The development of sexual reproduction and its great evolutionary importance was emphasized in Chapter 8. The Rotifera were undoubtedly evolved from bisexual ancestors, and some species are still bisexual. Nonetheless there are species in which the males are unknown. These still reproduce sexually, for the ova develop parthenogenetically, but the major advantages of sexual reproduction are lost. The development of the parasitic habit almost always involves degenerative evolution. In the tapeworms, this is extreme. Although derived from free-living flatworms, with well-developed digestive, nervous, reproductive, muscular, and other systems, the tapeworm is reduced substantially to an absorptive sac containing gonads. But, while such degenerative evolution is characteristic of parasites, it is by no means confined to them. As already pointed out, the class Archiannelida was most probably derived from polychaete ancestors by a process of simplification and loss of parts Thus it is clear that evolution can be retrogressive as well as progressive

Origin of New Groups from Primitive Ancestors. A second very important generalization relates to the form of the tree of life. It is more properly a shrub than a tree, for new groups do not arise from the most advanced and specialized members of their parent groups, but from the primitive, unspecialized ones. Thus the primitive flagellates have given rise to many additional plant and animal groups, but the more specialized protozoan and algal groups are generally terminal. Again, if indeed the Hemichordata and Chordata were derived from echinoderms, it seems certain that the more advanced phyla must have arisen from the ancestral Dipleurula in a very primitive stage before radial symmetry developed. One more qualification relative to the shape of the phylogenetic tree is

151

necessary for plants. Because of the phenomenon of allopolyploidy—to be discussed in detail below—hybridization of related species may result in new species. Hence branches may form, then fuse again, so that a network results. This phenomenon is of great importance for plants, less so for animals.

Rates of Evolution. The rates of evolution have not always been uniform. In general, periods of major geological change have been periods of rapid evolution, while periods of geological uniformity have been periods of slow evolution. This might be expected *a priori*, for the selective forces themselves should be in a state of flux during geologically unstable periods, while they should be quite stable during geologically uniform periods. Thus in a period of uniformity, organisms generally should tend to become well adapted to their environments. In such a situation, almost any change would be disadvantageous, and would tend to be eliminated by natural selection. But in periods of geological change, only those organisms which also changed could possibly have a selective advantage. Thus there was a rapid burst of evolution in the Silurian and Devonian when plants and animals were colonizing the land. Not only were vast new habitats thus opened up, but this was a time of mountain-building and of great changes in sea level. In the Mississippian, physical conditions were again quite stable, and evolution was slow. Many such alternations of periods of rapid and slow evolution have occurred. It appears that we are at present in the midst of a period of rapid evolution.

Two special situations favor rapid evolution. The first of these is adaptation of a group to a new mode of life, as invasion of land by aquatic plants and animals. During the transition, the organisms must be rather ill-adapted, hence strong selection pressure favors rapid change, and only those that can respond will leave descendants. This will be greatly facilitated if, in the earlier environment, characteristics developed which are *preadaptive* to the later habitat. Thus many Devonian fishes developed lungs, which were adaptive to life in stagnant, drying ponds. This preadaptation later speeded the adaptation to life on land. Simpson has called this "quantum evolution," as it involves sudden change from one "adaptive orbit" to another. He regards the term as unsatisfactory, because quantum events in physics are very small, while quantum evolution is on a large scale. Perhaps *macroadaptive* evolution would be a better term.

Second, when a group has invaded a new major mode of life, many variations of habitat are open to it with a minimum of competition, and hence rapid diversification is favored until these empty ecological niches are largely filled. This is the adaptive radiation which was discussed above.

Entirely apart from changes in the general rate of evolution, the rates in different lines of descent have not been equal. In some, change has been extremely slow for long periods of time. The brachiopod *Lingula*, for example, has been substantially unchanged since the Ordovician. In this connection, it should be pointed out that the warm, shallow seas which it inhabits constitute perhaps the most stable environment on earth. On the other hand, the mammals have evolved very rapidly during the Cenozoic Era.

Trends in Size. A very common trend in evolution, sometimes called Cope's Law, is one toward increasing size of individuals. The original studies of the phenomenon were made upon vertebrates, but comparable studies have shown the same tendency in many groups of invertebrates and plants. A review of the paleontology of almost any group shows that its largest representatives are not its earliest ones, though not necessarily its latest ones either. Newell has pointed out that species now living are the largest known representatives of the vertebrates, crustaceans, echinoderms, pelecypods, gastropods, cephalopods, and annelids. Yet the tendency toward size increase has been by no means universal. As already mentioned, the rise of herbs and shrubs is a recent thing, and they have been derived from trees and other large plants. Hooijer has pointed out that progressive size decrease has been characteristic of many vertebrate groups during the Quaternary period, which is now in progress.

Complexity and Efficiency. It is obvious that the general progress of evolution has involved the development of new organ systems and increasing complexity. Yet development of increased efficiency often involves reduction in number and complexity of parts. Vestigial organs in general could perhaps be considered in this light, but it is equally true of actively functional structures. Thus the teeth of fishes are very numerous and usually indefinitely replaceable. They are less numerous in amphibians and reptiles, still less so in mammals, where they reach their maximum degree of specialization and efficiency. Much the same thing is true of the vertebrae, and of the bones of the skull. This tendency can also be exemplified by plants, for example by the reduction of numbers of stamens in specialized plants.

Increased efficiency is also often obtained by fusion of originally separate parts. Thus the sacrum of mammals is formed by the fusion of three to five originally separate vertebrae, thus making a much stronger attachment of the hind limb to the vertebral column than would otherwise be possible. Another good example is afforded by the pectoralis muscles which, in tetrapods, arise near the midline of the chest region and insert upon the humerus. Muscle slips from many adjacent body segments join to form these muscles. Among plants, the corollas of flowers such as the cucurbits or petunia are formed by the fusion of originally separate petals.

Dollo's Law. Many times during the long history of life, advanced organisms have returned to ancestral habitats and modes of life. This gives selective value to adaptations similar to those of the ancestral species, and raises the question whether evolution might be reversible. Study of such cases shows that always a gross similarity between ancestral and descended structures is achieved without any genuine reversal at all. Thus many reptiles and mammals have reverted to an aquatic mode of life. They have assumed a generally stream-lined, fish-like form, and the limbs have become shortened, webbed, and fin-like. Yet the skeleton of such flippers is always distinctly that of the class to which the animal belongs rather than that of a fish fin. Similarly, many angiosperms have returned to the water and assumed alga-like appearances, but their morphology is still that of flowering, vascular plants. The evidence indicates that major evolutionary steps, once taken, are never reversed. This is known as Dollo's Law. It even might be expected *a priori*, for major evolutionary steps are compounded of many smaller steps, each preserved by natural selection. That such a sequence, occurring by chance once, should by chance be exactly reversed would be a most extraordinary thing. If not impossible, it is at least most improbable for whole organisms. Attempts to apply Dollo's Law to individual characters have failed, for these are, indeed, reversible by mutation.

The Significance of Extinction. Some closing remarks should be made upon the subject of extinction, for this has been the fate of most species since the origin of life. Extinction may have completely different significance in different instances. The dinosaurs were a highly specialized line of Mesozoic reptiles which dominated the earth for a long time. But when conditions became unsuited to their survival, they became extinct and left no descendants. They were succeeded by other unrelated forms. The cynodont reptiles also became extinct, but they were succeeded by their own descendants, the adaptively superior mammals. Thus extinction may mark the end of a line of descent or it may be the accompaniment of the origin of new and superior types. It ought to be added that the major adaptive types—phyla and classes—very rarely become extinct.

These, then, are some of the conclusions which may be drawn from a study of the course of evolution. The causative factors which have led to these results will be taken up in parts III and IV.

REFERENCES

BORRADAILE, L. A., and F. A. POTTS, 1958. "The Invertebrata," 3rd Ed., Macmillan Co., New York, N.Y. An authoritative and comprehensive text on the invertebrates.

HADZI, J., 1953. "An Attempt to Reconstruct the System of Animal Classification," *Systematic Zool.*, **2**, 145–154. A radical revision, yet to be adequately assessed by other zoologists.

HARDY, A. C., 1953. "On the Origin of the Metazoa," *Quart. J. Microscop. Sci.*, **94**, 441–443. A brief but stimulating paper urging the origin of the Metazoa from plants.

HYMAN, LIBBIE H., 1940. "The Invertebrates," McGraw-Hill Book Co., New York, N.Y. Five volumes of this series have been published, and at least three more are planned. As far as it goes, it is the best, most comprehensive, and most profound treatment of the invertebrates in the English language. (Cleveland, Kofoid.)

IVANOV, A. V., 1955. "The Main Features of the Organization of the Pogonophora," *Systematic Zool.*, **4**, 170–178. New phyla are rarely described, but this is such a description.

MARCUS, E., 1958. "On the Evolution of Animal Phyla," *Quart. Rev. Biol.*, **33**, 24–58. A scholarly presentation of a phylogeny similar to the present one in general, but differing in important respects.

SIMPSON, G. G., 1953. "The Major Features of Evolution," Columbia University Press, New York, N.Y. The viewpoint of a genetically minded paleontologist. This book includes much material applicable to the last part of this chapter. (Cope, Dollo.)

TIEGS, O. W., and S. M. MANTON, 1958. "The Evolution of the Arthropoda," *Biol. Rev.*, **33**, 255–337. A penetrating review for mature students.

CHAPTER ELEVEN

The Phylum Chordata

THE MOST PROBABLE RELATIONSHIP of the chordates to the invertebrate phyla has already been outlined in the preceding chapter. But evidence in favor of the echinoderm theory is not conclusive, and almost every invertebrate group has been suggested at one time or another as a possible ancestor of the chordates. Only a few of the more plausible suggestions will be discussed here.

THEORIES OF CHORDATE ORIGIN

The Nemertean Theory. Kofoid and Hubrecht suggested that the chordates originated from nemerteans because of the arrangement of the nemertean nervous system in eight longitudinal cords. They proposed that the development of the two dorsal cords at the expense of the ventral and lateral cords could explain the origin of the dorsal nerve tube of the vertebrates. While this is plausible, there is no positive evidence in favor of it. As to other systems, the most that can be said is that the nemertean is sufficiently generalized to permit the formation of chordate structures or any others. But there is no evidence that they have specialized in a chordate direction, or that they ever have done so in the past. Much the same can be said for the turbellarian origin of the chordates which was urged by Haeckel. The theory can neither be proved nor disproved, because there is a complete lack of evidence. Similarly, the evidence in favor of the origin of the chordates from the Coelenterata is negative in character. The coelenterates are so primitive that they could conceivably have given rise to any of the more advanced phyla, including the Chordata. But there is no positive evidence that they did, and this theory has no adherents today.

The Arachnid Theory. It has also been suggested that an arachnid similar to *Limulus* might be the ancestor of the chordates. The principal evidence for this rather surprising theory is the superficial resemblance between *Limulus* (together with its fossil relatives) and the ostracoderms, the earliest known vertebrate fossils. Yet every detailed study has shown that this resemblance is illusory, and the theory no longer has any advocates.

The Annelid Theory. With greater reason, it has been suggested that the annelids may have been the source of the chordates. This is based upon the resemblance of primitive chordates, such as the Ammocoetes larva of the lamprey, to an inverted annelid. This may best be visualized by studying Figure 51, first upright, then with the book inverted. The digestive system in either case is a simple tube with a ventral mouth at one end and an anus at the other. The nervous system of the annelid consists of a pair of ventral cords and cerebral ganglia which are connected to the ventral cords by a pair of commissures which form a ring around the esophagus. If the worm were to form a chordate by turning over on its back, it would be necessary for the original mouth to close and a new one to form on the new ventral (former dorsal) surface. The nervous system would then be entirely dorsal, with the cerebral ganglia forming the brain. The circulatory system of the annelid is based upon a dorsal vessel in which the blood flows anteriorly and a ventral vessel in which the blood flows posteriorly. At the anterior end, there is a series of pulsating vessels in which the blood flows from the dorsal vessel to the ventral. If inverted, this would approximate the primitive chordate system in which a dorsal aorta carries the blood posteriorly, the posterior cardinal veins and ventral aorta carry it anteriorly, and the aortic arches connect the ventral and dorsal vessels at the anterior end. Further, the annelids and the chordates are the two most conspicuously segmented phyla.

The resemblance is thus extensive, but there are some serious difficulties. First, there is the difficulty of forming a new mouth. Then there is no structure in any annelid which even remotely suggests a notochord. And although the "hearts" which connect the dorsal and ventral vessels might be homologized with the aortic arches, there is no suggestion of gill slits in the annelids. Further, the annelids show the typical protostome embryological characters, while the primitive vertebrates show the typical deu-

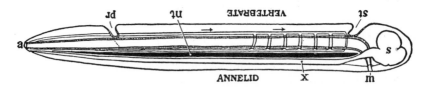

FIGURE 51. DIAGRAM TO ILLUSTRATE THE SUPPOSED TRANSFORMATION OF AN ANNELID WORM INTO A VERTEBRATE. In normal position this represents the annelid with a "brain" (s) at the front end and a nerve cord (x) running along the underside of the body. The mouth (m) is on the underside of the animal, the anus (a) at the end of the tail; the blood stream (indicated by arrows) flows forward on the upper side of the body, back on the underside. Turn the book upside down and now we have the vertebrate, with nerve cord and blood streams reversed. But it is necessary to build a new mouth (st) and close the old one; the worm really had no notochord (nt); and the supposed change is not as simple as it seems. (From Wilder, "Pedigree of the Human Body," (1927), by permission of Henry Holt & Co., Inc., publishers.)

FIGURE 52. DIAGRAMS OF THE TORNARIA LARVA OF *Balanoglossus* (A), THE BIPINNARIA LARVA OF A STARFISH (B), AND THE AURICULARIA LARVA OF A SEA CUCUMBER (C). (After Delage and Hérouard, from Romer, "The Vertebrate Body," 2nd Ed., W. B. Saunders Co., 1955.)

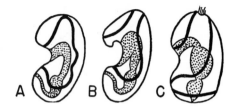

terostome characters. Thus it appears that the objections to the annelid theory outweigh its merits.

The Echinoderm Theory. And so we are left with the echinoderm theory, based as it is upon conformity of both echinoderms and primitive chordates to that set of embryological characters described for Deuterostomia generally. The theory received its major impetus from the discovery that the Tornaria larva, originally described as a starfish larva, was actually that of *Balanoglossus,* a hemichordate (Figure 52). At that time (the last quarter of the nineteenth century), the hemichordates were generally regarded as the most primitive subphylum of the Chordata. This is reinforced by serological and other biochemical evidence. Perhaps the majority of zoologists still accord them that position, but specialists on the group are more inclined to feel that a separate phylum, Hemichordata, should be recognized. Whether this will decrease the prestige of the echinoderm theory of descent remains to be seen. It is generally agreed that the phylum Hemichordata must be closely related to the phylum Chordata, even if it be conceded that the two phyla are distinct. At any rate the echinoderm theory now has more support than does any other theory of chordate origin, but few would care to claim that it is securely established.

MAJOR DIVISIONS OF THE PHYLUM CHORDATA

The phylum Chordata includes three subphyla, the Urochordata or tunicates; the Cephalochordata, including Amphioxus, the favorite of most elementary zoology texts; and the Vertebrata, much the most important subphylum, including as it does the dominant animals of land, sea, and air. The first two, together with the Hemichordata, have been exhaustively studied for evidence of the origin of the vertebrates. Morphologically, the tadpole larva of the tunicates is well described by the three fundamental characters of the chordates, that is, it has a dorsal nerve cord, a notochord, and a pharynx perforated by gill slits, and Berrill believes that the vertebrates arose from ancient tunicates. In the adult, however, the nerve cord and the notochord degenerate. The gills, which are a feeding mechanism rather than respiratory structures, become extremely highly developed. Adult tunicates are generally sessile and there are three well marked classes in the subphylum. Amphioxus was long regarded as an ideal "ancestor" because of its beautiful simplicity and its organization around the basic chordate characters. It has since become apparent, however,

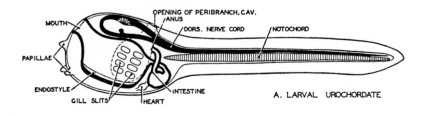

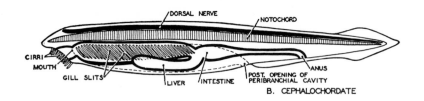

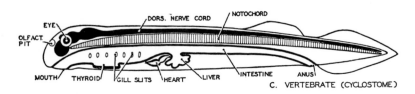

FIGURE 53. PRIMITIVE CHORDATES. (From Neal and Rand, "Comparative Anatomy," The Blakiston Co., 1939.)

that Amphioxus is quite specialized in some characters, such as the inclusion of the gills within an atrium, the fact that the notochord runs the entire length of the body, and the inexplicable fact that the kidneys of Amphioxus have more in common with those of annelids than they do with those of vertebrates. These primitive chordates (Figure 53) must have branched off from the main chordate stock very early, and their exact relationship to the vertebrates is no more clear today than it was when the question was first raised. The fossil record is of no help, for none of the prochordates are known as fossils at all (unless Jamoytius qualifies).

The vertebrates themselves are divided into eight classes, of which one is extinct. Four of these are entirely aquatic, the classes Agnatha, Placodermi, Chondrichthyes, and Osteichthyes. They may be grouped together as the superclass Pisces, corresponding to the common term "fish," yet they differ from one another more fundamentally than do the remaining four classes of land animals, which everybody recognizes as distinct. These are the classes Amphibia, Reptilia, Aves, and Mammalia, and they comprise the superclass Tetrapoda.

SUPERCLASS PISCES

Class Agnatha. The ostracoderms, primitive, armored members of the class Agnatha, are the first vertebrates to appear in the fossil record, and they are probably ancestral to the remaining classes. The earliest ostracoderm fossils are of Ordovician age, but they are neither numerous nor well preserved until the Silurian. They reach a peak of expansion during the Silurian and Devonian, then disappear from the record. Yet they did not actually become extinct, for unarmored representatives of the class have survived to the present as a minor part of our fish fauna, the cyclostomes. But, bereft of their bony armor, these have not been fossilized.

The Agnatha are extraordinarily primitive vertebrates. Morphologists have long agreed that the Ammocoetes larva of the lamprey (one of the living agnaths) approximates the archetypal vertebrate more closely than does any other living form. While there is no doubt that the living lampreys could not have been ancestral to the other vertebrate classes, nonetheless, a lamprey may be regarded as morphologically just an ostracoderm stripped of armor. Agnaths have no paired fins or limbs, in contrast to all other vertebrates. The mouth is suctorial and without jaws (hence the name of the class). The gills are well developed, but unlike those of higher vertebrates. The vertebrae are extremely simple, consisting only of the dorsal arcualia, the notochord still being the major element of the axial skeleton. There are only one or two semicircular canals in the ear. The kidney is pronephric in some (hagfishes), but it is mesonephric in others (lampreys). This brief sketch should suffice to show that the group is extremely primitive, and could potentially be a source for the higher vertebrates. This, coupled with the facts that the ostracoderms are the first vertebrates to appear in the fossil record and that the next class of fishes, the Placodermi, appeared soon afterward, lend weight to the hypothesis that the ostracoderms actually were ancestral to the higher vertebrates.

Class Placodermi. The class Placodermi shows significant advances over its ostracoderm progenitors. It first appeared late in the Silurian, then rose rapidly to dominance during the Devonian. During the Devonian, it gave rise to the remaining two classes of Pisces, and as these increased in importance, the Placodermi dwindled. By the beginning of the Mississippian they had been reduced to a minor place in the fauna, and they finally became extinct in the Permian. Like their predecessors, the ostracoderms, the placoderms were mainly a fresh-water group, but some of them did invade the seas.

Perhaps the most important advance of the placoderms was the acquisition of jaws, making possible a predatory mode of life. These were quite unlike those of modern fishes, and in some the lower jaw was fixed while the upper jaw and the entire head were movable. There was extensive bony armor, and the head armor was movably jointed to that of the thorax. The remainder of the skeleton was largely cartilaginous. Paired fins were also present, and these were quite variable. In some species, they resembled those of modern fishes, while others had bizarre fins. In some, the

159

fins were broad at the base and pointed at the tip. In many, smaller accessory fins extended along a line between the main pectoral and pelvic pairs. These and other data indicate that paired fins may have originated from a pair of longitudinal fin folds. During their period of expansion, the placoderms produced a wide variety of adaptive types. Most of these were not successful for long, and it is probable that none of the known placoderm fossils was in the direct line of descent leading to the higher vertebrates. But the Chondrichthyes were almost certainly derived from some placoderm line, and it is quite probable that the Osteichthyes were also derived from placoderms.

Class Chondrichthyes. At about the same time, in the late Devonian, both the Chondrichthyes and the Osteichthyes appear in the fossil record, each arising from a different stock of placoderms. The Chondrichthyes originated in the sea, while the Osteichthyes originated in fresh water. As the history of the Chondrichthyes, the sharks and their allies, is much simpler than that of the Osteichthyes, it may be sketched first. Processes involved in the origin of the sharks from placoderms were many, but a few may be singled out for discussion. One of these was the loss of armor, thus permitting a very much more active existence than was possible for their predecessors. Of the many jaw types with which nature experimented in the placoderms, the one carried over into the new class was (and still is) characterized by an upper jaw which was rather firmly jointed to the skull, but not fused to it (except in the chimaeras); and a lower jaw which was freely movable. Where the placoderms generally had bony plates for biting, the sharks, like some placoderms, developed true teeth with a core of dentine and an enamel surface. These are identical in structure and in mode of development with the placoid scales which cover the shark skin, and it may be that the teeth of sharks are derived from such scales by simple enlargement. Further, while their forebears possessed considerable bone, this was lost in the sharks, leaving a skeleton of cartilage.

The sharks first appeared in the late Devonian and reached a climax in the Mississippian. Their numbers were reduced in the Permian and Triassic, but they recovered and reached a new climax in the Cretaceous. This was accompanied by the formation of a new adaptive type, the rays. These are essentially flattened sharks, the teeth of which are modified to form plates for crushing the shells of the molluscs on which they feed. One more group of allied fishes, the chimaeras, comprising the order Holocephali, forms a part of the modern class Chondrichthyes. The chimaeras are not well known in the fossil record. The class as a whole has been reduced somewhat since the Cretaceous, but it continues to form a fairly important part of the marine fauna of the world, and a few sharks have even invaded fresh water.

Class Osteichthyes. The dominant fishes of the world are, and long have been, the Osteichthyes, or bony fishes. Following Haeckel, it was long believed that these took their origin from primitive sharks. But this no longer seems likely, for the fossil evidence indicates that the Osteichthyes arose earlier than did the Chondrichthyes. Further, the bony fishes originated in fresh water and remained there for long ages, while the sharks originated in the sea. Thus it seems most probable that both groups were produced independently from placoderm ancestors. The name of the class is based upon its possession of a bony skeleton. While this is a general character, yet it is variable, and some "bony fishes" have largely cartilaginous skeletons. Because of the extensive bony skeletons of ostracoderms and placoderms, that of the Osteichthyes is now regarded as a primitive character, while the cartilaginous skeleton of the sharks and the partially cartilaginous skeletons of some "bony fishes" are regarded as based upon secondary retention of an essentially embryonic character.

Even at their first appearance in the fossil record, the bony fishes were divided into two subclasses. One of these, the Choanichthyes (meaning nose breathers), is of especial interest because it appears to have given rise to the land vertebrates, while the other, the Actinopterygii, is of interest because it includes the most successful of fishes. The Actinopterygii take their name from the structure of their fins, which consist of a web supported by more or less parallel skeletal rays. The basal bones of these parallel fin rays are imbedded in the body wall, and the fin itself includes little or no musculature. This is in contrast to the Choanichthyes, in which the basal bones extend out longitudinally into the fin, with the fin rays arranged radially around it. The fin musculature forms a mass over the basals, and hence the Choanichthyes are often called the "lobe-finned" fishes.

Throughout the Paleozoic, the only actinopterygians were members of the superorder Chondrostei. In the Devonian, they were very much outnumbered by the Choanichthyes, but they expanded rapidly in the Mississippian, and soon they were the dominant fishes of the fresh-water lakes and streams. The skeleton of these fishes was largely bony. The external armor was reduced to a coat of ganoid scales, bony plates covered with a shiny, enamel-like substance called ganoine. The tail was shark-like. In common with the Choanichthyes, these fishes had lungs. Embryological evidence based upon living relatives of both groups of fishes and upon land vertebrates indicates that the lungs originated as a modification of the sixth gill pouch. The pouch first failed to break through to the outside, thus forming a moist internal chamber lined with a respiratory membrane. Swallowed air could thus be utilized for respiration in this incomplete gill pouch. Subsequently, these modified gills migrated back into the body cavity, thus becoming lungs. The lungs of tetrapods follow this course of development, and the known facts seem to indicate that a phylogenetic interpretation of the embryological data is justified in this case.

The Chondrostei reigned as the dominant fresh-water fishes until well into the Triassic, when they were superseded by another superorder, the Holostei. The chondrosteans contracted almost to the point of extinction during the Mesozoic, but a few genera have persisted to the present. They are represented at present in the Nile Valley by *Polypterus* and *Calamoichthys,* and in the United States by the sturgeons and the spoonbill of the Mississippi Valley. The sturgeons are widespread in the northern hemi-

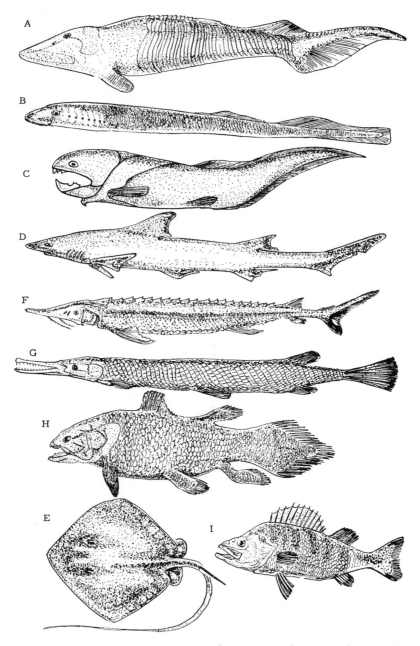

FIGURE 54. A PAGE OF FISHES. A, an ostracoderm, *Hemicyclaspis; B*, a lamprey, *Entosphenus; C*, a placoderm, *Dinichthys; D*, a shark, *Borborodes; E*, a ray, *Dasyatis; F*, a sturgeon, *Scaphirhynchus; G*, a garpike, *Lepisosteus; H*, the living crossopterygian, *Latimeria; I*, a teleost, *Chaenobryttus*. (A, C, and H redrawn from Romer.)

sphere, while the spoonbill is found elsewhere only in the Yangtze River of China. With the exception of the African species, these living fossils are rather degenerate. They have largely lost the ganoid scales of their ancient progenitors. The internal skeleton is largely cartilaginous, in contrast to the bony skeletons of their ancestors. Only the African forms retain the lungs.

The Holostei appears in the record in the Triassic, but probably took origin from chondrostean progenitors in the Permian. They rapidly rose to dominance, reaching a maximum in the late Jurassic. In these fishes, the tail shortened and became symmetrical, thus losing its shark-like aspect. The scales tended to lose their ganoine covering, thus leaving thin, simple, flexible bony scales, similar to those of the familiar fishes of today. But heavy scales were retained by some holosteans, including *Lepisosteus*. Furthermore, the holosteans invaded the seas, and their greatest diversification took place there. Yet it is a curious fact that the two genera which have persisted to the present are both fresh-water fishes. These are *Amia*, the bowfin, and *Lepisosteus*, the garpike, both of which occur in the United States. The "lungs" have fused to form a single sac, which functions as a hydrostatic organ or swim bladder. Yet its respiratory function is still important, for they come to the surface frequently to gulp air, and a garpike can be drowned by holding it under water.

Finally, the superorder Teleostei, in which the skeleton is almost entirely bony, appears to have originated from holostean ancestors early in the Mesozoic (Figure 54). They remained unimportant until the Cretaceous, when they began a rapid expansion which is still in progress. By the end of the Cretaceous, they were the dominant fishes both of the seas and of the fresh water of the world. Today they comprise upward of 95 per cent of the world's fishes. In the oceans, only the Chondrichthyes compete with them. In fresh water, only the few stragglers of the chondrosteans and holosteans and a very few unusual sharks are their competitors. The skeleton is always entirely bony in the teleosts. The scales are always thin, flexible chips of bone; the tail is invariably symmetrical; and the lungs, if present, are fused to form a swim bladder. As all of the familiar fishes of today are teleosts, it is obvious that they have undergone an immense adaptive radiation. The taxonomy of the teleosts is still quite controversial, but one widely accepted classification lists no less than twenty-eight orders. The teleosts, then, are one of the most successful and varied groups of vertebrates.

CROSSOPTERYGIANS, DIPNOANS, AND THE ORIGIN OF THE SUPERCLASS TETRAPODA

Soon after their origin, the subclass Choanichthyes became the dominant fishes, although they quickly yielded this position to the Chondrostei. The subclass included two orders, the Crossopterygii and the Dipnoi. The Dipnoi, or lungfishes, are represented by three living genera, one each in Australia, Africa, and South America. In the early evolutionary studies, they were given great theoretical importance as the probable little-changed descendants of the progenitors of the Amphibia, and through them, of the higher tetrapods. The Dipnoi have certain characters which lend themselves to this interpretation. Most important, they are lungfishes, and their lungs are developed in much the fashion described above. In degree of subdivision, their lungs surpass those of the Amphibia. Their fins have an arrangement somewhat like an elm leaf. There is a single row of basals running the length of the fin, and around this the fin rays are arranged like the veins in an elm leaf. A muscular lobe extends along the row of basals. Gegenbaur regarded this type of fin as the probable source of tetrapod appendages, and he called it an archipterygium (= primitive limb) to indicate this. He supposed that the tetrapod limb was formed from the archipterygium by the suppression of all of the fin rays except the terminal five. Finally, the aortic arches of the Dipnoi are closely similar to those of the tailed Amphibia (Urodela).

The theory of the dipnoan ancestry of the Amphibia enjoyed wide acceptance for a time, but it has some serious faults. The bones of the skull of the Dipnoi show a peculiar pattern, and efforts to homologize them with bones of amphibian skulls have been futile. And it is difficult to see how the archipterygium could have given rise to a limb in which two parallel bones form the second segment. In spite of these difficulties, the dipnoan theory might have continued to enjoy favor were it not that the Crossopterygii provide a much more plausible solution to the problems of amphibian ancestry.

The Crossopterygii, like other Devonian bony fishes, had lungs. The question now arises, why were lungs so general a feature of these fishes? Paradoxically, it appears that lungs were originally an adaptation to permit fishes to remain in the water. The fresh-water streams of Devonian

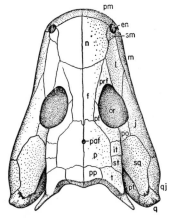
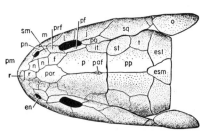

Figure 55. Dorsal Views of the Skulls of a Crossopterygian, *Eusthenopteron* (A), and of a Labyrinthodont Amphibian, *Palaeogyrinus* (B). Homologous bones are indicated by the same letters. (From Romer, "The Vertebrate Body," 2nd Ed., W. B. Saunders Co., 1955.)

times were subject to alternating periods of flooding and of stagnation and drying. As a result, only those fish which could breathe air could survive the periods of stagnation. But the air-breathing habit also made possible excursions over the land to reach larger and more favorable lakes or streams. The lungfishes of today also live in habitats in which seasonal drying gives selective value to the ability to breathe air.

The crossopterygian skull also has much in common with the skulls of primitive amphibians (Figure 55). They differ mainly in that a few bones of the crossopterygian skull are no longer present in the amphibian skull, and that certain bones have fused. In both, there is an opening for a pineal eye. In both, the internal choanae or nostrils are present. And in both, labyrinthodont teeth are present. This is a peculiar type of tooth, known only in the crossopterygians and in a very primitive type of amphibian, in which the enamel of the teeth forms deep ridges extending into the dentine. The crossopterygian fin is also of a dichotomous type which can be much more readily homologized with the limbs of amphibians than can the archipterygial fin. In fact, the similarity between primitive amphibian limbs and the dichotomous fins of crossopterygians is close, as can be seen in Figure 56. Both are characterized by a single, heavy piece, the humerus, which articulates with the shoulder girdle; by two parallel members, the radius and ulna, distal to the humerus; and by the less exactly homologized radial bones at the distal end of the appendage.

Finally, the vertebrae of both the Crossopterygii and the primitive Amphibia were diplospondylous, that is, in each body segment there were two centra, one developed from the pleurocentrum, the other from the

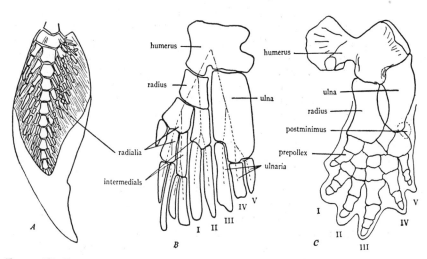

FIGURE 56. PRIMITIVE FORE LIMBS. A, archipterygium of *Ceratodus; B,* dichotomous fin of *Sauripterus,* a crossopterygian; *C,* limb of *Eryops,* a labyrinthodont. Note the similarity of B and C, and the complete dissimilarity of both of these to A. (From Hyman, "Comparative Vertebrate Anatomy," 2nd Ed., University of Chicago Press, 1942.)

hypocentrum. There are other parallels between these groups, but this may be sufficient to indicate the probability that the Crossopterygii were the vertebrates which led to the conquest of the land.

The discussion above is based upon only one of the two suborders of the Crossopterygii, namely the Rhipidistia. These were the dominant fishes of the Devonian, but they became extinct early in the Permian. At the height of their development in the Devonian, they gave rise to a marine suborder, the Coelacanthini, which appears in the fossil record as a minor group until the Cretaceous. But, by the end of the Cretaceous, some 75,000,000 years ago they had disappeared from the known fossil record, and all paleontologists regarded them as completely extinct. It was, therefore, an event of great scientific importance when a strange fish, brought into a South African port by commercial fishermen in 1939, was identified as an extant coelacanth crossopterygian. Unfortunately, the soft parts of the body were already extensively deteriorated, and little new knowledge was gained from this fish beyond the important fact that the coelacanths are not extinct. The species was described under the name of *Latimeria chalumnae*. Since 1952, many specimens have been taken near the Comoro Islands, north of the Mozambique Channel. As this is French territory, they are being studied in Paris.

Class Amphibia. The first amphibians which crawled out upon the mud banks of late Devonian streams and lakes were little more than fishes with fins sufficiently modified to support the weight of the body. These limbs were longer than typical crossopterygian fins, and it is probable that the muscular lobes were more highly developed. The radials were simplified to form a five-fingered—pentadactyl—hand which could be turned palm down to support the body. These early amphibians appear to have been aquatic animals in competition with their near relatives, the crossopterygians. As long as they remained in water, they probably were the poorer competitors, for legs are less efficient swimming organs than are fins. But seasonal droughts were the rule, and hence those animals which could leave a stagnating pond and go over land to a more favorable one had a selective advantage. In the long run, it was these which survived and gave rise to the land vertebrates, while those crossopterygians which failed to make adaptations to land living became extinct. The single known exception has been noted above.

The Amphibia quickly broke up into several orders. Three more orders are living today. The problem of how these orders are related to one another, and how they should be classified is a very difficult one, and the scheme to be outlined here must be regarded as tentative. All of the primitive, extinct Amphibia resembled our present-day Urodela (newts and salamanders) in general bodily form, but differed from them greatly in the details of anatomy. The most primitive group, including those forms most similar to the crossopterygians, comprises the superorder Labyrinthodontia. This name is based upon the fact that, in common with the crossopterygians, they possessed labyrinthodont teeth (Figure 57). They continued into the Triassic, but were extinct by the end of that period. The labyrinthodont orders were generally characterized by the possession

FIGURE 57. CROSS-SECTION OF A PART OF A LABYRINTHODONT TOOTH. The sinuous lines represent the complex infolding of the enamel layer of the tooth. (From Colbert, "Evolution of the Vertebrates," John Wiley & Sons, Inc., 1955.)

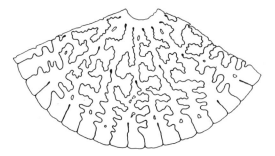

of diplospondylous vertebrae based upon a hypocentrum and a pleurocentrum. The relative sizes and spatial relations of the parts of the vertebrae were characteristic for each order. Taking their origin in the Devonian, they were a major part of the fauna of the Carboniferous swamps.

A second type of amphibian, the Lepospondyli, also appears among the Devonian fossils. They lack labyrinthodont teeth, and their vertebrae were formed by direct ossification around the notochord rather than from arcualia. These vertebrae are perforated longitudinally, permitting passage of the notochord. Because they are structurally more remote from the crossopterygians, it is believed that the Lepospondyli must have been derived from labyrinthodont ancestors, but their separation must have occurred before either group actually appears in the fossil record. And at least one competent anatomist has come to the conclusion that they arose independently, with the Lepospondyli coming from Dipnoan ancestors. The Lepospondyli were fairly prominent during the Pennsylvanian, but they disappear from the record in mid-Permian times.

By the end of the Triassic, the last of the ancient orders of Amphibia had disappeared. As early as the Pennsylvanian, some labyrinthodonts showed reductions of the skull and vertebrae similar to those of the Anura (frogs and toads). The hypocentrum developed at the expense of the pleurocentrum. A more advanced anuran-like labyrinthodont is known from the Triassic of Madagascar, and scattered remains af anurans have been found in Jurassic deposits. Numerous amphibian fossils, both of Anura and of Urodela (tailed Amphibia, newts and salamanders) occur in the Cretaceous, and these orders have continued up to the present as minor groups. Intermediates in the formation of the urodeles are unknown, but they have lepospondylous vertebrae, and so may have been derived from the Lepospondyli. The obscure, worm-like Apoda do not appear in the record until the beginning of the Cenozoic. Like the Urodela, they have lepospondylous vertebrae, so it is plausible that they have a similar origin.

Class Reptilia. The Amphibia play a minor role in the vertebrate fauna of today, and perhaps their greatest importance lies in their role as the source of the class Reptilia. The origin of the reptiles from primitive labyrinthodonts is unusually well attested, for there are many transi-

167

tional genera. Thus *Eryops*, which is now usually classified as an amphibian, and *Seymouria*, which is now usually classified as a reptile, have, with much justification, been placed in both classes by different authorities, or even by the same authority at different times. The reptiles first appear in the fossil record in the Pennsylvanian. By the Permian, they had already begun a great diversification which led to the formation of six orders in the Permian and ten more in the Triassic. Throughout the Mesozoic, they were the dominant vertebrates, and hence the ordinary designation of this era as the "Age of Reptiles."

The original reptiles were substantially just amphibians adapted for permanent land life. Like modern reptiles, the skin was probably thickened and cornified to prevent drying of the animal. It had four short limbs set more or less at right angles to the body, so that it could lift its weight only clumsily. It had a large number of undifferentiated conical teeth. And, perhaps most important of all, the developing embryo was enclosed in embryonic membranes including the amnion and chorion, and it respired by means of a third membrane, the allantois. Thus the reptiles were freed of the necessity of returning to water for reproductive purposes. The eggs, once laid, were unattended.

Adaptive radiation within the class Reptilia has been extremely varied (Figure 58). Three orders returned to the water and again developed specializations appropriate to that habitat. Most extreme of these was the order Ichthyosauria, in which the external form became completely fishlike, but the skeletons prove that these animals were reptiles. The plesiosaurs were less extremely modified. They had turtle-like bodies, and large flippers. The neck was often very long and serpentine in appearance. The third aquatic order is the Chelonia, including the turtles, which are generally adapted to an amphibious mode of life. But, as is well known, some turtles have become exclusively terrestrial and others have become exclusively marine. Representatives of most of the terrestrial orders have also invaded the water. The best known modern reptiles are the snakes and lizards, and these occur in both terrestrial and aquatic forms, and are adapted to predation upon almost every type of animal. The greatest range of adaptive radiation occurred within the several orders of dinosaurs, the ruling reptiles of the Mesozoic era. Their principal types are illustrated in Figure 59.

Why the Ruling Reptiles became extinct at the end of the Cretaceous is not known, but a number of plausible theories have been suggested. The disappearance of the great reptiles coincides roughly with the rise of the birds and mammals, and it has been suggested that the reptiles were simply unable to compete with these progressive newcomers. But this is improbable, because these groups had been present since the Jurassic, perhaps since the Triassic in the case of the mammals, and they had not been able to achieve a place of importance in competition with the dominant reptiles. It is more probable that the rise of the birds and mammals in the Cenozoic era occurred because their reptilian competitors disappeared. Another suggestion which has been given more credence is that the world climate became more severe, and that the great reptiles were

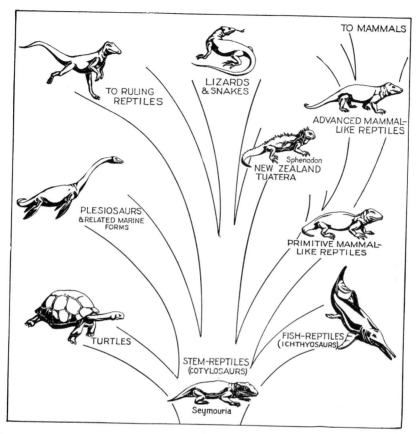

FIGURE 58. ADAPTIVE RADIATION IN REPTILES. (From Romer, "Man and the Verte-brates," University of Chicago Press, 1946.)

not able to adapt themselves to conditions of cold. Just the opposite sug-gestion has been made by Cowles, largely on the basis of studies on the reproductive physiology of living reptiles. It is well known that the testis is heat-sensitive. Mammals are sterilized by a temperature only slightly above the normal temperature of the scrotum. Birds, which characteristi-cally have a higher body temperature than mammals, show spermatogenic activity principally in the early morning hours, when the body tempera-ture is lowest. Cowles has demonstrated that the optimum temperature for normal activity of living reptiles is only a shade below the sterilizing temperature. Reptiles, of course, are "cold blooded," meaning that they do not maintain a constant body temperature. But large bodies cool off much more slowly than do small ones, and Cowles has suggested that, in an increasingly hot climate, reptiles so large as many of the dinosaurs might only rarely cool off sufficiently to permit spermatogenesis. Thus the great size of the Ruling Reptiles, in combination with an increasingly hot climate, could lead to their extinction by sterilization of the males.

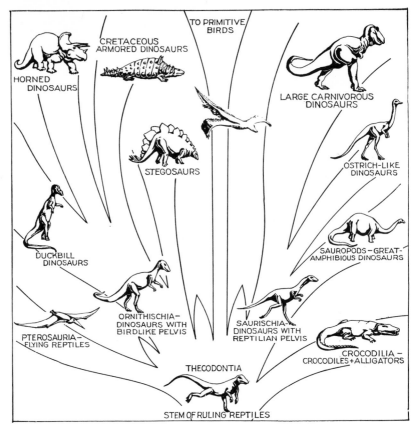

Figure 59. Ruling Reptiles. (From Romer, "Man and the Vertebrates," University of Chicago Press, 1946.)

But as yet no final decision on this problem is possible, and other factors, such as drought and scarcity of food, deserve consideration.

While the reptiles dominated the entire Mesozoic era, only four of the sixteen orders have survived into Cenozoic times. One of these, the order Rhyncocephalia, is represented today only by a single species, *Sphenodon punctatum*, a lizard-like reptile which is restricted to a few small islands off the coast of New Zealand. Although this animal superficially resembles the lizards, its skeleton is very much more primitive. For example, in the earliest reptiles, there was a progressive reduction of the hypocentrum, with the pleurocentrum finally forming the entire body of the vertebra. In most other living reptiles, this process is complete, but in *Sphenodon* the hypocentra still form small wedges between the successive vertebrae. It possesses many such archaic characters, and is often referred to as a "living fossil," because its morphological affinities are with long extinct types rather than with living types.

A second living order of reptiles is the Chelonia, the turtles, an ancient order of which Permian records are known. Although the turtles are less varied than many orders, yet they have been able to occupy habitats ranging from desert to marine. In terms of numbers of genera, they are second only to the Squamata today.

The order Crocodilia is represented today only by a few genera, remnants of a once important line. It includes the crocodiles and alligators. They arose in Triassic times from the order Thecodontia, the same group which gave rise to the great Ruling Reptiles of the Mesozoic Era. They reached a peak in numbers in the Cretaceous, but have since been only a minor part of the reptilian fauna of the world.

The final order of living reptiles is the Squamata, the snakes and lizards, popularly the best known reptiles today, but by no means the most typical. The order appears to have arisen in the Jurassic, reached a peak in the Cretaceous, and continued to the present on a more restricted scale. Yet they are found in all parts of the world except the arctic and antarctic zones. All are carnivorous, but they are adapted to predation upon animals ranging from insects to large mammals. Because there is a tendency to think of most reptiles as being poisonous, it is worth mentioning that all poisonous reptiles are confined to the order Squamata, and that only a few of the many families in this order have developed poison mechanisms.

Class Aves. The oldest known bird fossils are of late Jurassic age, and they are most instructive with respect to the probable origin of the Aves. Their skeletal characteristics are largely those of the primitive dinosaurian order Thecodontia. Like these, they had many simple, conical teeth; a skull of similar pattern; vertebrae unfused; a long tail composed of many unmodified vertebrae; and bipedal locomotion. These are all reptilian characters, but the fossils also include feathers, a characteristic known only in birds. Hence they have been assigned to the genus *Archeopteryx* (Figure 60), the oldest known genus of birds. But, had the feathers not been preserved in the fossils, it is quite probable that the specimens would have been assigned to the Thecodontia on the basis of the skeleton. Thus the origin of the birds from thecodont ancestors is quite probable. It should be noted that the birds are not the only flyers developed from the Ruling Reptiles, for the pterodactyls also belong in this series. But they were remotely related to the birds, and they became extinct without leaving any descendants.

The features in which birds differ from reptiles are almost all adaptive to flight. At the outset, they do have feathers, which form a planing surface for flight. The early embryology of feathers is quite similar to that of reptilian scales, and it is commonly believed that feathers developed as modifications of scales. The legs of birds are still covered with reptilian-type scales. The feathers also provide insulation, thus aiding in the maintenance of the high body temperature of birds, so necessary for the maintenance of the high metabolic rate required for flight. The light, hollow bones of birds and the air sacs associated with the respiratory system are probably also best interpreted as adaptations to flight, for they reduce the weight of the bird. Other modifications affect practically every organ system.

171

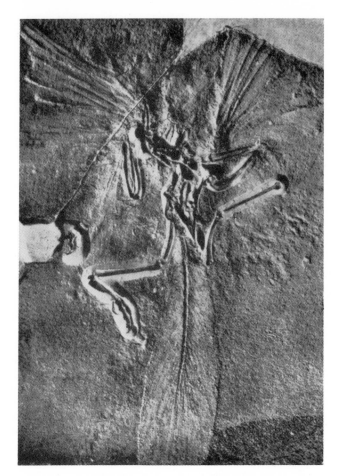

FIGURE 60. PHOTO-GRAPH OF THE ORIG-INAL ARCHAEOP-TERYX SPECIMEN. Found in a quarry of lithographic stone at Solenhofen, Germany in 1861 and purchased by the British Museum in 1862. (From De-Beer, *Archaeop-teryx lithographica*, British Museum of Natural History, 1954.)

Although birds appeared in the fossil record in the Jurassic, they remained rare and insignificant until the Cenozoic era, when they began a great expansion. A few orders, all located on the southern continents, have lost the power of flight and developed powerful running legs, or other adaptations to a purely terrestrial existence. Several orders have become adapted to various aquatic niches. Still others have become adapted to a wide variety of land habitats. Altogether, there are twenty-eight orders of extant birds, but it is worth bearing in mind that the birds are taxonomically the best known class of animals, and hence there is a tendency to subdivide them more finely than other classes.

Class Mammalia. Reptiles which diverged in a mammal-like direction occurred in the Permian or even the Pennsylvanian, comprising the order Pelycosauria. These gave rise in the late Permian to a more advanced order, the Therapsida, which in many skeletal characteristics (the only ones which are preserved) approach quite closely to primitive mammals.

FIGURE 61. A THE-RAPSID, *Lycaenops.* (After Colbert, from Romer, "The Vertebrate Body," 2nd Ed., W. B. Saunders Co., 1955.)

The therapsids (Figure 61) expanded rapidly and were among the most common Permian vertebrates. But apparently they were unable to compete with the Ruling Reptiles which rose to prominence in the Triassic, for all but the smallest therapsids became extinct. In other words, they were able to hold only those ecological niches for which the Ruling Reptiles did not compete.

The major mammal-like trends include the rotation of the limbs so that the elbows and knees were brought in under the body. This made it possible to lift the weight of the body without working against a leverage, a very important advance which made possible rapid and long sustained running. The rotation of the limbs also requires modification of the girdles and strengthening of the vertebral column. The skull was simplified somewhat, but the major changes in the skull involve the jaws and the articulation of the lower jaw to the skull. In typical reptiles, the lower jaw is composed of many bones. In the therapsids one of these, the dentary, which bears the teeth, tends to replace the others. In typical reptiles, the articular bone of the lower jaw articulates with the movable quadrate bone, which in turn articulates with the temporal region of the skull. In the therapsids, the dentary forms a second joint with the temporal, and the articular-quadrate-temporal joint becomes reduced in size and in functional importance. The articular and the quadrate become loosely attached and tend to become associated with the columella, an ossicle of the adjacent ear. Finally, a certain amount of regional differentiation of the teeth occurred in the therapsids.

A few mammalian remains of Jurassic age are available. It is probable that the remote ancestors of the order Monotremata were separated from the general mammalian stock as early as this. Of all the orders of mammals (thirty-two, including fourteen extinct orders), this is the only one which lays eggs. The young are fed by milk which is secreted into shallow depressions on the abdomen of the mother. As there are no nipples, the young must lap up the milk. The order is represented by only two living species, the duckbilled platypus and the spiny anteater, both of which are confined to Australia. Like all mammals, they are covered by a coat of hair, an insulating material which assists in the maintenance of a relatively constant body temperature.

A few fossils from the late Cretaceous show that the order Marsupialia had appeared by that time, but it is only in the Tertiary that they became numerous. These are mammals which no longer lay eggs, but rather the young undergo a brief development in the uterus (modified oviduct), then are born in a very incomplete state of development. They are transferred to an abdominal pouch (marsupium), where each of the young becomes permanently attached to a nipple and is thus nourished until it is sufficiently developed to leave the maternal pouch and fend for itself. The kangaroo is the best known marsupial, but it is also one of the most highly specialized. The opossum of the United States is a much more primitive marsupial. In general, the order has been rather unsuccessful in competition with the placental mammals, but they have been extremely successful in Australia, where they have filled almost every possible adaptive niche. But Australia has been separated from the northern centers of diversity since the Cretaceous, and it is consequently almost devoid of placental mammals. South America was similarly isolated during much of the Tertiary, and hence a rich marsupial fauna developed there, only to become extinct when placental mammals invaded from North America. Today, the Australian region is the home of the only rich marsupial fauna, but unfortunately very little is known of its fossil history.

The placental mammals also appear in the fossil record in the late Cretaceous for the first time, and they underwent an explosive expansion at the beginning of the Tertiary, so that almost all of the orders are present from the beginning. The major feature differentiating them from the marsupials is that the embryos develop an efficient placenta for obtaining nourishment from the mother's bloodstream. This also serves as an organ of respiration and excretion. These are by far the dominant animals of the world today, and this is in no small measure due to the prolonged period of embryonic development which is made possible by the placenta, as well as by parental care and the enlargement of the cerebral hemispheres, diminutive in all lower classes of vertebrates. This last trend is already observable in the lower mammalian groups, but is carried much further by the placental mammals.

The ancestral placentals were small animals, probably because only such could compete with the reptiles of the Cretaceous. Although they were potentially meat eaters, their small size restricted them largely to a diet of insects, worms, and other small invertebrates, with perhaps a small amount of vegetable matter. This is about what the living members of the order Insectivora eat. Some of the members of this order, which includes shrews, moles, hedgehogs, and tenrecs, are very primitive anatomically. It seems probable that, of all the living orders of mammals, the Insectivora is closest to the primitive placental stock.

Because the extant (and extinct) orders appear so rapidly at the beginning of the Tertiary, it is very difficult to trace probable relationships. But, on the basis of comparative anatomy, comparative serology, and paleontology, certain probable relationships have been drawn up, some of which are much better supported by evidence than others. Because of their antiquity and primitiveness, the order Insectivora is generally regarded as the probable source from which other orders of placentals have been derived. The derivation of one series of orders, the cohort Unguiculata, from the Insectivora is fairly clear, but others vary considerably in

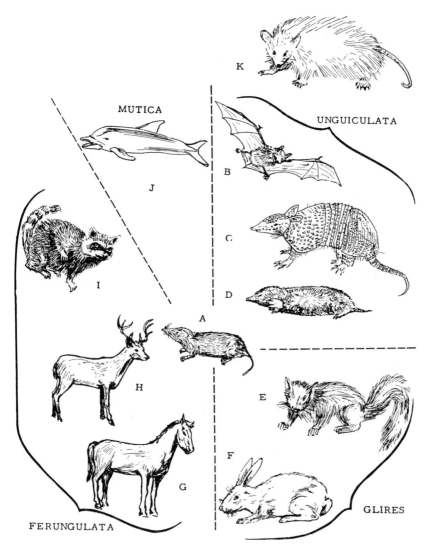

FIGURE 62. REPRESENTATIVE MAMMALIAN TYPES. *A*, shrew, similar to the primitive placental progenitor; *B*, bat, the only mammal capable of true flight; *C*, armadillo; *D*, mole; *E*, squirrel; *F*, rabbit; *G*, horse; *H*, deer; *I*, racoon; *J*, porpoise; *K*, opossum. *A–D*, cohort Unguiculata; *E* and *F*, cohort Glires; *G–I*, cohort Ferungulata; *J*, cohort Mutica. The opossum, being a marsupial, is of remote and uncertain relationship to any of the others.

the degree of assurance with which such an ancestry can be postulated. (Because arrays of orders within the class are quite distinct from one another, Simpson, whose classification of the mammals is generally accepted, advocates the use of the *cohort* as a category intermediate between order and class.)

The cohort Unguiculata (Latin: having claws or nails) comprises eight orders, of which two are extinct. Three of the six extant orders are quite familiar: the order Insectivora itself; the order Chiroptera, or bats, which are substantially just flying insectivores; and the order Primates, including lemurs, monkeys, apes, and man. Three less well known orders are the Dermoptera, including only the so called "flying lemur," which is not a lemur and glides rather than flies; the Edentata, including such animals as sloths, anteaters, and armadillos, all of which have rather few, oddly developed teeth; and the Pholidota, including only the pangolin, an Old World anteater which is not at all closely related to the Edentata. Of the other cohorts, the Glires is most easily related to the Insectivora. This cohort includes two very well known orders, the Rodentia, including a great host of widely varied gnawing animals, and the Lagomorpha, including rabbits, hares, and pikas. These were formerly united in one order, but as long ago as the turn of the century, it was proposed that they ought to be separated on the basis of characters of the skull and teeth. Subsequent studies have consistently supported this decision, and the paleontological record shows that the two orders do not converge as one traces it back further. Hence Simpson has concluded that the inclusion of the Rodentia and the Lagomorpha in a single cohort "is permitted by our ignorance rather than sustained by our knowledge."

The third cohort, the Mutica, includes only the single order Cetacea, including whales and porpoises. This is perhaps the most aberrant of mammalian orders, and its relationships are much befogged. Yet, as has been pointed out in a previous chapter, its skeleton bears clear testimony of descent from land mammals.

The last cohort, Ferungulata (Latin: Ferae, beasts plus Ungulata, hoofed), includes a large number of diverse orders, most of which are extinct. The living orders of this series are the Carnivora, the Tubulidentata, the Proboscidea, the Hyracoidea, the Sirenia, the Perissodactyla, and the Artiodactyla. The first of these and the last two are among the most successful of animals. One of the major characteristics of early mammalian evolution was the development of large herbivores and of carnivores adapted to prey upon these and upon other types of mammals. The latter type is the order Carnivora, comprising such diverse forms as cats, weasels, wolves, bears, and seals. The herbivores have diversified into a great many more orders, of which the dominant ones are the Perissodactyla, which includes the horses and their allies with an odd number of toes; and the Artiodactyla including the pigs, camels, deer, cattle, and other hoofed animals with an even number of toes.

The remaining living orders of ferungulates are represented by only a few living forms. The Tubulidentata includes only the African aardvark, an anteater which is unrelated to the several other animals called anteaters. The Proboscidea includes the elephants and their extinct allies, the mammoths. The Hyracoidea includes only the conies, small rabbit-like ungulates of Africa and Asia. Finally, the Sirenia includes only the sea cows.

At first, the inclusion of the principal carnivores and herbivores within a single series may seem surprising; however, they do converge as one traces the fossil record back, and many an early Tertiary genus could with equal justification be assigned to either type. Thus their common origin is quite probable.

REFERENCES

BERRILL, N. J., 1955. "The Origin of the Vertebrates," Oxford University Press, New York, N.Y. An interesting analysis of evidence for the tunicate theory.

COLBERT, E. H., 1955. "Evolution of the Vertebrates," John Wiley & Sons, Inc., New York, N.Y. An interesting and well-illustrated account, from the viewpoint of a paleontologist.

MILLOT, J., and J. ANTHONY, 1958. "Anatomie de *Latimeria chalumnae*," Centre National de la Recherche Scientifique. A thorough and beautifully illustrated study of this living fossil, for those who read French with facility.

ROMER, A. S., 1958. "Tetrapod Limbs and Early Tetrapod Life," *Evolution*, 12, 365–369. The most recent in a series of papers debating the factors which led to the origin of the Amphibia.

ROMER, A. S., 1945. "Vertebrate Paleontogy," 2nd Ed., University of Chicago Press. A classic in its field.

ROMER, A. S., 1959. "The Vertebrate Story," University of Chicago Press. A well-rounded, phylogenetic introduction to vertebrate zoology.

SMITH, J. L. B., 1956. "The Story of the Coelacanth," Longmans, Green & Co., New York, N.Y. The discoverer's account of *Latimeria*.

CHAPTER TWELVE

The History of the Primates

WE COME NOW to the climax of this phylogenetic history—the order Primates, including tree shrews, lemurs, tarsiers, monkeys, apes, and man. But before discussing the history of this group, it may be well to review the classification and the major characteristics of the living members of this order.

CLASSIFICATION OF THE PRIMATES

The order Primates is singularly difficult to define because of the absence of salient distinguishing characters, comparable to the chisel-like incisors of rodents, or the hooves with an odd number of toes of the Perissodactyla. Mivart long ago defined the Primates as placental mammals with nails (or claws in some); with clavicles; with orbits encircled by bone; with three kinds of teeth; possessing a brain with a posterior lobe having a fold called the calcarine fissure; a thumb and great toe having a flat nail or none; a large intestine with a blind pouch, the cecum; with penis pendulous, and the testes descended into a scrotum; and with two pectoral mammary glands.

These are largely primitive mammalian characters, and it may be said that, except for a tendency toward expansion of the brain, the Primates are relatively unspecialized mammals. The teeth are adapted to a generalized diet, but this is itself lack of specialization. Thumb and great toe are usually opposable to the other digits, which gives efficiency in grasping objects. While the eyes of most mammals are on the sides of the head, so that each eye sees a different field, those of Primates are placed toward the front, thus permitting binocular, stereoscopic vision. Vision is generally more highly developed than in other mammals, the sense of smell less so. Finally, although enlargement of the brain is a general mammalian characteristic, this is most marked in the Primates.

Tree Shrews. The tree shrews are certainly the most primitive of Primates; indeed, many zoologists prefer to class them with the Insectivora, from which they are derived. However, they vary from ground shrews in several ways which tend to associate them with Primates. The digits are more mobile and the thumb and great toe are somewhat opposable. These digits, however, are capped by typical claws. The eyes are larger than in typical ground shrews, and the nasal apparatus is somewhat less developed. Thus a good case can be made for the inclusion of the tree shrews with the order Primates, but they are exceedingly primitive. They are at present known only in the Oriental Region, where they are widely distributed.

Lemurs. There are two groups of lemurs: the lemuriforms, now confined to Madagascar; and the lorisiforms, found both in Africa and in Asia. They vary in size from that of a mouse to that of a small monkey. They are arboreal, primitive animals, some being scarcely more advanced than the tree shrews. They exhibit the basic primate character of well-developed hands and feet, with opposable thumbs and great toes. The snout is usually long and projecting. The ears are large and mobile, but there is little mobility of facial expression. The lemurs are generally nocturnal, hence their eyes are large.

Tarsiers. The living tarsier of the Philippines and other Oriental islands is the last survivor of an old and important group of Primates which was probably derived from lemuroid ancestors. The tarsier (*Tarsius spectrum*) is about the size of a young kitten. An exclusively nocturnal animal, its eyes are immense relative to the size of its head, and they look forward, thus permitting binocular vision. The snout is correspondingly reduced, so that it has a monkey-like appearance. The hind legs are modified for jumping, and it can leap from branch to branch with considerable accuracy. The ears are large. The tail is long and naked, except for a terminal hairy segment. Although tarsiers resemble lemurs in many details, the structure of the brain and of the reproductive organs is essentially simian. Hence some students group them with the monkeys, anthropoids, and man rather than with lemurs and tree shrews.

Monkeys. The monkeys are generally larger than the primitive Primates which have just been discussed, and they are generally diurnal. The eyes are set well forward, and the nasal apparatus is reduced. Generally they are arboreal, but some are largely or entirely terrestrial. They are divided into two contrasting groups, both of which are quite varied. These are the platyrrhine monkeys of the New World and the catarrhine monkeys of the Old World. The terms refer to the character of the nasal septum, which is broad in the platyrrhines and narrow in the catarrhines. But there are many other features distinguishing these two groups of monkeys. The platyrrhines, which are restricted to South America, have generally been regarded as the more primitive. Some have prehensile tails, and they are the only Primates that do. They are well typified by the spider monkey (*Ateles*) and the capuchin monkey (*Cebus*), the well-known organ grinder's beggar. The catarrhines are widely distributed in the Old World, and include such diverse types as the macaque (among which is the rhesus monkey of medical research), the guereza, the guenon, the baboons, and the mandrills.

The Anthropoid Apes. Of all the extant Primates, those which resemble man most closely are the anthropoid apes. There are only five genera living, the gibbon of Asia, the siamang of Sumatra, the orangutan of Borneo and Sumatra, and the gorilla and chimpanzee of equatorial

FIGURE 63. REPRESENTATIVE PRIMATES. *A*, tree shrew, *Tupaia; B*, lemur, *Galago; C*, tarsier, *Tarsius; D*, a macaque monkey, *Macaca; E*, gibbon, *Hylobates; F*, chimpanzee, *Pan*. (From Clark, "The History of the Primates," 3rd Ed., British Museum of Natural History, 1953.)

Africa. These most nearly resemble man in structure of skull and skeleton, dentition, physiology, blood groups, parasitic susceptibilities, and other characteristics. They are, however, highly specialized for an arboreal mode of life. Their arms are greatly elongated, and they swing through the trees by a method called brachiation, that is, they swing from branch to branch with their arms alone, their bodies and legs playing only an indirect role. The thumbs have accordingly become reduced, so that the hand can function largely as a hook in grasping branches. The legs are much shorter than the arms, in contrast to the Old World monkeys and man.

The gibbon is the smallest of the anthropoid apes, and in many respects it is the most primitive. It is completely arboreal, and, while it is capable of remarkably swift and accurate brachiation, it is more adroit on foot than most of the apes, for it can run along the branches quite skillfully. The siamang is closely related to the gibbon, and differs from it only in details. The orangutan is a much larger ape, often weighing well over 100 pounds, but it is still primarily arboreal, and moves quite successfully by brachiation. It rarely descends to the ground. The chimpanzee is somewhat larger and is arboreal. The gorilla is much the largest of the apes, reaching weights in excess of 600 pounds. Although gorillas arc brachiators morphologically, the huge size of the adults confines them to the ground, where they use a peculiar type of quadripedal locomotion.

Summary of Classification. This, then, is the array of organisms which makes up the order Primates, of which man is the dominant member. In the most recent taxonomic revision of the mammals, the Primates are divided into two suborders, the Prosimii, including tree shrews, lemurs, and tarsiers; and the Anthropoidea, including monkeys, the anthropoid apes, and man. The Anthropoidea are further subdivided into three superfamilies, the Ceboidea, including the platyrrhine monkeys; the Cercopithecoidea, or catarrhine monkeys; and the Hominoidea, including man and the anthropoid (man-like) apes. The Hominoidea is divided into two families, the Hominidae including only man, and the Pongidae including all of the anthropoid apes.

Each group of Primates discussed above shows some significant advances over the one preceding it, and there is a temptation to treat them as an evolutionary series, as has been done in the past. Yet a little reflection will make it clear that each of these groups is the more or less specialized end product of a long evolution of its own, and the living tree shrews, for example, could hardly be ancestral to the lemurs. But very primitive ancestors of the tree shrews of today may well have been ancestral also to the lemurs. It is important to bear this in mind, for there used to be much argument over which of the living apes was the progenitor of man, and this viewpoint has been revived in a recent book. But most authorities now regard the question as quite absurd.

PROSIMIANS IN THE FOSSIL RECORD

The paleontological record of the Primates is very incomplete but it is ancient, for primate remains have been found from rocks of mid-Paleocene age, the oldest epoch of the Cenozoic Era. From this remote time, near the beginning of the Age of Mammals, the skulls of small mammals suggesting affinities with tree shrews have been discovered. The structure of the molar teeth of these fossils is primate in character, and there is a tendency toward expansion of the brain. The family Plesiadapidae has been erected for these early Primates.

Lemurs first appear in the record in the Eocene epoch, from which age records have been found both in Europe (genus *Adapis*) and in America (genus *Notharctus*). These were comparable in size to modern lemurs, but the brain was smaller, and they had not yet formed certain specializa-

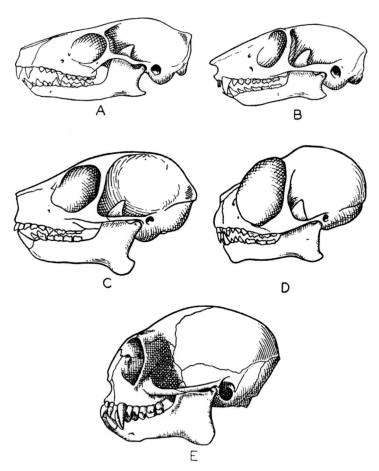

FIGURE 64. SKULLS OF PRIMITIVE PRIMATES, SHOWING A MORPHOLOGICAL (BUT NOT PHYLOGENETIC) TRANSITION FROM TREE SHREW THROUGH LEMUR AND TARSIER TO MONKEY. *A*, tree shrew, *Ptilocercus; B*, mouse lemur, *Microcebus; C*, an Eocene tarsier, *Necrolemur; D*, the modern tarsier, *Tarsius; E*, marmoset, a platyrrhine monkey, *Callithrix*. (From Clark, "The History of the Primates," 3rd Ed., British Museum of Natural History, 1953.)

tions of the teeth which characterize modern lemurs. Of somewhat later date (Oligocene), is the tree shrew *Anagale*, of especial interest because it is in many respects intermediate between tree shrews and lemurs. While its general features are those of the tree shrews, certain features of the skull and of the molar teeth are lemuroid. Also, the digits were capped by nails rather than claws. While this animal could not be ancestral to *Adapis* (because it occurred later), it indicates the probability that similar tree shrews of an earlier time may have been ancestral to lemurs.

Tarsiers were present in great abundance in the Eocene, for no less than twenty-five genera are known from Europe and America. Some of these were quite similar to the modern tarsier (Figure 64). Others were decidedly more primitive in skull pattern, brain, and limbs. Some retained the primitive insectivore dentition of forty-four teeth, while others had the dentition reduced to thirty-two, the number characteristic of the higher Primates. Other changes in the teeth of these Eocene tarsiers also tended in the direction of the higher Primates. Thus they were the first Primates to develop bicuspid premolars. The molars of the tree shrews and insectivores have three cusps, but some of the Eocene tarsiers, in common with all of the higher Primates, have molars with four cusps. Thus many of these primitive tarsiers had decidedly monkey-like features. It is a much debated question whether the tarsiers were derived directly from the tree shrews or whether they were derived from very primitive lemurs.

APES IN THE FOSSIL RECORD

Parapithecus and Propliopithecus. Anthropoid apes first appear in the fossil record in the Oligocene epoch, but these remains are very scant indeed, being known only from two lower jaws found in Egypt. The first of these was described under the name *Parapithecus*. A single jaw gives a very scant basis for judgment, but some conclusions can be drawn. The first of these is the fact that anthropoids did exist as long ago as the Oligocene (although the anthropoid and even primate status of *Parapithecus* has been seriously contested). *Parapithecus* was a small animal, about the size of the little squirrel monkey, and it was very primitive. It had the dental formula characteristic of Old World monkeys and anthropoids, that is, there were in each half jaw two incisors, one canine, two premolars, and three molars. The canines were no larger than the adjacent incisors and premolars, unlike the modern apes in which the canines form powerful tusks. Also, there was no simian shelf, the shelf of bone connecting the two sides of the lower jaw of modern apes. The premolars were tarsioid in character, a fact which has been interpreted to mean that the apes were derived from tarsiers independent of the monkeys. In other words, the three superfamilies of the suborder Anthropoidea may have been separate since their origin in the late Eocene or early Oligocene. The molars, however, are anthropoid in character. The second genus, *Propliopithecus*, was larger, and more specialized. It is regarded as being already on a collateral line of descent leading toward the gibbons and siamangs of today.

Proconsul. Anthropoid remains from the Miocene are very numerous, especially in east Africa, where a variety of fossil apes has been found. Some of these, belonging to the genus *Proconsul,* range in size from rather small animals up to some approaching the size of a gorilla. The limb bones indicate that these Miocene apes were of much lighter build than modern apes. Specializations for brachiation are either lacking, or very moderate. The teeth of *Proconsul* show some moderate specializations in the direction of the great apes. The skull, however, is more generalized, and resembles those of the smaller Old World monkeys. Moreover the brain, as shown by endocranial casts, resembled those of monkeys rather than anthropoid apes. An animal similar to *Proconsul,* but lacking its dental specializations, could have been ancestral to both the Hominidae and the Pongidae.

Dryopithecines. While the early Miocene development of the apes appears to have occurred in central Africa, they soon spread. In the late Miocene and early Pliocene, apes are known from several localities in Europe and India. The several genera are collectively called the subfamily Dryopithecinae, but it is doubtful that this heterogeneous array is a natural taxonomic group. One of the Pliocene genera, *Dryopithecus,* was very widespread, very common, and represented by many species. This was a fairly large ape, comparable to the chimpanzee of today. Variations of the teeth in different species suggest in some cases the chimpanzee, in others the orangutan, and in still others, the gorilla. It has been suggested that *Dryopithecines* may be ancestral to all of these modern types. (The gibbon and siamang appear to have been derived from another Pliocene genus, *Pliopithecus.*) Only two limb bones of *Dryopithecus* have been recovered, one humerus and one femur. They indicate that *Dryopithecus* was more slender than its modern descendants, and that it was not yet specialized for brachiation.

It has been much debated whether man could have been derived from a dryopithecine. On the positive side, the limb structure of these apes, as far as it is known, was still primitive, and could conceivably have given rise either to the limbs of man or to those of the apes. But the teeth show specializations of a type not found in man, including enlarged, pointed, tusk-like canines. Those who oppose the dryopithecine ancestry of man suggest that the ancestors of man must have branched off before the dental specializations had developed. Much more evidence is needed, especially with regard to parts of the skeleton other than the skull and teeth. Straus believes that the line of descent leading to man probably branched off from the primitive catarrhine stock when the latter was more monkey-like than ape-like, and hence before the development of actual anthropoid apes.

Oreopithecus. More promising as a Pliocene ancestor of man is *Oreopithecus,* a primate from the early Pliocene of Tuscany, which was described in 1872 as a fossil monkey, and then largely forgotten. Re-examination by Hürzeler has revealed many hominid characters, and other students have concurred in his judgment. Materials studied include many fragments of skulls, jaws (Figure 65), and teeth, as well as some vertebrae

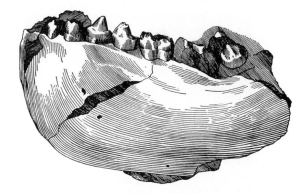

FIGURE 65. MANDIBLE OF
Oreopithecus bambolii.
(From Hürzeler, *Verh.
Naturf. Ges.,* Basel, V. 69,
1958.)

and limb bones. In 1958, a nearly complete skeleton was recovered, but its thorough study is still in the future. The dental formula is the typical anthropoid 2-1-2-3, and the proportions of the teeth tend to ally them to man rather than to the great apes. Unlike the apes, there is no diastema between the canines and the premolars. The details of the molars and of patterns of wear also tend to ally them with man. The limb bones and vertebrae have been studied less completely, but they too seem to support the conclusion that *Oreopithecus* was hominid. Hürzeler considers that the hominid trends of this animal were sufficiently advanced that they must have been in progress for a considerable time, and hence that the separation of the earliest hominids from primitive anthropoids must be sought even earlier than the Miocene. Yet it is possible that *Oreopithecus* represents a new family, allied to the Hominidae, but distinct from it.

The Australopithecines. Most of the later hominoid fossils can be clearly assigned either to the genus *Homo* or to the anthropoid apes, but one important group of fossils which has been discovered in South Africa is controversial in this respect. The original find was made in 1925 at Taungs by R. Dart, and described by him under the name of *Australopithecus* (southern ape). This was the skull of a child of about six years, and it showed a curious mixture of human and simian characters. The difficulties of study in this case were increased by the fact that most comparisons are based upon adult specimens. But a considerable number of additional skeletons, some nearly complete, have since been found by Broom, Dart, and others. Three genera have now been described, but all are included in a single subfamily of the Hominidae, the Australopithecinae.

The skull of the australopithecines resembles that of a modern chimpanzee, but the differences are significant. First, the brain case is larger in the fossil, having a capacity of about 600 cubic centimeters. This is somewhat larger than that of the gorilla, but is much larger compared to body size (about four feet tall). The forehead is more rounded out than in the chimpanzee, possibly indicating greater development of the highest centers of the brain. The eyebrow ridges are very prominent, but less so than in the chimpanzee. The jaws protrude quite prominently, but again less so than in the modern ape (Figure 66). The dentition is quite human in

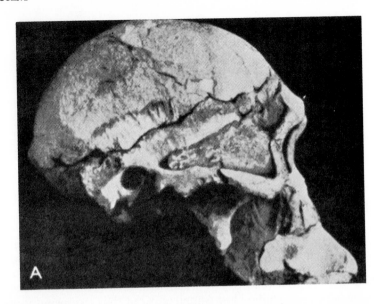

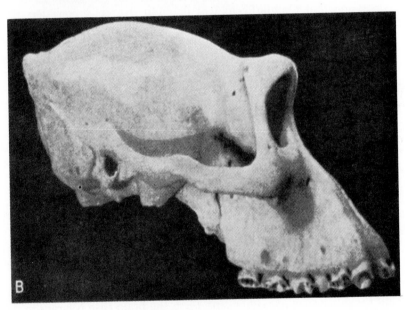

FIGURE 66. A RECONSTRUCTED AUSTRALOPITHECINE SKULL (A) COMPARED WITH THAT OF A CHIMPANZEE (B). (From Clark, "The History of the Primates," 3rd Ed., British Museum of Natural History, 1953.)

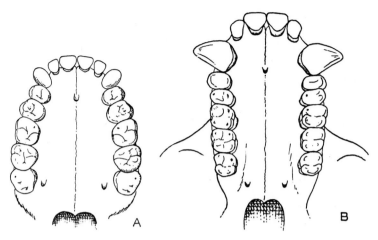

FIGURE 67. PALATE AND UPPER TEETH OF AN AUSTRALOPITHECINE (A) AND OF A
GORILLA (B). (From Clark, "The History of the Primates," 3rd Ed., British Museum
of Natural History, 1953.)

character. The canines are larger than in modern man, but much smaller
than in any modern ape. Further, the shape of the tooth rows is altogether
different. In the apes, the canines, premolars, and molars form parallel
rows, with the incisors being set at right angles to them at the front of
the jaws. In man and in the australopithecines, the entire tooth row is
more evenly curved (Figure 67). Finally, the occipital condyles, by
which the skull articulates with the spinal column, are set much farther
forward on the ventral surface of the skull in the australopithecines than
in any living ape. This suggests a relatively erect posture.

The rest of the skeleton gives evidence which corroborates that of the
skulls. The limb bones also indicate that this southern ape may have been
erect or nearly so in posture. There is no indication of the overdevelop-
ment of the arms which goes with brachiation. The hip bone is character-
istically long and narrow in apes, but in man and in the authralopithecines
it is broad and flat, an anatomical feature which is associated with erect
posture.

These, then, are some of the major facts relative to the Australopithecinae. Regarding the factual findings, there is no disagreement, but there
is much disagreement regarding their interpretation. Some highly competent anthropologists, including Broom and Robinson, believe that the
details in which they agree with man are too extensive and exact to be
explainable on any basis other than that they were in the direct line of
descent leading to man. Others equally distinguished, such as Weidenreich, believe that the australopithecines belong to a line of descent which
preserved some of the primitive characters of the original anthropoid
stock, but did not lead to any modern group. LeGros Clark takes an intermediate position, considering that they are certainly closely related to the
ancestors of man, but possibly not old enough to be themselves our an-

cestors. At the center of the problem is the unsatisfactory dating of these fossils. They have been found in deposits which are very difficult to date. Broom believed that they are at least of late Pliocene age, which would make them more than a million years old, older than any certainly identified human fossil. If this is correct, they would at least be old enough to qualify as ancestors of man. Current opinion favors early Pleistocene, but perhaps not early enough for the Australopithecines to be ancestral to man. Much more exact data are necessary on this important subject.*

MAN IN THE FOSSIL RECORD

Pithecanthropus. The known history of man begins with a fossil discovered by Dubois in 1891 in Pleistocene deposits of central Java. The fossil consisted of a single skull cap (Figure 68), a jaw fragment, and a femur. This was sufficient to show that its possessor had heavy, ape-like eyebrow ridges, but a much larger brain case (about 900 cubic centimeters) than any known ape, and occipital condyles apparently set sufficiently far forward to permit erect posture. Dubois regarded this as the "missing link" which was then so much under discussion, and he named it *Pithecanthropus erectus* (erect ape-man) to indicate this belief.

For nearly fifty years, the true nature of the Java fossils was controversial, but in 1938, von Koenigswald found a second and more complete skull. Additional finds have raised the total to five skulls, several jaw frag-

---

* Mention must be made of one more fossil of this group, found in July, 1959, by L. S. B. Leakey in the Oldovai Gorge of Tanganyika. A fairly complete skull and tibia, of late Lower Pleistocene age, were found in association with stone chopping tools and bones of the fauna of the time. All were sealed between two layers of rock in what Leakey considers to have been an actual living site of early Pleistocene man. Leakey named this fossil *Zinjanthropus*. While the general characters of this fossil are those of other Australopithecines, the differences are all *closer* to man. More material and more study will be necessary to assess the significance of this find, but it now appears to strengthen the claim of the Australopithecines to a place in the ancestry of modern man.

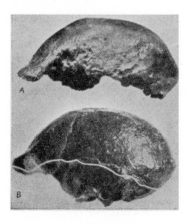

FIGURE 68. PITHECANTHROPUS SKULLS. The second specimen is broken off along the white line. (From Weidenreich, "Apes, Giants, and Man," University of Chicago Press, 1946.)

ments, and a femur. On the basis of these fossils, anthropologists are now generally agreed that *Pithecanthropus* was in fact true man. The age of the fossils is estimated to be on the order of 500,000 years.

Meanwhile, an important group of fossils was discovered in China. In 1927, Black found a single tooth, which he identified as probably human, while excavating a cave near Choukoutien. Two years later, he found a nearly complete skull, including parts of the lower jaw and teeth. Subsequent finds by Black, Weidenreich, and their collaborators have raised the total to fifteen skulls and other bones representing a total of some forty individuals. The Chinese fossils were originally described under the name *Sinanthropus pekinensis,* but detailed study by Weidenreich and von Koenigswald has shown that these do not differ significantly from the Javanese fossils, and so the name should be *Pithecanthropus pekinensis.*

Collectively, the Chinese and Javanese fossils give a fair picture of this most primitive of known men. He was of moderate stature, and the straight limb bones, broad hip bone, and position of the occipital condyles all show that he stood erect or nearly so. Proportionate lengths of arms and legs were much as in modern man, suggesting that this may be a primitive character, while the elongated arms of the apes are specialized. His forehead was retreating, and his jaws projecting, but much less so than those of any ape. His jaws and teeth were rather large, and there was no chin, a structure usually regarded as specifically human. The teeth, while larger than usual in man, agree with those of man rather than with those of apes in all specific differences. The size of the brain case was quite variable. In the Javanese skulls, it varies from 775 to 900 cubic centimeters, with an average at 860; while in the Chinese skulls it varies from 850 to 1300, with an average of 1075 cubic centimeters. This compares with an average of some 500 cubic centimeters in the gorilla and 1350 cubic centimeters in modern man. Intelligence is only very loosely correlated with brain size at best, and measurements on badly damaged and incomplete skulls are very crude estimates, yet is seems probable that *Pithecanthropus* was very clever by ape standards, yet very dull by human standards. No cultural remains have been found with the Javanese fossils, but the Chinese fossils are associated with crude tools of chipped stone and bone. Peking man used fire, and charred deer bones indicate that he had learned to cook. Thus his cultural attainments clearly indicate some human intelligence.

Two more fossils in this series deserve passing mention. An extraordinarily large jaw fragment was found in Java and described as *Meganthropus;* while three teeth, which dwarf even those of a large gorilla, were found in native drug stores of Hong Kong. These afforded the basis of a new genus *Gigantopithecus.* Weidenreich considered these to be ancestral to *Pithecanthropus,* and on this basis he concluded that man was descended from gigantic ancestors. This conclusion was never well received by anthropologists, but it was virtually demolished by the recent discovery of a lower jaw of *Gigantopithecus,* with teeth in *situ.* This jaw is clearly not that of a hominid, but rather it is that of a pongid, and not an extraordinarily large one, in spite of its prodigous teeth.

Finally, two lower jaws of mid-Pleistocene age, found in Algeria in 1954, have been described under the name *Atlanthropus*. While one of these is larger than the typical *Pithecanthropus* jaw, they are otherwise quite similar to the latter; and so it appears probable that *Pithecanthropus* was widely distributed in the Old World in mid-Pleistocene times.

These fossils also illustrate the major differences between man and the great apes. These may be summarized in two categories: adaptations to erect posture, and the development of the brain and associated changes in the skull. Of these, enlargement of the brain has lagged behind in human evolution, as shown by the study of australopithecines and *Pithecanthropus*.

The adaptations to erect posture may have begun as early as *Oreopithecus*, were quite far advanced in *Australopithecus*, and were largely complete in *Pithecanthropus*. In order to show how many lines of evidence may bear upon seemingly so simple a character, some detail may be given on this subject. The ankle bones bear the main weight of the body while erect. In man, these are large, while in the apes they are small. Correspondingly, the toes of the apes are long and freely movable, the great toe being opposable to the others. In man, the toes are short, and the great toe is held in line with the others, thus making a more rigid support for the weight of the body. As already pointed out, the leg bones are longer than the arm bones in man, while the reverse is true in the apes. The trunk of man is short relative to the legs, while the opposite is true in the apes. The human situation is obviously mechanically more stable for erect posture. The hip bone of the apes is long and narrow, while that of man is broad, thus giving optimum support to the viscera when in the erect position. The curvature of the spinal column of the apes is a single, sweeping, outward curve, like that of any four-footed mammal. This tends to throw the animal off balance when standing erect. But the curvatures of the spinal column of man alternate in direction, thus averaging out to a straight line. In the upright position, the human knee and hip joints are held straight, while those of the apes are flexed slightly. Finally, the occipital condyles of the apes are near the posterior end of the skull and are directed backwards, while those of man are near the center of the base of the skull and are directed downward. Thus the human adaptations to erect posture affect every part of the skeleton, as well as the viscera in many ways which have not been discussed. These adaptations appear to be largely completed in the most primitive men known.

The most important change in the skull has been the increase in size of the brain case. The largest brain case known in apes has a capacity of 685 cubic centimeters. The average of the Javanese fossils is about 900 cubic centimeters, while that of modern man is about 1350 cubic centimeters. This has been accomplished by an increase in the height of the vault of the skull, and by an increase in the diameter of the skull above the ear line. In the apes and in *Pithecanthropus* the greatest width of the skull is at the level of the ears. In later human races, the skull has become wider in the parietal region. As a result, the shape of the skull has become more nearly globular. This change started later than the adaptation to erect posture, and it has proceeded more slowly. As the brain case has enlarged, the jaws have decreased in size, with the result that the face has gradually receded to a position *under* the brain case, rather than in front of it as in all other mammals. As already pointed out, the premolar and molar teeth form parallel rows in the apes, while in man, all of the teeth form a gently curved arch (Figure 67). Several differences in the teeth may be mentioned, apart from size. The canines of man are no larger than the adjacent teeth, while those of apes are powerfully developed tusks. The first lower premolar of the apes is modified as a shearing tooth to work against the upper canine, whereas in man the first lower premolar is a typical bicuspid grinding tooth. Also, the surface of the molars of the apes is rather elaborately etched, in contrast to that of man. Finally, man has recently developed the chin, a protuberance on the lower jaw which is unknown in any other mammal.

Pleistocene deposits in Europe are dated according to the alternating periods of glaciation and warmer times which characterized the Pleistocene. There were four major advances of the glaciers separated by three interglacial periods. The fourth postglacial period has now been in progress for something like 25,000 years, and it constitutes the Recent epoch. Dating of the glacial periods is quite uncertain, but we may provisionally accept the following estimates. The first glaciation began about 600,000 years ago, and lasted perhaps 75,000 years. The second glaciation began about 500,000 years ago, and lasted about as long as the first. The second interglacial period was much longer, for the third glaciation began only about 250,000 years ago. Its duration was again about the same as the first two. The final glaciation began about 120,000 years ago, and has been receding for about 25,000 years.

The Heidelberg Jaw. One European fossil appears to be of an age comparable to *Pithecanthropus*. This is the massive Heidelberg jaw, found in 1907 near Heidelberg, Germany, in a gravel pit which also included bones of known early Pleistocene mammals. It is rather human in general aspect, except that it lacks a chin, and the teeth are quite human, but the entire structure is on a size scale unknown in modern man. It has been described under the name of *Homo heidelbergensis*. No subsequent finds have yet come to light.

The Piltdown Fraud. In 1911 and 1912, Charles Dawson, an amateur collector, recovered parts of a skull and lower jaw from a gravel pit near Piltdown, England. The skull was thick but remarkably human, the jaw quite anthropoid, and it soon became the center of controversy. Its proponents regarded it as the earliest known fossil man, claiming that such a combination of human and simian characters was to be expected in the most primitive man. Their opponents considered it to be a spurious association of human and ape bones. Stone and bone implements were found in the same deposit, and were claimed (or disclaimed) as evidence of a simple culture. In 1949, a reinvestigation of this fossil was begun at the British Museum, to which Dawson had presented the material. Tests of fluorine content (fluorine accumulates in buried bone) showed that whereas the skull was perhaps 50,000 years old, hardly a relic of the early

Pleistocene, the jaw was modern! Chemical tests showed that it had been stained to simulate a fossil of great age. The teeth showed atypical wear, and microscopic examination revealed file marks. X-rays showed that the roots of the teeth were too long for the crowns, being actually of the size of those of chimpanzees. In short, it was proven that a chimpanzee or orang jaw had been deliberately modified to simulate a transitional stage between ape and man. Thus the Piltdown man was proven to be a very clever fraud. The story of its exposure, principally by K. P. Oakley, J. S. Weiner, and W. E. LeGros Clark, is one of fascinating scientific detective work.

The Swanscombe Skull. If man was not in England 500,000 years ago, the Swanscombe skull, discovered in 1935, leaves no doubt that he was there 250,000 years ago. This skull consists only of the parietal (left found in 1935, right in 1955!) and occipital bones, and is thus very incomplete. It was found in deposits from the second interglacial period in association with crude flint implements and the bones of elephants, rhinoceroses and deer. The size and curvatures of the bones recovered closely resemble those of modern man, but nothing is known of the face, nor of other parts of the skeleton. On the basis of the bones available, one cannot point out significant differences between the Swanscombe man and modern man. But much more evidence is needed before the significance of Swanscombe man can be properly assessed.

The Steinheim and Galley Hill Skulls. A few more finds consist only of single skulls or parts of skulls. The Steinheim skull was found near a town of that name in Germany in 1933, and it is probably of the same age as the Swanscombe skull. This skull suggests an intermediate between *Pithecanthropus* and modern man. The general shape of the vault of the skull is more modern, but the capacity is somewhat greater. The eyebrow ridges are still very prominent, but the jaws are less projecting. At Ehringsdorf, Germany, a much more advanced skull was found in deposits indicating an age of about 120,000 years. This skull has a very large capacity, 1450 cubic centimeters, which is in excess of the average for modern man. On the other hand, the chin was not well developed. Two more skull fragments of very modern appearance, found at Fontéchevade, France, in 1947 have been assigned to the third interglacial period. Thus *H. Sapiens* may have been in western Europe 150,000 years ago. The Galley Hill skull, discovered in 1888 in middle Pleistocene deposits of England, shows no simian characteristics whatever, and it has been used to support a claim to the great antiquity of modern man. But many anthropologists are gravely doubtful about the correctness of the dating of this skull, and it now appears to be of very much later origin.

Neanderthal Man. The first human fossil to be found was a skull fragment found in a cave at Gibraltar in 1848. The bones of this skull were very thick. The eyebrow ridges were very prominent, the nose was broad, and the jaws were massive. This skull did not attract much attention, but eight years later, a similar skull cap together with a few ribs and limb bones were recovered from a cave in the Neanderthal valley of Germany. The remains became very well known under the name of *Homo neander-*

thalensis, or the Neanderthal man, which was popularly regarded as "the prehistoric man." The dating of the Neanderthal man was first established in 1886, when two skeletons were found at Namur, Belgium, in association with bones of the mammoth and the woolly rhinoceros, which were characteristic European animals of the last ice age. Since then, a large number of Neanderthal fossils, some of them quite complete, have been found in localities ranging throughout the Palearctic Region. Dating of deposits indicates that the Neanderthal man arose during the last interglacial period and did not become extinct until about 25,000 years ago.

A rather complete picture of the appearance of the Neanderthal man can be constructed on the basis of available skeletons (Figure 69). The skull was large and thick-boned. The eyebrow ridges were very prominent and the forehead receding. Although the cranial capacity was greater than that of modern man (the average was about 1450 cubic centimeters), the roof of the skull was rather flat. While the brain was large, no inferences regarding its quality are possible. However, the quality of his stone tools and the fact that he buried his dead indicate a high order of intelligence. The eyes were large and the nose was broad. The teeth and jaws were very large and heavy by modern standards, and the chin was receding. The occipital condyles are not quite so far forward as in modern man. It has often been said that the posture of Neanderthal man was stooped, but it now appears that this was a misinterpretation based upon the study of an arthritic skeleton. The spinous processes of the cervical vertebrae were exceptionally large, indicating that the neck musculature was powerfully developed. The standing height of the Neanderthal man could not have been much in excess of five feet. The hands and feet were dispro-

FIGURE 69. NEANDERTHAL FAMILY GROUP. (Courtesy of the Chicago Natural History Museum. Frederick Blaschke, sculptor, and Charles A. Corwin, artist.)

portionately large. As the Steinheim and Ehringsdorf skulls also show some of these features, some anthropologists regard them as the earliest Neanderthal men.

Fossils which have sometimes been considered Neanderthalian have also been found in South Africa and in Java. The South African fossils were found in Rhodesia, in 1921, and have been described under the name of *Homo rhodesiensis.* The find includes a nearly complete skull (Figure 70), part of another upper jaw, portions of limb bones and of the pelvic girdle. The limb bones are not distinguishable from those of modern man, but the skull is somewhat Neanderthaloid in appearance. The brain capacity is about 1250 cubic centimeters. Geological evidence on the age of Rhodesian man is inconclusive, but bones of species of mammals now living were found in the same cave, and it seems probable that the fossils are of relatively recent origin. In 1953, a skull cap and jaw fragment were found near Saldanha Bay in South Africa, and were described as Saldanha man. These bones resemble those of Rhodesian man. They appear to be older, and may have been ancestral to Rhodesian man. The Javanese find consists of eleven incomplete skulls and a tibia. The tibia is indistinguishable from that of modern man, but the skulls (all lacking the facial skeleton) are Neanderthaloid. The cranial capacity, however, is small, varying from 1150 to 1300 cubic centimeters. As the fossils were found on the Solo river in 1931 and 1932, they have been described under the name *Homo soloensis.* Both the Rhodesian and Solo men have been considered as late Neanderthal survivors, but perhaps more anthropologists now regard them as of uncertain relationship to other men.

The problem of the relationship of Neanderthal man to modern man is a much vexed question. It was originally assumed that the relationship was one of direct descent from Neanderthal to modern man. But the discovery of the Swanscombe skull in 1935 showed that a much more modern type of man was already present in Europe long before the date of the earliest known Neanderthal remains. Further, Neanderthal man was at least partly contemporaneous with Cro-Magnon man (see below), and this would appear to disqualify the Neanderthal as an ancestor, for it is generally agreed that Cro-Magnon man was ancestral to modern man. Some anthropologists now favor the idea that Neanderthal man is, like modern man, descended from *Pithecanthropus,* but by a collateral line which died out without leaving any descendants. Thus the relationship of Neanderthal man to modern man is that of an uncle rather than that of a parent. But so distinguished an anthropologist as Weidenreich, in reviewing the extensive fossils of Neanderthal and modern man, as well as many intermediates at Mount Carmel in Syria, has concluded that "No matter how the occurrence of such a mixture of forms may be explained, this find proves that the Neanderthalians did not die out but survived somewhere by continuing in *Homo sapiens.*" * The classical Neanderthal remains were all found in western Europe and were of rather late date. However, many skeletons of earlier date and of less marked Neanderthal-

---

* From Weidenreich, "Apes, Giants, and Man," 1946, University of Chicago Press.

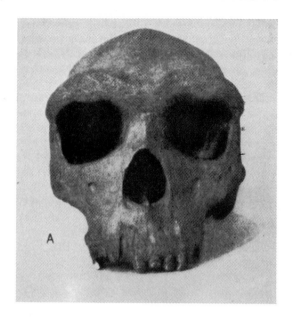

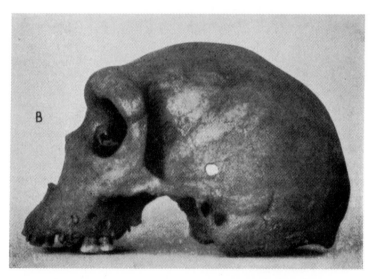

FIGURE 70. RHODESIAN SKULL. A, face view; B, side view. (From Clark, "The History of the Primates," 3rd Ed., British Museum of Natural History, 1953.)

oid type have been found in eastern Europe and in Asia. It is probable that, as early as the second interglacial period, a modern type, but highly variable, race of man inhabited much of the Old World. As the fourth glaciation set in, those with more extreme Neanderthaloid traits were isolated from the rest in southwestern Europe. Developing in isolation, and probably aided by considerable inbreeding, they developed the classical Neanderthaloid habitus. Meanwhile, the main population to the east developed along different lines, and produced *Homo sapiens*. Subsequently, this more progressive man replaced his Neanderthal cousins in western Europe.

Cro-Magnon Man. A considerable number of skeletons of modern man (*Homo sapiens*) have been found in late Pleistocene deposits of Europe. They make their advent around 40,000 years ago, during the last glaciation, and have dominated the European (and world) scene ever since. Their origin is not known, but it is commonly believed that they developed their physical and cultural characteristics elsewhere, probably in Asia, and then invaded Europe. If Weidenreich is correct they are simply the racially differentiated descendants of the early, generalized Neanderthals. Around a hundred of these fossils, called Cro-Magnon after the French cave in which they were first found, are now known. Their characteristics are well established and it may be said that they did not differ significantly from many modern Europeans. The cranial capacity appears to have been somewhat greater. There is considerable variation among the fossils and they appear to have blended into modern man.

INTERPRETATIONS OF THE FOSSIL RECORD

These, then, are the main outlines of the fossil record of man. These data can be agreed upon by every one, but they are subject to a wide variety of interpretations. Practically every fossil type was originally described as a distinct species, either of *Homo* or of a presumed ancestral genus. With these as data, varying phylogenies have been constructed. Hooton, for example, visualized two different lines of descent. One line leads from the Piltdown to modern man, with Swanscombe, Galley Hill, and Cro-Magnon men as intermediate steps. The other line leads from *Pithecanthropus* through *Sinanthropus* to Neanderthal man. The Mount Carmel fossils he accounts for on the basis of the assumption that some interbreeding occurred between Cro-Magnon and Neanderthal men. Gates believes that each of the major races of modern man should be regarded as a distinct species, and he believes that they have developed independently over the entire span of time covered by known human fossils. At the opposite pole is the opinion of Weidenreich that the anatomical evidence offers no alternative but to unite all of the known human fossils and modern man in a single species, *Homo sapiens*. Mayr has tried to simplify the interpretation by the application of the ordinary standards of zoological taxonomy. He feels that this requires that one genus, *Homo*, include all of the fossils from the australopithecines up through Neanderthal and Cro-Magnon to modern man. In this array, he finds no evidence that there has ever been more than one species at any one time, although subspecies have been contemporaneous. The australopithecines he designates as *H. transvaalensis*, with a possibility of several subspecies. Java and Peking man represent two subspecies of a second and later species, *H. erectus*. Finally, all of the later types comprise a single species, *H. sapiens*, with the extreme

Neanderthals of the last glacial period being simply a well-marked subspecies. The suggestion is inherent that each of these species is the progenitor of the succeeding one in temporal sequence. Finally, the possibility should be mentioned that *Oreopithecus* may be an intermediate between one of the many known Miocene apes and a man of the australopithecine type.

At the turn of the century and before, it was commonly charged that only vacuous theories could be presented with respect to human ancestry, because fossil evidence was entirely inadequate. This viewpoint was justified at that time, for only Neanderthal man was then well established as a fossil man. The same viewpoint can no longer be maintained without ignoring known facts, for available human fossils number in the hundreds. Yet such great discrepancies of opinion among qualified authorities as those cited above could hardly exist if really adequate data were at hand. Thus the greatest need for progress in anthropology is still for the accumulation and careful study of additional fossils.

In considering the problems of human phylogeny it is also well to bear in mind the warning of Zuckerman: "No one who has paid attention to the history of the study of fossil primates and its bearing on human evolution will doubt the need for the most critical of attitudes in dealing with these far-reaching conclusions. The difficulty is not only that stories of human phylogeny can never be more than a series of probabilities largely based on guesswork. We also have to consider the fact that speculation clouds almost every single stage in the treatment of the physical evidence itself. It begins with decisions as to which fragments found in a deposit are to be individually associated with each other. It continues into the stage where their anatomical characters, and the influence of slight variations in the manner of reconstruction, are considered. It ends with a variety of individual views about the theoretical framework of evolutionary change to which the facts can be fitted. When to all of this we add the uncertainties associated with the geological dating of fossil remains, and the fact that only rarely have those who have written on the subject had any understanding of the discipline of quantitative biology, we have all the ingredients necessary to produce endless speculation and controversy. It is true that several generally accepted conclusions about our fossil ancestry, based upon just such speculative grounds, have stood the test of time and argument—if these can be regarded as adequate scientific tests. But just as many have had to be considerably modified, while at least one, like the believed primate genus *Hesperopithecus,* the single fossil tooth of which is now known to be that of a peccary, has in the end been abandoned in spite of the fact that at one point in their histories they were supported by the selfsame leading authorities who to-day press the hominid claims of the new South African fossils...." Zuckerman is not an unconditional opponent of anthropological studies: he is a distinguished anatomist who has himself made valuable contributions to anthropology. His strictures may be rather extreme, but they are not illconsidered.

REFERENCES

BROWN, W. L., JR., 1958. "Some Zoological Concepts Applied to Problems in Evolution of the Hominid Lineage," *American Scientist*, **46**, 151–158. A very stimulating paper in which evolution is considered from a causative viewpoint.

CLARK, W. E. LEGROS, 1953. "History of the Primates," 3rd Ed., British Museum of Natural History, London. Brief but authoritative, clear, and well illustrated.

CLARK, W. E., LEGROS, 1955. "The Fossil Evidence for Human Evolution," University of Chicago Press, Chicago, Ill. A very scholarly and penetrating analysis. (Broom, Dart, Dubois, von Koenigswald, and Robinson.)

Cold Spring Harbor Symposium on Quantitative Biology, V. 15, 1950. "Origin and Evolution of Man," Long Island Biological Assn., N.Y. A valuable collection of papers by many authorities, including a paper by Mayr.

GATES, R. R., 1948. "Human Ancestry," Harvard University Press. A severely criticized book which presents a radical interpretation of its subject.

GAVAN, J. A., Ed., 1955. "The Non-Human Primates and Human Evolution," Wayne University Press, Detroit, Mich. A valuable collection of papers by many authorities.

HOOTON, E. A., 1945. "Up from the Ape," 2nd Ed., Macmillan Co., New York, N.Y. A classic, now badly out of date.

HOWELLS, WILLIAM, 1959. "Mankind in the Making," Doubleday & Co., Inc., Garden City, N.Y. A well-written, up-to-date, popular book on human evolution.

HÜRZELER, J., 1958. "Oreopithecus bambolii Gervais. A Preliminary Report," *Verh. Naturf. Ges. Basel*, **69**, 1–48. The principal English-language report to date on this important find.

MONTAGU, M. F. ASHLEY, 1952. "Introduction to Physical Anthropology," 2nd Ed., Chas. C Thomas, Springfield, Ill. A reliable text. (Black, Zuckerman.)

WEIDENREICH, FRANZ, 1946. "Apes, Giants, and Man," University of Chicago Press, Chicago, Ill. A systematic presentation, in readable form, of the viewpoint of an excellent anatomist.

WEINER, J. S., 1955. "The Piltdown Forgery," Oxford University Press, New York, N.Y. A fascinating scientific detective story, told by one of the principal sleuths. (Oakley.)

PART THREE

The Origin of Variation

"Descent with Modification"

CHAPTER THIRTEEN

Gene Mutation

THE UNIT of Mendelian heredity is the gene. The genes are parts of the chromosomes, and hence they ordinarily exist in pairs, just as the chromosomes do. Whenever the two genes of a pair are identical, there can be no doubt as to what trait they will determine. Whenever the two genes of a pair are unlike, however, several possibilities arise. It may be that the effect of one gene shows up to the exclusion of the other, in which case the one which is expressed is referred to as dominant, while its allele (alternative gene) is referred to as recessive. Dominant and recessive genes are symbolized by capital and small letters respectively. Or the two genes may collaborate to cause a trait intermediate between the two pure types. Such intermediates are called compounds. Lastly, the two unlike genes could collaborate to produce a character unlike either pure type.

ELEMENTARY MENDELIAN CONCEPTS

The Mendelian Laws. These genes are inherited in a statistically predictable manner (Figure 71). When gametes are formed, the two genes of a pair are separated into sister gametes, so that each gamete contains only one gene for each character (Mendel's first law—the Law of Segregation). As a result, whenever an organism is heterozygous for a particular gene (that is, the two members of the pair are unlike), two types of gametes will be formed in equal numbers. These unlike genes are not in any way diluted or modified in the direction of an intermediate because of their association in the hybrid. If either gene again becomes homozygous (both members of the pair alike) in a zygote, the original character will reappear unmodified. This is Mendel's second law, the Law of the Purity of Gametes. With the exception of certain special cases, when hybrid organisms interbreed, any type of sperm or pollen has an equal probability of fertilizing any type of egg or ovule (the principle of random fertilization). As a result, the offspring of two hybrids (Aa) include 25 per cent homozygous dominant (AA), 25 per cent homozygous recessive (aa), and 50 per cent heterozygous (Aa). As the heterozygotes show the dominant character, this results in 75 per cent dominant and 25 per cent recessive offspring, the famous 3 to 1 ratio. Such ratios are usually not obtained exactly, but rather, because they depend upon the laws of

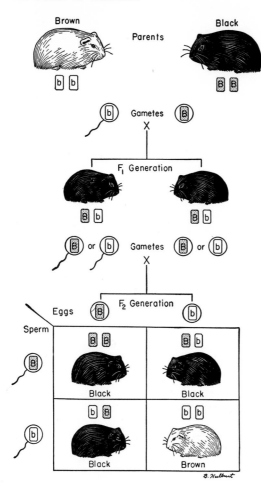

FIGURE 71. A MONOHY-
BRID CROSS BETWEEN
BROWN AND BLACK-
COATED GUINEA PIGS.
(From Villee, "Biology,
The Human Approach,"
3rd Ed., W. B. Saunders
Co., 1957.)

chance, deviations from them within the limits of statistical probability are obtained.

An example may be discussed with the aid of Figure 71. In guinea pigs, black coat color (B) is dominant over brown (b). If homozygous strains are crossed, all of the offspring (F_1, or first filial generation) must be heterozygous (Bb), and black because of dominance. If these are then interbred, each parent will produce two kinds of gametes (B and b) in equal numbers. Random fertilization of the two kinds of eggs by the two kinds of sperm results in offspring of which 25 per cent are homozygous black (BB), 50 per cent are heterozygous (Bb) and black because of dominance, and 25 per cent are brown (bb). Thus the F_2 generation consists of 3 blacks to 1 brown. An example without dominance is afforded by flower color in Four-o'clocks. If red flowering plants (RR) are crossed with whites (rr), then the F_1 plants are all hybrids (Rr) and are pink-flowering. Again, these produce pollen and ovules of two types (R and r)

in equal numbers. Hence random fertilization leads to an F_2 generation consisting of 25 per cent red flowering (RR), 50 per cent pink flowering (Rr), and 25 per cent white flowering plants (rr).

Such pairs of alleles exist because the original gene has mutated, that is, it has undergone a reproducible change which results in a modified character. There is nothing, however, to restrict the number of alternative forms of a gene to two, and actually large series of *multiple alleles* are known. Thus the white eye gene of the fruit fly, *Drosophila*, is represented by at least 14 alleles, and the self-sterility genes of many plants are represented by large numbers of alleles, up to around 200 in some cases. Only *two* members of any such series can be represented in any individual (although any number can be present in a population), and these are inherited in the usual Mendelian fashion, as described above.

If a cross is made between organisms differing in two pairs of genes, then each segregates as though the other were not there. This is shown by the fact that the ratio obtained is nine dominant for both genes, to three dominant for the first but recessive for the second, to three recessive for the first but dominant for the second, to one recessive for both (9:3:3:1). But this is simply the algebraic expansion of the binomial $(3 + 1)^2$.

An example is shown in Figure 72. In the guinea pig, short hair (S) is dominant over long (s). If a black, short variety ($BBSS$) is crossed to a brown, long variety, the F_1 are all dihybrids ($BbSs$), and they are black and short because of dominance. When gametes are formed, one gene of each pair must be included in each gamete, but these include all possible combinations in equal numbers. In this example, gametes are formed of types BS, Bs, bS, and bs. Random fertilization yields an F_2 consisting of 9B?S?, 3B?ss, 3bbS?, and 1 bbss (the question mark after a dominant gene indicates that the second member of the pair might be either dominant or recessive); or, restating it, 9 black, short; 3 black, long; 3 brown, short; and 1 brown, long.

If three pairs of genes, all on different chromosomes, differ in a cross, then the ratio of phenotypes (appearances) obtained is the expansion of $(3 + 1)^3$, or 27:9:9:9:3:3:3:1. The same principle is indefinitely applicable, and is called the principle of Independent Assortment. But if a cross involves gene differences for two or more pairs of genes located on the same pair of chromosomes, then one would expect that the number of recombinations of the genes would be like that of a monohybrid cross. That is, if a cross were made between two homozygous individuals, $AABB$ and $aabb$, then one of the chromosomes of the offspring should contain the two dominant genes, AB, its homologue should contain the two recessive genes, ab, and these two combinations should be maintained indefinitely. This is the phenomenon of linkage, and it is a simple consequence of the fact that the genes are more numerous than the chromosomes. But linkage is not absolute, for blocks of material may be exchanged between homologous chromosomes. Thus, in the example discussed above, the combination Ab and aB could be formed, though only in a minority of the gametes. This is called crossing over.

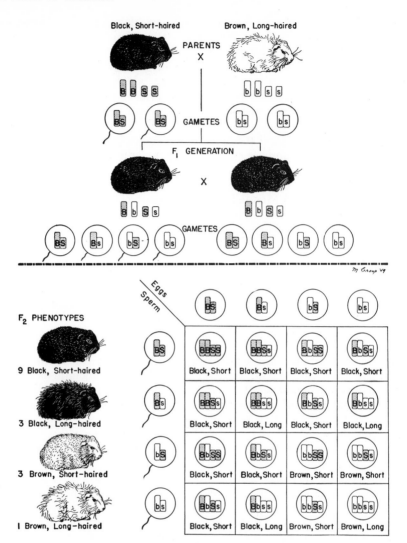

FIGURE 72. A DIHYBRID CROSS BETWEEN A BLACK, SHORT-HAIRED GUINEA PIG AND A BROWN, LONG-HAIRED ONE, SHOWING INDEPENDENT ASSORTMENT. (From Villee, "Biology, The Human Approach," 3rd Ed., W. B. Saunders Co., 1957.)

Sex Linkage. Sex was the first character to be successfully interpreted in terms of the chromosomes. It was found that many species of grasshoppers have 23 chromosomes in males and 24 in females. Thus 11 pairs of chromosomes (autosomes—A) are identical in the two sexes, but the twelfth pair is complete in the female yet represented by only one chromosome in the male. This is the *sex differential* pair, or the X chromosomes. Consequently, meiosis (maturation divisions) results in eggs all of which are similar, having $11A + X$. Sperm, however, are of two classes:

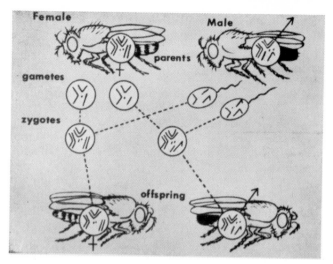

FIGURE 73. SEX DE-TERMINATION IN *Drosophila*. Note that all chromosomes of the female are perfectly paired, while the *X* chromosome of the male has an unlike mate, the *Y* chromosome, which is hooked in this species. (From Dodson, "Genetics," W. B. Saunders Co., 1956.)

$11A + X$ and $11A + O$ (no sex differential chromosome). When these two kinds of sperm fertilize the eggs, the parental conditions are re-established, that is, half of the zygotes have $22A + 2X$ and become females, while the other half have $22A + X + O$ and become males. More commonly, as in *Drosophila* and man, the *X* chromosome of the male has an unlike mate, the *Y* chromosome. Thus in *Drosophila* (Figure 73), females have $6A + 2X$, while males have $6A + X + Y$. Meiosis results in eggs with $3A + X$ and sperm of two types, $3A + X$ and $3A + Y$. Fertilization then leads to two equal classes of progeny, $6A + 2X$ (females) and $6A + X + Y$ (males).

In general, very few mutant genes are known on *Y* chromosomes, but the *X* chromosomes have many. The pattern of inheritance of such *sex-linked* genes necessarily follows that of the *X* chromosomes, and so is readily identified. Because such genes are unpaired in males, sex-linked recessive genes inherited from the mother show up in male progeny.

Quantitative Inheritance. Inheritance of quantitative traits also deserves special mention. These include all of those individual differences which must be defined by measurement rather than qualitatively. Examples include size, proportions of parts, intensity of color, rate of production of a vitamin, concentration of a protein, and many others. The genes which influence these traits are inherited in the usual Mendelian fashion, but swarms of pairs of genes collaborate in the determination of each character. The effect of each such gene is small, so that variation in a heterozygous population appears to be continuous. Also, dominance is often lacking, so that there is a simple additive effect of the genes present. Because of these characteristics, quantitative inheritance can only be studied with the tools of statistics.

All of the basic phenomena of inheritance can be understood in terms of the above principles. They should therefore be thoroughly reviewed, as should the entire field of elementary genetics, before going any further in the study of evolution.

THE GENE THEORY

The name *gene* was applied to the hereditary unit by Johannsen with the intention that it should be just a convenient term to designate the units of heredity, which Mendel had referred to as "factors" or "elements." He expressly stated that he proposed this term without intention of implying any particular theory of the nature of the hereditary units. The name has become closely associated, however, with the theory of T. H. Morgan that the genes are corpuscular bodies in the chromosomes, arranged in linear order like beads on a string, each gene separated from all of the others and different from all of the others in substance, while all of the genes of a particular chromosome are held together by an indifferent substance. This theory is based upon three types of evidence: the fact of mutation; the fact that a linear order of genes in the chromosome can be established by cross-over tests (see Chapter 14); and the fact that, once this order is established, it can be reshuffled by subsequent crossing over.

This *morphological* concept of the gene dominated genetic thinking for many years. Its boundaries were soon blurred by position effects (Chapter 14); and more recently investigations of the biochemistry and physiology of the gene have increased the difficulty of setting exact boundaries to it, so that physiology rather than morphology dominates current thinking about the gene.

Gene Number and Gene Size. Attempts have been made to estimate the number of pairs of genes for various organisms. Obviously, a direct estimate of the number of genes is not available for any organism, both because a complete study of all hereditary characters of a single organism has never been made, and because a particular gene is identifiable by genetic methods only when it is available in more than one form—in other words, when it has mutated to form two or more alleles. Estimates for the fruit fly, *Drosophila melanogaster*, genetically the best known of all organisms, range from 5000 to 15,000. Belling has estimated about 2200 genes for *Lilium*. And Curt Stern estimated not less than 5000 nor more than 120,000 for man. Estimates of the size of the gene are based upon the volume of the chromosomes (especially the euchromatic portions) divided by the estimated number of genes. On this basis, Gowen and Gay calculated the average size of the genes of *Drosophila* to be 1×10^{-18} cubic centimeters. More recently, Pease and Baker, using the electron microscope, have observed in the salivary gland chromosomes of *Drosophila* leaf-shaped bodies which they believe to be the genes (Figure 74). These vary in size by a factor of about 3, but the average is about 1×10^{-17} cubic centimeters. As the salivary gland chromosomes are giant chromosomes to begin with, the mere fact that these bodies are ten times the size estimated by Gowen and Gay cannot be regarded as conflicting with their estimate. But neither is there any proof that these bodies actually are the

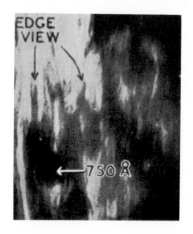

FIGURE 74. THE GENES? This is an electron micrograph of a portion of the salivary gland chromosomes of *Drosophila*, and the leaf-shaped bodies were once thought to be genes. (From Pease and Baker, *Science*, V. 109.)

genes. If the boundaries of the gene are actually indefinite, then the significance of such estimates is considerably reduced.

Whatever their nature, then, there is very abundant evidence that the genes do mutate to produce permanently inheritable alleles. And it is equally evident that this is a source of variability which must be studied for its bearing on evolution. A mutation may be defined as a permanent change in a gene. Alleles exist only because the original wild-type gene has mutated at some time. Like the original gene, the mutant is recognized by the character which it causes.

Mutation in Nature. Critics of genetics long held that the mutants with which geneticists worked could not be significant, because such mutants were not a really natural phenomenon, but a sort of degenerative effect of laboratory environment. This has been abundantly disproved, for many investigators have found in nature many mutants identical with or similar to those dealt with by laboratory zoologists. The original *Drosophila* mutant (white eyes) was, as a matter of record, captured in nature by the entomologist Lutz at about the same time that it arose in the laboratory stocks of T. H. Morgan. Goldschmidt found many mutants in the wild populations of *Drosophila* near Berlin. Tschetverikoff inbred the offspring of 239 wild *Drosophila melanogaster* from southern Russia, and 32 recessive mutants segregated out. Dubinin collected the same species in several localities in the Caucasian Mountains. On inbreeding, he found that the incidence of lethal genes varied from 0 to 21.4 per cent in different localities, while visible mutations varied from 3.9 to 33.1 per cent. Many of these had minor effects, such as a slight reduction in size of bristles. Baur studied wild snapdragons (*Antirrhinum*) and found that about 10 per cent of the plants showed at least one mutant. Dobzhansky showed that in wild populations of *Drosophila pseudoobscura* 75 per cent of the chromosomes showed at least one mutant. Dice has found that wild populations of the deer mouse *Peromyscus* always have many mutants. It may then be regarded as established that mutation is a normal phenomenon in nature.

Rate of Mutation. The rate at which mutation occurs has also been the subject of many studies. Spencer has shown that the rate of mutation in *Drosophila* differs from time to time. These differences were attributed to environmental influences. Demerec found that the rate differed in different genetic strains of *Drosophila,* and hence proposed that the rate of mutation is itself controlled by specific genes. He found that the rate of lethal mutation varied from one in every 100 chromosomes to one in every 1000 chromosomes. Ives has identified such "mutator" genes in wild populations of *Drosophila,* genes which increase the mutation rate as much as ten fold. Baur found that 5 to 7 per cent of the progeny of normal *Antirrhinum majus* showed at least one mutant, while none was observed in *A. siculum* during twenty years of breeding. These data were explained on the basis of a great difference in mutation rate in these two closely related species. Stadler measured the frequency of the appearance of eight mutants in maize. When expressed as the number of times each occurred per million gametes, the following series was obtained: 492, 106, 11, 2.4, 2.3, 2.2, 1.2, and 0. It is clear, then, that different genes mutate at different rates even within the same strain. This is supported by laboratory studies of *Drosophila.* Some mutants have appeared many times in laboratory stocks, some rather rarely, and some are known only from single records.

The explanation of variations in mutation rates has been tentatively considered in thermodynamic terms. Change from one allelic state to another involves physicochemical change, and this must require free energy. The amount of energy required depends upon the specific mutation. The rate of any mutation, then, will generally be inversely proportional to the amount of energy required. Thus in Stadler's study, I mutated to i 106 times per million gametes, while Pr mutated to pr only 11 times per million, so it is probable that the latter mutation requires much more free energy than the former. The same considerations apply to different mutational steps in a series of multiple alleles. Thus in the eye color series in *Drosophila,* W (red) mutates to its various alleles far more readily than does w (white).

One may ask where mutator genes fit into this picture. They could act by increasing the amount of available energy or by lowering the energy threshold for mutation. No data are available which permit a choice between these hypotheses. Many data, however, show that mutation requires time. Thus increased pressure reduces the mutagenic effects of radiation (see below), even if the pressure is applied as much as 20 minutes after radiation. Hence it appears that the mutating gene first passes to a labile state of some duration, from which it may go to a mutant state or to its initial state.

Direction of Mutation. It is commonly said that the direction of mutation is random, meaning that chance alone determines in which of an infinity of possible ways a particular gene will actually mutate. This is true in the sense that the environment does not cause the appearance of mutants which are appropriate to it. In all environments, both natural and experimental, the majority of the mutants are disadvantageous: their prospective fate is elimination by natural selection. Yet *a priori* one would expect that there must be a limited number of ways in which a particular gene can mutate. For any known chemical substance, there are certain classes of reaction which are possible and certain classes which are not possible. Whatever theory of the gene one may hold, the genetic material must have definite physical and chemical properties which will restrict the potential scope of mutation. That this is actually the case is indicated by the frequency with which many mutations recur, for example albinism, which is common in most groups of vertebrates.

Experimental Production of Mutation. As early as 1927, H. J. Muller showed that mutations were produced in *Drosophila* at several hundred times the normal frequency if the gonads were X-rayed. He selected sex-linked lethal mutations for special study because of the ease with which they can be identified. It has since been shown by Muller and others that any high-energy radiation will produce mutations. In 1947, Muller was awarded the Nobel Prize in Medicine on the basis of this work. With any type of radiation, the rate of induced mutation is directly proportional to the dosage, but is independent of the rate of administration of the radiation. Chemical mutagens were sought unsuccessfully for many years, but during World War II, Charlotte Auerbach demonstrated that mustard gas is as effective as radiation. Subsequently, mutagenic activity has been found in such diverse substances as urethane, formaldehyde, peroxides, manganese chloride, aluminum chloride, purines, and pyrimidines, and the list may well become indefinitely large.

When Muller's work was first reported, the question was naturally raised, is cosmic radiation, or some other naturally occurring radiation, responsible for mutation in nature? Turel presented evidence that the mutation rate is higher in alpine areas than in lowlands, and he suggested that this might be caused by ultraviolet radiation from the sun, which is known to be more intense at higher altitudes. But it is also possible that the great diversity of alpine plants has been caused by selection of different mutants occurring in isolated areas, without the need of a high mutation rate. Babcock tried raising *Drosophila* in areas of high and low natural radiation. Significant differences in mutation rate were not observed. Hence it must be admitted that studies of experimental mutation have not yet led to the understanding of naturally occurring mutation. Chemical mutagens may well be of great importance here.

GEOGRAPHIC SUBSPECIES AND NEO-DARWINIAN EVOLUTION

An important aspect of natural species is the fact that the various types of variability occurring within a species are not scattered evenly over its entire range, but rather that local populations, commonly more or less isolated from their neighbors, show distinctive patterns of the variable characters of their species, with the result that they may be defined as subspecies. They are fertile in crosses, and intergrades between adjacent subspecies are commonly found wherever their ranges meet. But the

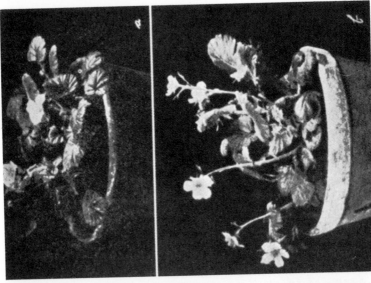

FIGURE 75. ALTITUDINAL ECOTYPES OF *Geum montanum*. *a* from 2800 meters, *b* from 2400 meters, and *c* from 2000 meters altitude in the Tyrol. (After von Wettstein, from Goldschmidt, "The Material Basis of Evolution," Yale University Press, 1940.)

several subspecies of any one species do inhabit ranges which are mutually exclusive for the most part. Closely related species may have identical ranges, but subspecies replace one another geographically. The whole series of geographic subspecies for any species is called a Rassenkreis (German—group, or circle of races).

Wherever subspecies are distributed over an area which presents a progressive change in some physical feature, such as mean annual temperature, some of the differences in the subspecies are likely to show a progressive change also. Such series are called *clines*. One can scarcely doubt that such characters, which vary progressively along with a progressive change in climate or topography, must be of adaptive value to the various subspecies concerned. Thus, Alpatov found that the average size of bees increased progressively from southern to northern Europe, while the lengths of legs and tongue decreased along the same cline. Similarly, Turesson has shown that plants commonly show definite character complexes according to the type of habitat in which they live. Thus in *Primula* (primrose) there are *ecotypes* adapted to alpine habitats and there are ecotypes adapted to meadowlands, and to as many more habitat types as the plant is naturally found in (Figure 75). Here again, one can scarcely doubt the adaptational value of a character complex which is always found in plants which live in a particular type of environment. Wherever such ecotypes form a cline, they coincide with the subspecies. Sometimes factors leading to such cline formation can be summarized in concise rules. Thus Bergmann's rule states that, in a given species or genus, northern populations have a larger mean body size than do southern populations. This is believed to depend upon the fact that a large body has a smaller surface area per unit volume than does a small body. Thus conformity to the rule promotes heat conservation in the north and heat conduction in the south. A corollary to this is Allen's rule, which states that extremities (limbs and tails) tend to be smaller in the north than in the south. Again, this is interpreted in terms of restriction (north) or extension (south) of radiating surfaces. Scholander has recently explained adaptation on quite a different basis. Arctic mammals have very effective insulation of fur or feathers, as well as circulatory control of heat loss, which permits them to maintain the body temperature against severe cold by basal metabolism alone. The less protected extremities are physiologically adapted to function well even when thoroughly chilled. He believes that these findings invalidate Bergmann's and Allen's rules. Still, the number of analyzed cases in which the rules appear to apply is so great, and their explanation on any other basis would be so difficult, that zoologists generally have not been convinced of Scholander's thesis. Probably the physiological adaptations are of major importance, yet under conditions of severe competition even the small additional advantage gained by size variation in accordance with these rules is sufficient to give a selective advantage to those populations which conform. Gloger's rule states that animals in cool, dry regions tend to be lighter in color than those in warm, humid regions. Exceptions to these rules can be found, but they have considerable validity.

Subspecies need not occur in clines. That is, if a series of subspecies, A, B, C, D, and E occur in that order over the geographic range of the species, A might resemble B most closely in size, while it resembles D most closely in color. In a third character, it might resemble C most closely, and so on. In many cases, subspecies can be defined only by a statistical analysis of several variable characters in the various populations of a species. Subspecies of North Sea herring, for example, have been defined by means of statistical studies of such characters as number of vertebrae, number of scales in the lateral line, number of rows of scales above and below the lateral line, and many other variable characters. The resulting statistics show that the averages and standard deviations for each school are typical, and differ from all others, with the result that they may be described as subspecies.

Now Darwin had already spoken of subspecies as "incipient species ... (which) ... become ultimately converted into good and distinct species ... by ... natural selection." As already mentioned, Darwin realized that the weakest point in his theory was the lack of knowledge of heredity. The dominant school of evolutionary thinking today is based upon the Darwinian principles of the prodigality of nature, variation, and natural selection. But to this is added the interpretation of variation in terms of the theory of the gene, and the study of populations and changes in gene frequency with the tools of statistics which, like genetics, did not exist in Darwin's time. This school is therefore known as the neo-Darwinian school, and it is exemplified by most of the outstanding students of evolution of today, as Sewall Wright, Th. Dobzhansky, G. L. Stebbins, and Julian Huxley.

For the neo-Darwinians, then, subspecies are, at least potentially, incipient species. Each subspecies of a particular species is characterized by a particular complex of gene-determined characters, and these various genes are derived by mutation from an originally more uniform progenitor. The genes which any particular subspecies possesses are not peculiar to it, but are found in varying percentages in different subspecies. It is the constellation of genes which is unique for each subspecies, not the individual genes. Such gene combinations are the materials upon which selection acts. Within such subspecies, mutation continues at random, with the result that different alleles arise in different subspecies. Now subspecies are usually at least partially isolated from each other geographically. Other types of isolation (to be discussed below) may also arise, for example physiological or ecological isolation. Within such isolated subspecies, new mutations, as they arise, cannot spread throughout the species. Thus, by the gradual accumulation of mutations, such isolated subspecies may "become ultimately good and distinct species." The spectrum of mutation extends from gross defects such as lethality or crippling through moderate effects like color change or change in bodily proportions to very minor changes detectable only by special methods. Neo-Darwinians generally favor the latter, for several reasons. First, most of the larger mutations are plainly abnormalities which would be eliminated by natural selection. This may be a consequence of the fact that the more favorable ones, which must have recurred many times in the past because of the high natural mutation rate, have already been incorporated into the normal genotype of the species. Second, although unproven, it is generally accepted that a species can harmoniously incorporate into its genotype a series of small mutations more readily than it can a single, equivalent, larger mutation. Most important, however, is the fact that subspecies and closely related species generally differ from each other in a series of quantitative traits, such as size, proportions, intensity of color, or extent of a pigmented area. As explained above, such traits are generally inherited upon the basis of swarms of genes, each with a very small effect individually.

This theory is plausible in every detail. It has the advantage of incorporating the greater part of historical Darwinism with some of the main trends of modern genetic research. It is backed up by a great body of experimental data. It is hardly to be wondered at then, that this theory has achieved very nearly universal acceptance. But there is another possibility, also based upon modern genetic research, and this will be taken up in the next chapter.

REFERENCES

BABCOCK, E. B., 1947. "The Genus Crepis," Part 1, Univ. Calif. Publs. Botany, 21, 1–199. This paper summarizes an extraordinarily comprehensive study of one genus, with the neo-Darwinian theory providing the direction. (Turesson.)

BLUM, H. F., 1955. "Time's Arrow and Evolution," 2nd Ed., Princeton University Press. The application of the laws of thermodynamics to mutation and evolution is here explored.

CLAUSEN, JENS, 1951. "Stages in the Evolution of Plant Species," Cornell University Press, Ithaca, N.Y. A brilliant exposition of evolution in several genera of plants, from a neo-Darwinian viewpoint. (Baur.)

DOBZHANSKY, TH., 1951. "Genetics and the Origin of Species," 3rd Ed., Columbia University Press, New York, N.Y. A very readable book which is the cornerstone of the neo-Darwinian theory. (Auerbach, Demerec, Dice, Dubinin, Ives, Spencer, Stern, Tschetverikoff.)

DODSON, EDWARD O., 1956. "Genetics: the Modern Science of Heredity," W. B. Saunders Co., Philadelphia, Pa. If you have enjoyed your evolution book, you might like this one too.

MORGAN, THOMAS HUNT, 1928. "The Theory of the Gene," 2nd Ed., Yale University Press, New Haven, Conn. A classic of modern biology. (Belling.)

STADLER, L. J., 1954. "The Gene," Science, 120, 811–819. A very penetrating study of the theory of the gene by one of its major architects.

CHAPTER FOURTEEN

Chromosomal Mutations

ONE OF THE MAJOR ACHIEVEMENTS of classical genetics was the demonstration, principally by T. H. Morgan and his associates, that each gene can be assigned to a definite locus, or position, in its chromosome by means of cross-over tests. If a wild-type female *Drosophila* is mated to a yellow-bodied, white-eyed male, the F_1 females should all be heterozygous, with the two dominant genes in the X chromosome derived from the mother and the two recessive genes in the X chromosome derived from the father. If no crossing over occurs, and if these F_1 females are backcrossed to the recessive type, then all of the backcross generation should be of one parental type or the other. Actually these types (gray-bodied, red-eyed, and yellow-bodied, white-eyed) make up 98.5 per cent of the offspring, but the remaining 1.5 per cent comprises flies which are gray-bodied but white-eyed, or yellow-bodied but red-eyed.

LONGITUDINAL DIFFERENTIATION OF THE CHROMOSOMES

Crossing Over. These exceptional types can be accounted for only by the exchange of genes between the two X chromosomes of the female parent. The frequency of the exchange is typical for any two pairs of genes which may be studied, and is the same in reciprocal crosses. If the cross involves yellow and cut (a wing mutant), the exceptional types will always make up 20 per cent of the backcross generation. If white and cut are tested, then the exceptional types comprise 18.5 per cent of the progeny. These data give a basis for mapping the relative positions of the genes in the chromosome, if it be assumed that the frequency of crossing over is a function of the distance between the genes. Thus the white and yellow genes, which show only 1.5 per cent of crossing over, should be fairly close together in the chromosome, while white and cut should be twelve times as far apart. Such crossover experiments between sets of three pairs of genes have made it possible to map the chromosomes of the genetically better known plants and animals. They leave no room for doubt that the chromosomes must be differentiated longitudinally.

Specificity of Synapsis. Cytological evidence of the longitudinal differentiation of the chromosomes is also available. The synapsis of homolo-

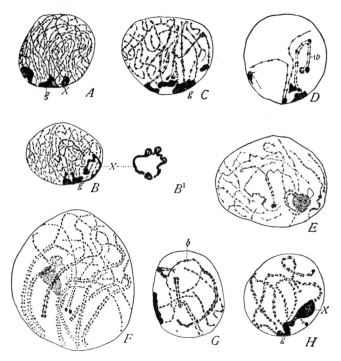

FIGURE 76. PROPHASE OF THE FIRST MEIOTIC DIVISION IN GRASS-HOPPERS, SHOWING THE REGULAR PATTERN OF THE CHROMOMERES, AND THE SPECIFICITY OF SYNAPSIS. (From Wilson, "The Cell in Development and Heredity," 3d Ed., The Macmillan Co., 1925.)

gous chromosomes is specific not only for the whole chromosomes, but for the chromomeres which are the smallest visible components of the lepto-tene (greatly extended preparatory to synapsis) chromosomes (Figure 76). Whenever the chromomeres are individually identifiable because of differences of size or shape, it turns out that only like chromomeres synapse, never unlike ones. If a group of chromomeres has been lost from one chromosome of a pair, then at synapsis the corresponding chromomeres of its mate form an unpaired bulge projecting to one side of the chromosome. Other gross changes in the chromosomes, which will be discussed below, produce changes in the synaptic behavior of the chromosomes which are understandable only on the principle that synapsis is specific for each point along the length of the chromosome.

The Salivary Gland Chromosomes. The most impressive cytological evidence for the longitudinal differentiation of the chromosomes is derived from the study of the salivary gland chromosomes of Diptera, especially *Drosophila*, in which they have been most intensely studied. These are giant chromosomes (Figure 77), up to half a millimeter in length, so that they would be visible to the naked eye if they were opaque. They are very closely synapsed pairs, in which reorganization on the molecular

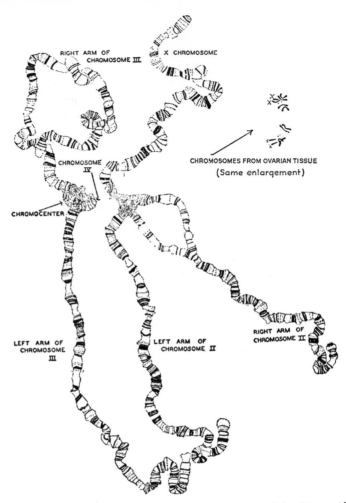

FIGURE 77. THE SALIVARY GLAND CHROMOSOMES OF *Drosophila*. (From Altenburg, "Genetics," 2nd Ed., Henry Holt & Co., Inc., 1957.)

level appears to be involved in the great size increase. These chromosomes have a cross-banded appearance, and the order, shape, and intensity of the bands are all perfectly regular, so that the salivary gland chromosomes can be mapped, and even small segments can be identified by reference to a standard map.

ARCHITECTURAL CHANGES IN THE CHROMOSOMES

The above data lead to the conclusion that, beneath the homogeneous appearance so often presented by the chromosomes, there is a longitudinal physicochemical differentiation, the "architecture of the chromosomes."

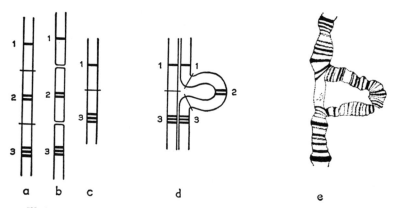

FIGURE 78. HETEROZYGOUS DEFICIENCY IN THE SALIVARY GLAND CHROMOSOMES. *a–d,* diagrammatic; *e,* actual specimen. (From Altenburg, "Genetics," 2nd Ed., Henry Holt & Co., Inc., 1957.)

Once a standard architecture for any chromosome is established, it ought to be possible to rearrange it in various ways. Four such types of rearrangement are known. A *deletion* or *deficiency* constitutes the loss of a segment of a chromosome (Figure 78). Such losses can be produced experimentally by high-energy radiation, and they may occur in this way naturally also. At synapsis, in an organism which is heterozygous for a deficiency, the unpaired portion of the normal chromosome projects to one side as a loop, while those points which are present in both chromosomes of the pair are synapsed normally.

Just the opposite of a deletion is a *duplication,* in which a segment of a chromosome is repeated. In a heterozygote, synapsis looks very much like it does in a deficiency heterozygote, this time because a region which is present only once in the normal chromosome is present twice (or more) in the modified chromosome. In the salivary gland chromosomes, the difference between a deficiency and a duplication can be easily detected. It is believed that the production of duplications is based upon unequal crossing over, that is, crossing over in which the breaks in the homologous chromosomes occur at somewhat different points. This unequal crossing over would produce simultaneously two gametes with a duplication and two gametes with a deficiency.

A third type of chromosomal rearrangement is an *inversion,* and is simply a reversal of a segment of a chromosome. Thus, a chromosome in which the order of parts runs ABCDEFGHIJK might undergo an inversion of the segment D to H, with the result that the order of parts would run ABCHGFEDIJK. In synapsis in the heterozygote, homologous point still synapses with homologous point, with the result that one chromosome of the pair must form a twisted loop, while the other continues around it without twisting (Figure 79). An inversion can be produced only if a chromosome is broken at two points, and if it reheals in the reversed relationship. They are produced frequently in radiation experiments, and

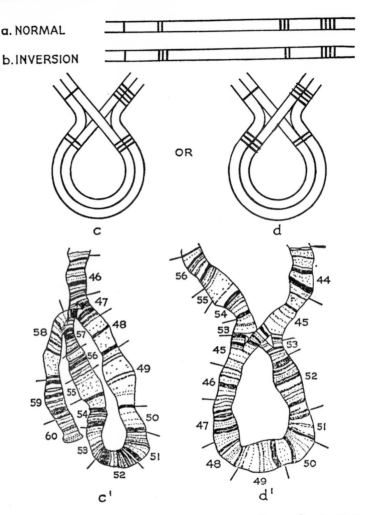

FIGURE 79. HETEROZYGOUS INVERSION IN THE SALIVARY GLAND CHROMOSOMES. *a–d*, diagrammatic; *c'* and *d'*, actual chromosomes. (From Altenburg, "Genetics," 2nd Ed., Henry Holt & Co., Inc., 1957.)

it seems probable that naturally occurring radiation plays a role in the production of naturally occurring inversions. In any case, inversions appear to be rather common in nature.

The fourth and final type of architectural rearrangement of the chromosomes is called a *translocation*. This means that a segment of one chromosome has been transferred to a nonhomologous chromosome (Figure 80). Typically, translocations are reciprocal, that is segments are exchanged between two nonhomologous chromosomes, so that one may speak of "illegitimate crossing over." In synapsis of the heterozygote, the

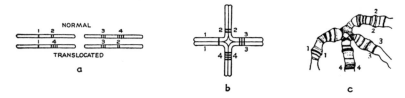

FIGURE 80. TRANSLOCATION HETEROZYGOTE. *a* and *b* diagrammatic, *c* actual chromosomes. (From Altenburg, "Genetics," 2nd Ed., Henry Holt & Co., Inc., 1957.)

exchanged portions retain their original synaptic specificity, with the result that two different tetrads are bound together.

Position Effects. In accordance with the classic gene theory (see the preceding chapter), it might be expected that such chromosomal rearrangements would alter linkage relations, but would never actually affect phenotypes. Yet in fact there are many examples in which each type of chromosomal rearrangement behaves as though it were a gene mutation: that is, it causes a definite phenotype, and is inherited according to the usual principles of Mendelian heredity. Many are lethal in the homozygous state, but some are not, and the fact that closely related species can often be shown to differ by a few homozygous rearrangements shows that these can play a definite role in evolution. An obvious explanation of the mutational effect of rearrangements would be that the same force which causes the rearrangement (breaking the chromosome) also causes mutation of the gene nearest the break. Everyone who has worked in this field agrees that most, if not all, of the mutational effects of chromosomal rearrangements cannot be explained on this basis. But the final disproof was obtained by Dubinin in the course of a study of a reciprocal translocation between the second and third chromosomes of *Drosophila*. The untranslocated third chromosome carried the recessive gene *hairy*, which causes excessive development of the bristles when homozygous. The translocated third chromosome, however, carried the normal allele of the *hairy* gene. When these two genes were interchanged by a crossover, *hairy* became dominant over normal. As crossing over has no such influence in the absence of the translocation it must be that change of dominance is a simple consequence of change of neighborhood: *hairy* gene in a normal chromosome is recessive, while *hairy* gene in a translocated chromosome is dominant. Because the rearrangement of the genes in the chromosome seems to have a mutant effect independent of the genes, such chromosomal mutations are called *position effects*.

The Bar Eye "Gene." The Bar Eye mutation of *Drosophila* was the first to be analyzed in terms of position effect. The eyes of normal flies are oval shaped. The Bar gene, however, results in a smaller number of facets than usual, and so the eyes are narrower than those of normal flies. Zeleny showed that, in homozygous Bar-eyed stocks, about one fly in 1500 mutated to the more extreme Double-Bar, while an equal number mutated back to the wild type. By experiments involving crossing over between other

219

genes near Bar, Sturtevant and Morgan showed that this simultaneous mutation from Bar to Double-Bar and to normal was based upon unequal crossing over. Thus, while each chromosome of the pair ought to contain the Bar gene, after the unequal crossing over, one chromosome had *two* Bar genes in tandem, and the other chromosome had none at all. Those zygotes which got the chromosome with no Bar gene were wild type, while those having the chromosome with the two Bar genes showed the Double-Bar phenotype. It appeared then, that two Bar genes in the same chromosome caused a different phenotype than did two Bar genes located one in each of a pair of chromosomes (Double-Bar and Bar, respectively). To describe this unexpected phenomenon, it was Sturtevant and Morgan who introduced the term *position effect*.

The analysis of the Bar "gene" could be completed only after the introduction of the salivary gland chromosome technique. It turned out that there is, in the region of the X chromosome which is indicated by crossover tests to include this gene, a series of about six bands which is present only once in wild-type flies, but twice and three times in Bar and Double-Bar flies, respectively (Figure 81). It appears, then, that the Bar "gene" is actually a duplication for a small segment of the X chromosome. If

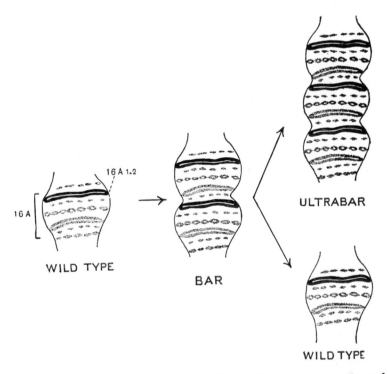

WILD TYPE

16 A 1.2

16 A

BAR

ULTRABAR

WILD TYPE

FIGURE 81. THE BAR "GENE" IN THE SALIVARY GLAND CHROMOSOMES OF *Drosophila*. (After Bridges and Sutton in White, "Animal Cytology and Evolution," 2nd Ed., Cambridge University Press, 1954.)

crossing over is unequal in a fly homozygous for this duplication, then the resulting chromosomes will be a normal one in which this segment is present only once, and one in which the segment is present three times. The normal allele of Bar now becomes simply the unmodified chromosome.

Chromosomal Rearrangements Differentiating Species. Thus far, two position effects have been discussed, one resulting from a translocation, the other from a duplication. Both types play a role in evolution. The morphology and synaptic behavior of the chromosomes of many species suggest that duplications may be present. Duplications have been suggested as a possible source of "new" genes, the duplicated ones presumably becoming completely different from the original through repeated mutations. Evidence of this is found in the common occurrence of *pseudoalleles*. These are genes of very similar effects which are located so close together in the chromosome that they are very rarely separated by crossing over. That new genes might arise in this way is suggestive, but there are some difficulties. No examples of strongly divergent pseudoalleles are known, and some geneticists consider them simply as evidence that limited regions of the chromosomes are concerned with unified functions.

Translocations certainly play a role in producing changes of chromosome number. For example, Makino has studied the chromosomes of the Japanese loaches (teleost fishes of the family Cobitidae) *Misgurnus anguillicaudatus* and *Barbatula oreas*. These species have 26 and 24 pairs of chromosomes respectively, and these are, for the most part, rod-shaped (Figure 82). However, two pairs of the chromosomes of *B. oreas* are V-shaped, and Makino drew the unavoidable conclusion that the reduction in chromosome number has been accomplished by translocations, thus uniting originally distinct chromosome pairs. Similarly, in *Drosophila* there are six basic chromosome arms which in different species are combined in different ways to give haploid numbers varying from three to six (Figure 83). Thus in *D. melanogaster,* the X and fourth chromosomes consist of one arm each, while the second and third chromosomes consist of two arms each. But in *D. palustris* there are six pairs of chromosomes, all of which have only a single arm. *D. texana* has five pairs of chromosomes, of which four have a single arm while one has two arms. Finally, *D. pinicola* has only three pairs of chromosomes. Two of these have two arms, and the third a single arm. But undoubtedly the small chromosome

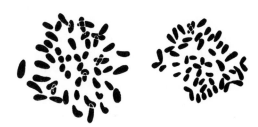

FIGURE 82. THE CHROMOSOMES OF *Misgurnus anguillicaudatus* AND *Barbatula oreas*. Note the two V-shaped pairs in the latter. (From Makino, *Cytologia*, V. 12, 1941.)

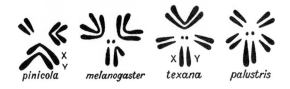

of species like *palustris* and *melanogaster* is incorporated into one of the three large chromosomes of *pinicola*.

In *Drosophila*, mutations based upon all four types of chromosomal rearrangements are known. Examples of mutant effects of duplications and of translocations have been discussed above. Notch wing is a well known sex-linked mutation based upon a deficiency near the left end of the X chromosome. Curly and dichaete wings are based upon inversions of the second and third chromosomes, respectively. Goldschmidt and others have found that about half of the classical mutant genes of *Drosophila* are in fact position effects. Of the other half, many have not been adequately examined, and it is to be expected that a good portion of them will also prove to be position effects when they are properly studied. This leads to the question: Are the remaining mutants really bona fide genes in the sense discussed in Chapter 13, or are they also position effects of rearrangements which are too small for detection by present methods? Goldschmidt has taken the latter position, while the majority of geneticists have tried to reconcile the facts of position effect with the classical theory of the gene.

That closely related species may differ by chromosomal rearrangements of the types discussed above is well established, as in Makino's loaches, which are distinguished by two translocations. *Sciara ocellaris* and *S. reynoldsi* (fungus-gnats) have been shown to differ by a number of small deficiencies and duplications. White and others have shown that in many insects for which salivary gland chromosomes are not available differences between the chromosome complements of related species are nonetheless most easily interpreted in terms of several types of rearrangements. This cannot, of course, preclude the possibility that gene mutation might also play a role, even a predominant role, in the differentiation of these species.

Overlapping Inversions and Phylogeny. By far the most complete study of chromosomal differences separating natural races and species is that of Dobzhansky and his collaborators on *Drosophila pseudoobscura* and its allies. As originally described, the pseudoobscura group comprised *D. pseudoobscura* and *D. miranda*, two closely similar species. But *D. pseudoobscura* was subdivided into races A and B on the basis of a series of differences so refined that they can usually be distinguished only on the basis of statistical analysis of populations. These differences, difficult though they may be to diagnose, are nonetheless highly consistent, and they include intersterility. After years of detailed study, Dobzhansky and

FIGURE 84. PAIRING IN SALIVARY GLAND CHROMOSOMES HETEROZYGOUS FOR INVERSIONS. Top row, a single inversion; second row, two independent inversions; third row, two inversions of which the shorter is included within the larger; bottom row, two overlapping inversions. (From Dobzhansky, "Genetics and the Origin of Species," 3rd Ed., Columbia University Press, 1951.)

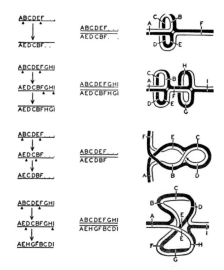

Epling have described race B as a distinct species, *D. persimilis*, reserving the original name for race A.

Phylogeny of the various local populations of this group has been studied by means of the analysis of inversions. When two inversions are present in the same chromosome, it is possible that they may be independent (ABEDCFGJIHKL), or the second may be included within the first (ABCJIHEFGDKL), or they may be overlapping (ABGFIHCDEJKL) (Figure 84). The last type is of especial interest because here the order of events can be determined. Thus, if there are three arrangements known for a particular chromosome, (1) ABCDEFGHIJKL, (2) ABGFEDCHIJKL, and (3) ABGFIHCDEJKL, it is obvious that either (1) or (3) could have been derived from the other only with (2) as an intermediate step.

Most of the chromosomal variability in the pseudoobscura group occurs in chromosome III. Actually, there are thirteen different arrangements of this chromosome which are known only in *D. pseudoobscura*, seven which are known only in *D. persimilis*, and one, called Standard, which is found in both species. By detailed analysis of overlapping inversions, starting with the assumption that Standard was the primitive arrangement, it has been possible to work out a nearly complete phylogeny for this group (Figure 85). Only a single gap occurs, between Standard and Santa Cruz, a variety of *pseudoobscura*. But even this gap is not entirely unfilled, for the necessary intermediate chromosomal pattern is found in the closely related species *D. miranda*.

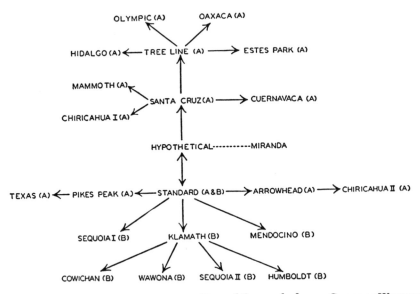

FIGURE 85. PHYLOGENETIC CHART OF THE *Drosophila pseudoobscura* GROUP AS WORKED OUT BY ANALYSIS OF OVERLAPPING INVERSIONS. Race A of this diagram is *D. pseudoobscura* and race B is *D. persimilis* of the newer classification. (From Dobzhansky, "Genetics and the Origin of Species," Columbia University Press, 1937.)

THE ARCHITECTURE OF THE CHROMOSOMES AND SYSTEMIC MUTATION

Goldschmidt urged that only two possibilities exist with respect to the significance of such chromosomal differences between closely related species. Either they are simply one among the many phenotypic characters which differentiate species, having no more causal significance than such characters as coat color, tooth specialization, or form of ovipositor; or else they constitute the actual genetic basis for the separation of species. He favored the latter viewpoint.

Goldschmidt's reasons cannot be treated in detail here, but some indications of the basis of his viewpoint must be given. At the outset, he rejected the corpuscular theory of the gene, which is basic to the neo-Darwinian theory. He did this because he believed that the principle of parsimony demands that the important facts of both gene mutation and position effect must have a single explanation. Very briefly, because gene mutations and chromosomal rearrangements respond to radiation similarly, because position effects resemble the effects of a gene near the break and may behave as an allele of it, and because all types of genic action also occur as position effects, Goldschmidt believed that these two phenomena were one. Further, because a position effect sometimes overlaps two or more genes, because breaks anywhere in a small region of the chromosome may give the same position effect, and because position effects may simulate genes at a considerable distance, he considered that the genic properties must actually reside in small segments of the chromosome, within which change of order is mutation.

With regard to this, Goldschmidt [*] said, "It might be stated first that there has been much misunderstanding of our conclusions. There is, of course, no doubt that the chromosome has a serial structure and that localized changes of this structure, the mutant loci, can be located by the cross-over method. There is no doubt either that these localized conditions of change can be handled descriptively as separate units, the mutant locus or gene, and that for all descriptive purposes the extrapolation can be made that at the normal locus a normal gene exists. Further, there can be no doubt that almost all genetical facts can be described in terms of corpuscular genes, and that a geneticist who is not interested in the question of what a gene is may work successfully all his life long without questioning the theory of the corpuscular gene. In the same way, . . . a chemist can describe and handle almost all the content of chemistry with valences represented as one or more dashes between atoms. But when he wants to know what valence is, he has to use the tool of quantum mechanics which the ordinary chemist does not need in his work. Similarly, the concept of the corpuscular gene comes under scrutiny only when the problem of the nature of the gene and the explanation of mutation and position effect is attacked."

Functional Organization of the Chromosome. Goldschmidt, then, rejected the idea of the corpuscular gene, with inactive spaces between the genes. Rather, he conceived of the chromosome as a more unified structure, a chemical continuum, in which the functional units (genes, in the traditional terminology) may be very small segments for some reactions, or larger segments for other reactions. Nor is there any reason why the functional segments for different genic reactions should not overlap. Finally, the whole chromosome may function as a unit in some instances, although this would be difficult to prove. There is some evidence that this may be the case for the sex-determining action of the X chromosomes. This may be restated in terms of the salivary gland chromosomes. Kodani has shown that the salivary gland chromosomes are basically tetrads. In each band the chromonemata are coiled, with 8 to 12 chromatic hairs or bulbs radiating from these coils in the plane of the band (Figure 86). Interstitial chromatin is packed between the chromatic hairs to complete the band. The interband spaces are greatly hydrated, and contain so-called matrix protein so as to expand to the diameter of the. bands. A rearrangement of the chromatic hairs within a band would be undetectable by present means, and hence, if it should have a mutant effect, it would be called a point mutation, although in fact it would be a position effect.

On the next level of function, a whole series of bands, corresponding to a single chromomere of the leptotene chromosome, might act as a unit.

---

[*] Goldschmidt, R. B., "Position Effect and the Theory of the Corpuscular Gene," *Experentia*, **2**, 24 (inclusive pp., 1–40), 1946.

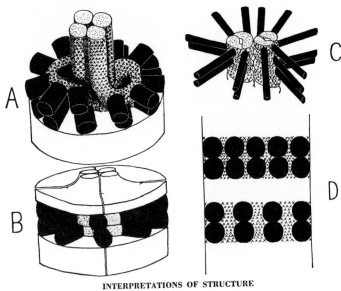

INTERPRETATIONS OF STRUCTURE

FIGURE 86. THE STRUCTURE OF THE SALIVARY GLAND CHROMOSOMES AS INTERPRETED BY KODANI. Cross-hatched strands are the chromonemata, coiled in the plane of the band. The radial bars are the chromatic hairs, which, with interstitial chromatin, form the bands. In the interband spaces, the chromonemata are hydrated to reach the diameter of the bands. (Kodani, M., *J. Heredity*, V. 33, 1942.)

Within such a unit, rearrangements of moderate size may be recognized in the salivary gland chromosomes, and they constitute the typical position effects whenever a mutant effect is associated with the rearrangement. The next larger sections are perhaps represented by the alternating blocks of euchromatin and heterochromatin, which may be equivalent to the large chromomeres of diplotene chromosomes (much thickened and visibly double, synapsed chromosomes). Rearrangements in which the mutant effect seems to be localized at a considerable distance from the break may well function in terms of these units. In such cases the second break is almost always in the heterochromatin. Finally, the chromosome as a whole may function as a unit.

In such a system, some rearrangements would be neutral; others, the ordinary mutations of genetics, would produce moderate effects; while still others might disturb the chromosomal functions so profoundly that, if viable, it might constitute a completely different reaction system, a new species or even higher group. Natural selection would now act directly upon the new species, usually destroying it, occasionally permitting its increase. Goldschmidt demonstrated that such so-called systemic mutations are consistent with the facts of embryology and physiological genetics. But he acknowledged that systemic mutations have not actually been demonstrated, just as the corpuscular genes have not yet been demonstrated.

Difficulties of Neo-Darwinism. Goldschmidt believed that the neo-Darwinian type of evolution, by accumulation of micromutations under the influence of natural selection, is largely restricted to subspecific differentiation within species, and that the decisive step in the formation of new species must involve an altogether different genetic process, the systemic mutation. Only a few of the reasons which led him to this conclusion can be indicated here, and these only briefly. If neo-Darwinian evolution gives rise to new species, then new species should come only from the terminal members of a Rassenkreis, and the Rassenkreise of closely related species should blend into one another. But actually this does not happen. He believed that good species are always separated from their nearest relatives by a bridgeless gap. Controversial cases he believed depend in part upon purely morphological definitions of species which do not take the genetic facts into account. Goldschmidt believed that interbreeding, or potentially interbreeding, populations should be treated as a single genetic unit, a species, from an evolutionary point of view, even if other factors may make it advisable for taxonomists to break it up into several species. On this basis (which is not acceptable to most geneticists and taxonomists), many difficult cases can be resolved in accordance with his ideas.

Another major neo-Darwinian tenet is that isolation of a subspecies is essential if it is to accumulate enough gene differences to become a good species, distinct from the parent species. But there are known instances in which long isolation has failed to produce more than subspecific differentiation. Thus there is a race of *Lymantria dispar* (Gypsy moth) which has been isolated on the island of Hokkaido (North Japan) since the early Tertiary period, yet in the intervening 60,000,000 years only subspecific differentiation has occurred. Again, the seasonal variation within a single race of *Papilio* (butterfly) may be greater than the variation between races at any one time.

Finally, Goldschmidt believed that the neo-Darwinian theory places too great a burden upon natural selection. The theory demands that only very minor mutants be significant for evolution, and these should be subject to very slight selection pressures. One may define selection pressure in terms of loss of survival value. Thus, if 1000 individuals of genotype AA survive to reproductive age while only 999 of genotype aa survive, the selection pressure against a is said to be 0.001. Haldane calculated the results where a selection pressure of this magnitude is operating in favor of a new gene which is present in a population to the extent of one allele in a million. If the favored gene is dominant, it would require 11,739 generations to raise the frequency to two in a million. But if the favored gene is recessive, 321,444 generations would be required! Selection operates more rapidly when the gene frequencies are higher. But it is not self-evident that a process operating as described above could produce the existing species of plants and animals, even given time on a geological scale. And there are many examples like the *Lymantria* case mentioned above in which selection has not produced more than subspecific differentiation even when aided by rather complete isolation.

There is no doubt that the majority of geneticists, as well as other students of evolution, do not agree with Goldschmidt with respect to some or all of his theses. But most will agree that he has amassed a body of highly pertinent evidence which cannot be lightly dismissed.

Perhaps these views are not so divergent as they seemed only a few years ago. At the center is the theory of the gene, which Goldschmidt conceived in physiological terms at a time when most geneticists were thinking in morphological terms. The implications of physiology for morphology and of morphology for physiology were not clear, and much acrimonious debate further clouded the issue. Several recent developments have tended to clarify it. First, the great development of biochemical genetics in recent years has focussed the attention of geneticists upon physiological aspects of the gene. Second, increased interest in pseudoalleles, whether interpreted as gradually diverging repeats or as different mutational sites in broad genic fields, has tended to blur the limits of the gene, and to leave the corpuscular theory as merely a first approximation. Finally, much recent evidence points to desoxyribonucleic acid (DNA) as the actual genic material, and Watson and Crick have devised a very clever model of its structure which readily relates genic action to a pattern in a biochemical continuum. Nucleic acids are made of nucleotides, each of which consists of a pentose (five-carbon sugar), phosphoric acid, a base, which may be a purine [adenine(A), or guanine(G)] or a pyrimidine [cytosine(C) or thymine(T)]. These are polymerized by sugar-phosphate linkages to form long chains, on which the bases appear as side

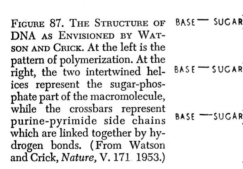

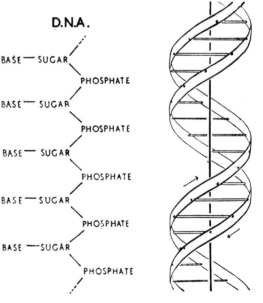

FIGURE 87. THE STRUCTURE OF DNA AS ENVISIONED BY WATSON AND CRICK. At the left is the pattern of polymerization. At the right, the two intertwined helices represent the sugar-phosphate part of the macromolecule, while the crossbars represent purine-pyrimide side chains which are linked together by hydrogen bonds. (From Watson and Crick, *Nature*, V. 171 1953.)

chains. X-ray diffraction studies showed that there are always two such chains spiraling about a common axis (Figure 87). The two chains are united by hydrogen bonds between the bases. Analysis showed that equal amounts of adenine and thymine or of cytosine and guanine were always present, although amounts of adenine and cytosine, for example, might be very unequal. They concluded that the bonds between DNA chains are always formed either by adenine and thymine or by cytosine and guanine. Thus there are four possible bondings between any two nucleotides: A·T, T·A, C·G, and G·C. It is suggested that the sequence of these bonds could encode unlimited genetic information, translatable by processes of development into the great array of possible phenotypes, just as the simple dots and dashes of Morse code can encode any verbal information. This theory is not without defects, but it is one of the most exciting and productive aspects of current genetics, and it tends to modify profoundly the older concepts of the gene.

One current theory of the gene envisions it as a more or less broad field of function with a point focus on the chromosome, comparable to the center of gravity of a physical object. The data of the corpuscular gene theory, such as crossing over, deal with these point foci, while position effects and pseudoalleles deal with the broader fields. These functional fields may vary in size from molecular dimensions to the entire chromosome, and they may overlap broadly so long as their foci remain separate, as required by crossover data. This theory embraces elements of both the divergent views of a few years ago, and it may well subserve species formation upon lines related to both theories, and perhaps according to others still unformulated.

REFERENCES

DARLINGTON, C. D., 1937. "Recent Advances in Cytology," McGraw-Hill Book Co., Inc., New York, N.Y. Badly out of date, but a classic of cytogenetics.

GOLDSCHMIDT, R. B., 1940. "The Material Basis of Evolution," Yale University Press, New Haven, Conn. The major statement of the evolutionary theory presented in this chapter.

GOLDSCHMIDT, R. B., 1955. "Theoretical Genetics," University of California Press, Berkeley and Los Angeles. The last major work by this author.

SWANSON, C. P., 1957. "Cytology and Cytogenetics," Prentice-Hall, Inc., New York, N.Y. A very penetrating analysis of many of the problems touched upon in this chapter.

WHITE, M. J. D., 1954. "Animal Cytology and Evolution," 2nd Ed., Cambridge University Press. The most complete presentation of this field now available.

PART FOUR

*The Origin of Species
and of Higher Categories*

CHAPTER FIFTEEN

Natural Selection

THE EARLY post-Darwinian students of evolution were inclined to think of natural selection as an all-or-none phenomenon: either an organism was favorably endowed and survived to reproductive age, or it was unfavorably endowed and died in the struggle for existence without leaving progeny (Figure 88). Among the factors that led to the agnostic reaction at the turn of the century was the fact that no such severe selection had been actually observed as a general phenomenon in nature. Yet it ought to be very obvious if it existed. Severe defectives were, of course, eliminated: crippled deer cannot long escape predators. But many minor defectives were observed in nature, and they did leave progeny. Coupled with the new mutation theory of De Vries, according to which a new species might be produced at a single step, these facts seemed to push natural selection out of the picture entirely.

SEVERE VERSUS MILD SELECTIVE FORCES

But there is another possibility, not considered at the time of the agnostic reaction. If a character confers only a slight disadvantage upon the organ-

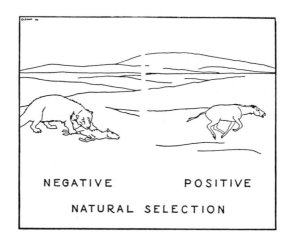

FIGURE 88. (From Simpson, *J. Washington Acad. Sci.*, V. 31, 1941.)

NEGATIVE   POSITIVE

NATURAL SELECTION

isms which show it, they may have a death rate *somewhat* higher than their more fortunate cousins, with the result that relatively fewer of them will reach reproductive age, and they will leave somewhat fewer progeny. Conversely, a variant which has even a slight advantage will have a somewhat higher survival rate and hence will leave somewhat more descendants. In either case, the net result will be a gradual change in the proportions of each variant in a natural species. Once it was realized that gradual changes, apparent only on statistical analysis, might be effected by selection rather than all-or-none changes, it became possible to study selective changes in natural populations, and, to some extent, to perform experiments in selection.

EXPERIMENTAL DEMONSTRATION OF NATURAL SELECTION

Sukatchew analyzed such changes experimentally in the dandelion, *Taraxacum officinale.* Strains tested were obtained from the Crimea, Leningrad, and Archangel. As the latitudes are respectively 45°, 60°, and 64°, the Crimea is comparable to Minneapolis, Leningrad to Seward, Alaska, and Archangel to Dawson, the gold rush city of the Yukon Valley. The dandelions tested were then adapted to temperate, subarctic, and arctic conditions, respectively. The three strains were planted in mixed plots at Leningrad at densities of every three centimeters, and every eighteen centimeters. The Crimean strain survived to about 60 per cent in the sparsely planted plots, but in the densely planted plots, where competition (and hence selection) was more severe, its survival rate dropped to about 1 per cent. At high densities, the Archangel strain had a greater survival rate than did the Leningrad strain (70 as against 11 per cent), but at low densities, there was a small but significant margin in favor of the local strain (96 as against 88 per cent). These observations are important for two reasons. First, they do demonstrate that the differences between populations of the same species result in differential survival when a mixed population is exposed to identical conditions. And second, they demonstrate that selection need not be an all-or-none phenomenon, but that it may operate by the statistical transformation of populations.

Timofeeff-Ressovsky, working at Berlin, similarly tested the relative viability of *Drosophila melanogaster* and *D. funebris* at temperatures of 15°, 22°, and 29° centigrade. *D. funebris* is an originally northern species, while *D. melanogaster* is originally tropical. His method was to put 150 eggs of each species together in a culture bottle containing insufficient food for 300 larvae. The numbers of adults of each species which emerged were then compared. As *D. melanogaster* usually showed the greater viability, results were expressed by stating the viability of *D. funebris* as a per cent of that of *D. melanogaster* $\left(\dfrac{\text{surviving funebris}}{\text{surviving melanogaster}} \times 100 \right)$.

As already mentioned, the survival rate of *D. melanogaster* was almost always higher than that of *D. funebris*. But northern races of *D. funebris* did much better relatively, when tested at 15°, than did southern races.

In fact, one northern race showed a slightly higher survival rate than did *D. melanogaster*. At the higher temperatures, however, the survival rate of *D. funebris* was never as much as half of that of *D. melanogaster*. The complete data are summarized in Table 2.

TABLE 2. THE RELATIVE VIABILITY OF THE STRAINS OF *Drosophila funebris* OF DIFFERENT GEOGRAPHICAL ORIGIN (from Timofeeff-Ressovsky) °

STRAINS OF D. FUNEBRIS	VIABILITY IN % OF THAT OF D. MELANOGASTER			VIABILITY IN % OF THAT OF BERLIN STRAINS OF D. FUNEBRIS		
	15°C	22°	29°	15°	22°	29°
Berlin	81	42	18	100	100	100
Sweden	88	40	21	108.6	95.2	116.6
Norway	80	41	21	98.7	97.6	116.6
Denmark	79	44	22	97.5	104.7	122.2
Scotland	84	43	20	103.7	102.4	111.1
England	78	42	21	96.3	100.0	116.6
France	80	44	25	98.7	104.7	138.8
Portugal	71	45	28	87.6	107.1	155.5
Spain	69	48	30	85.2	114.3	166.6
Italy	78	43	25	96.3	102.4	138.8
Gallipoli	75	44	26	92.6	104.7	144.4
Tripoli	64	47	31	79.0	111.9	172.2
Egypt	68	46	30	83.9	109.5	166.6
Leningrad	90	43	22	111.1	102.4	122.2
Kiev	91	44	28	112.3	104.7	155.5
Moscow	101	43	28	124.7	102.4	155.5
Saratov	92	42	30	113.6	100.0	166.6
Perm	98	41	26	121.0	97.6	144.4
Tomsk	96	42	28	118.5	100.0	155.5
Crimea	87	42	28	107.4	100.0	155.5
Caucasus I	89	43	31	109.9	102.4	172.2
Caucasus II	86	45	32	106.2	107.1	177.7
Turkestan	90	44	34	111.1	104.7	188.8
Semirechje	92	46	36	113.6	109.5	200.0

° By permission from Dobzhansky, "Genetics and the Origin of Species," 3rd Ed., Columbia University Press, 1951.

Goldschmidt studied adaptation to climatic conditions in the gypsy moth, *Lymantria dispar*. While the conditions of development were here determined by climate, rather than by experiment, they nonetheless permit an insight into the nature of the selective force. The moths lay their eggs in the fall, but the eggs remain dormant until the following spring. In the spring, when the sum of the daily temperatures has reached a certain minimum, development proceeds and the larvae emerge. Then they feed on the green foliage of certain plants. Now the hatching time of northern races is quite short, while that of Mediterranean races is much longer. It is evident that a rapid developmental cycle is necessary for the northern races because of the short summer. But if a southern race had a short hatching time, then after a mild winter the larvae would be likely to emerge before foliage appeared on their food-plants. Similarly, the Mediterranean plants have a long period of growth of vegetation; while the arctic plants have a very short period, corresponding to the short arctic growing season.

OBSERVED CHANGES IN NATURAL POPULATIONS

The observations reported above leave no room for doubt that selection can actually cause changes in the characteristics of species. Whether this has happened historically is, however, another question. Fortunately, there is evidence available that important changes have occurred in some species within historical times, and in response to known selective forces.

Industrial Melanism. Perhaps the best known example is the phenomenon of "industrial melanism," which has occurred in many species of moths during the progress of the industrial revolution. Dark (melanistic) forms have been known for hundreds of years as an occasional curiosity. But during the past 100 years they have become increasingly numerous, until now the lighter original forms are the rare ones, prized by collectors. The centers of distribution of the melanistic forms have been the large industrial cities. The melanism of the nun moth, *Lymantria monarcha*, was analyzed genetically by Goldschmidt. The original pattern is one of narrow zig-zag lines on a light background. Two independent, autosomal mutations cause an increase in the width of the zig-zag lines and some pigment deposition between them. A third gene is sex-linked, and causes an intensification of the pigments. The three pairs of genes, in various combinations, have additive effects, and so can cause the wide range of phenotypes which are illustrated (Figure 89). Calculation of the mutation rate necessary to cause the observed increase in melanism gave an absurdly high value, and so it was concluded that selection favored melanism.

The nature of the selective force which has produced industrial melanism is by no means agreed upon. The most obvious explanation is that the melanistic forms are less conspicuous to predators in smoky industrial areas than are the original light forms. Actual counts of moths taken by birds has in some cases substantiated this. Ford, however, believes that melanism is secondary to physiological changes, such as resistance to poisoning by lead salts of industrial smoke. In preindustrial times, melanism failed to spread because the melanistic moths were conspicuous to predators and resistance to lead poisoning was not of value. Both traits, however, proved to be preadaptive to the industrial environment, and hence the recent rapid spread of melanism. Whatever the selective force, here is a well documented case of transformation of species, at least to a subspecific degree, within historical times. As the transformation is associated with a known environmental change (industrialization), it is difficult to doubt that selection has been responsible for the change.

Additional Examples. Another often-quoted example is that of the crab, *Carcinus maenus,* reported by Weldon in 1899. The building of a breakwater in an English sound resulted in a higher silt content of the water. During a period of five years, the mean diameter of the carapace of the crabs in the sound was observed to decrease. Weldon captured both narrow-carapaced and broad-carapaced crabs and kept them in silty aquaria. The narrow-carapaced crabs survived, while the broad-carapaced ones died. Weldon attributed this to accumulation of silt on the gills of the broad-carapaced crabs. He assumed that the same thing happened in nature, and so constituted a selective force. The actual silting of the gills was not demonstrated, but it is plausible.

The relationship of wheat and wheat rust is also illustrative. One of the primary objectives of plant breeders is to develop disease-resistant varieties of commercially valuable plants. To a disease-producing parasite, however, scarcity of susceptible hosts is obviously a severe selective force. The history of rust-resistant wheats is monotonous in its repetition. When a new resistant wheat is introduced, it gives excellent, disease-free crops, and is widely adopted. After a few years, however, a few fields show some active rust infection. Then the successful rust spreads rapidly, and a new resistant wheat is again needed. It is evident that mutation has simply by chance produced a rust variety which is adapted to the new wheat, and it spreads rapidly because it is favored by a strong selection pressure. This has been the subject of much study by Stakman.

Quayle has reported an interesting case of selection in the scale insects which parasitize citrus trees. The standard method of combatting the insects is to cover each tree with a tent and fumigate with hydrocyanic acid. But in each of the three species concerned, cyanide-resistant varieties have appeared and have replaced the original cyanide-sensitive varieties. One of these resistant species has subsequently disappeared for unknown reasons. A similar development has more recently occurred with respect to DDT poisoning. When DDT first came into general use late in 1945, it gave promise of being an almost perfect insecticide for the control of household pests, such as flies. But soon DDT-resistant flies began to appear. Under the strong selective force of DDT poisoning campaigns, these resistant strains soon became well established, and in many localities they have largely replaced the original DDT-sensitive flies.

An interesting case in the fire-ant, *Solenopsis saevissima,* has been studied by Brown and Wilson. A large, dark variety of this ant was accidentally introduced into Alabama from South America about 1918. It spread very slowly and caused no alarm. During the 1930's, however, a smaller, lighter-colored form appeared, probably as a new immigrant from South America. This proved to be far more invasive and aggressive than the dark form, the nests of which it destroys. It is evidently favored by selection, for it is rapidly replacing the dark form and extending its range. It has become a major pest in the southeastern United States.

Polymorphism. Selection does not lead to uniformity: it leads to high frequencies of those genes which contribute to the most successful genotypes. However, heterozygosity itself has considerable value, for it pro-

FIGURE 89. See opposite page for legend.

vides a source of variability which permits a species to adapt rapidly to environmental changes. Most natural populations of sexually reproducing organisms are highly heterozygous. Whenever this heterozygosity results in obviously different phenotypes, one speaks of *polymorphism*. The most striking cases are those of mimetic polymorphism (see below). Specific selective forces may maintain a balanced polymorphism. The inversion types of *Drosophila pseudoobscura* were introduced above. Dobzhansky has shown that some chromosomal arrangements increase in frequency during the summer, others during the winter, hence they must be related to seasonal adaptations. A particularly instructive case is known in man. Sickle cell anemia is a fatal disease resulting from homozygosity of a certain gene. The same gene, when heterozygous, causes the harmless sickle cell trait. Strong selection pressure should tend to eliminate such a gene,

FIGURE 89. INDUSTRIAL MELANISM IN *Lymantria monacha*. Each row of three represents one combination of the two pairs of autosomal genes and the one pair of sex-linked genes which control melanism in this species. (From Goldschmidt, "The Material Basis of Evolution," Yale University Press, 1940.)

but it is nonetheless common in equatorial Africa. This was inexplicable until it was found that heterozygous persons are resistant to malaria, which is prevalent in the same region. Thus selection favors heterozygosity, in spite of the severe liability of homozygosity.

The fact of the effectiveness of selection in nature, originally postulated by Darwin as a necessary consequence of the prodigality of nature and of the variability of all species, has been thoroughly vindicated by observations such as those reported above. Some special aspects of natural selection have, however, fared less enviably when subjected to critical analysis. Two such aspects, adaptive resemblances or coloration, and sexual selection, will be discussed below.

ADAPTIVE RESEMBLANCES

Perhaps no single factor did more to cause the agnostic reaction than did the excessive speculations of nineteenth century evolutionists on adaptive resemblances. Every possible quirk of nature was interpreted by someone as having adaptive, and therefore selective, value to its possessor. Many speculations were published on the basis of preserved specimens, unsupported by field observations. And always the assumption was made that the sensory faculties of the potential predators were similar to those of man.

The principal types of adaptive coloration are cryptic resemblance, by which the animal blends into its background (Figure 90); aposematic, or warning coloration, by which an obnoxious or dangerous animal is made obvious to potential predators; and mimicry, by which one species resembles another, presumably taking advantage of aposematism.

Cryptic Coloration. Cryptic coloration is very general in the animal kingdom. It is useful not only for the protection of animals from their predators, but also to the predators themselves, permitting them to avoid detection by their prey long enough to permit a kill. Perhaps the simplest form is countershading, that is, dark pigmentation on the dorsal surface and light color on the ventral surface. Its protective function may be well illustrated with a fish. A bird looking down upon the fish will see the dark dorsal surface blending into the darkness of the depths. But a larger fish, looking up from below, sees the light ventral surface blending into the daylight. While the protective effect of such a pattern is plausible, even probable, it is by no means certain that its occurrence depends upon selective value. It has been suggested that the very universality of countershading indicates a simple physical cause. Often development of pigment requires exposure to light. As most organisms are exposed to light much more on the dorsal surface than on the ventral, it is to be expected that countershading would be the rule. Some fishes, however, normally swim with the ventral surface up: in these it is the ventral surface which is darkly pigmented and the dorsal surface which is light. Also, this pattern has been experimentally reversed in some fishes by raising them in aquaria lighted from below. But, while such facts demonstrate the *means* of development of countershading, they do not disprove its protective value. And that the sunlight mechanism is not the whole story is shown by the facts that small surface fishes are typically transparent, while abyssal fishes are typically dark-colored. Both would appear to be protective, but both run counter to the light gradient.

Many cases of cryptic resemblance, however, are less easily disputed. For example, the arctic hare, northern weasels, and ptarmigan molt and grow a coat of white fur or feathers (ptarmigan) in the fall, then revert to a brown coat in the spring. The protective utility is difficult to explain

FIGURE 90. CRYPTIC COLORATION OF POCKET MICE FROM ADJOINING AREAS. *Perognathus intermedius ater* on black lava and *P. apache gypsi* on white gypsum sand. (Paintings by Allan Brooks from Benson, 1933, by permission of University of California Press.)

away. Then there are numerous cases of stick and leaf insects (Figure 91), which, so long as they do not move, look like parts of the plants on which they are normally found. Many moths (Figure 92) have a close resemblance, when at rest, to the bark of the trees upon which they are usually found. In a few instances, resemblances to the environment are actively acquired. Thus the masking crabs, of which *Loxorhynchus crispatus* (Figure 93) is a good example, actually "plant" upon themselves algae, hydroids, sponges, and other sessile organisms from their environment. If a masking crab is moved from its original locale, it will seek, in its new environment, an area with the same kinds of organisms which it is carrying. If no such area is available, the crab will remove its riders and

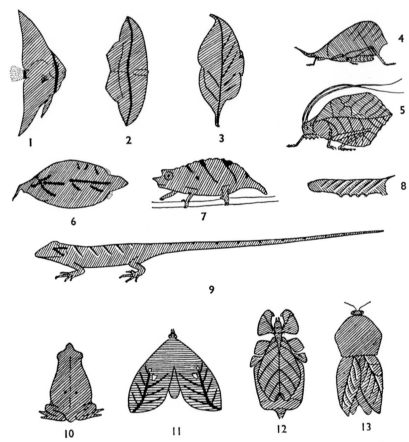

FIGURE 91. RESEMBLANCES OF ANIMALS TO LEAVES. 1, *Platax vespertilio*. 2, *Timandra amata*. 3, *Kallima paralekta*. 4, *Systella rafflesii*. 5, *Cycloptera* sp. 6, *Monocirrhus polyacanthus*. 7, *Rhampholeon boulengeri*. 8, *Smerinthus ocellatus*. 9, *Polychrus marmoratus*. 10, *Bufo typhonius*. 11, *Miniodes ornata*. 12, *Phyllium crurifolium*. 13, *Choeradodis rhomboidea*. Numbers 2, 3, 4, 5, 8, 11, 12 and 13 are insects; 1 and 6 are fish; 10 is an amphibian; and 7 and 9 are reptiles. (From Cott, "Adaptive Coloration in Animals," Methuen & Co., Ltd., London, 1940.)

FIGURE 92. PROTECTIVE RESEMBLANCES IN INSECTS. At left, a walking stick insect. At right, an underwing, *Catocala*, with wings spread (A) and at rest on bark (B). (From Folson and Wardle, "Entomology," P. Blakiston & Co., 1934.)

replace them with the sessile flora and fauna of the new locality. The denial of the protective value of so complicated an instinct leads to highly speculative and uncertain conjecture, or to complete default of a theory to explain the observed facts.

Warning Coloration. Warning coloration is also rather common in the Animal Kingdom. Here the object is just opposite to that of cryptic coloration: whereas a cryptic pattern tends to render the animal inconspicuous, a warning pattern is obtrusive, and advertises the presence of an otherwise well-protected animal. As Cott [*] has put it, "Their bite is worse than their bark." Such animals have formidable defense mechanisms, or they produce foul-smelling substances, or they have a disagreeable taste. The object seems to be to educate potential predators by presenting an easily recognized appearance which a predator will avoid after an initial bad experience. That the method can work is well known to every farm boy whose dog has attacked a skunk or a porcupine. A second encounter rarely occurs.

Examples of warning coloration are numerous. The skunk and the porcupine appear to be very clear examples among the mammals. Many

[*] Cott, "Adaptive Coloration in Animals," Methuen & Co., Ltd., London, 1940.

FIGURE 93. A MASKING CRAB, *Loxorhynchus crispatus*. (From Ricketts and Calvin, "Between Pacific Tides," 2nd Ed., Stanford University Press, 1952.)

poisonous snakes are brilliantly colored, and it may be that these are properly interpreted as aposematic. In general, the Amphibia are cryptically colored, but some have brilliant aposematic patterns (Figure 94). These, of which *Triturus torosus*, the western water-dog, is a good example, are well supplied with poisonous skin glands. Many of the most brilliantly colored insects are believed to have repugnant tastes (butterflies), while others have stings (bees) or emit foul-smelling fluids (coccinellid beetles).

Mimicry. Finally, there are the much disputed cases of mimicry. The basic situation here is aposematism in which the same pattern (or closely similar patterns) is shared by two or more species. In the original description, Bates assumed that one of the species, the model, is genuinely aposematic, and appropriately marked; while the other species, the mimic, is desirable prey, but shares the protection of the model by assuming its cloak. This is known as Batesian mimicry. Müller later described what appeared to be an exceptional case of mimicry in which *both* the model and the mimic appeared to be protected because of repugnance to predators. This type of mimicry, by which two or more species present a single aposematic pattern for their predators to learn, is called Müllerian mimicry. It now appears to be much more common than Batesian mimicry. It presumably operates as a kind of double insurance.

FIGURE 94. APOSEMATIC COLORATION IN AMPHIBIANS. Natural colors are very striking combinations of black, red, yellow, orange, and white.

Bombinator igneus

Hyperolius marmoratus

Atelopus stelzneri

Salamandra maculosa

Phrynomerus bifasciatus

Dendrobates tinctorius

Dendrobates tinctorius

Dendrobates tinctorius

(From Cott, "Adaptive Coloration in Animals," Methuen & Co., Ltd., London, 1940.)

Mimicry is widely scattered throughout the Animal Kingdom, but the best known examples occur among the Lepidoptera. The monarch and viceroy butterflies of this country are familiar examples, with the monarch as the model and the viceroy the mimic (Figure 95). That mimicry is effective among birds is shown by an experiment of Swynnerton. The African drongos, *Dicrurus afer* and *D. ludwigi*, are black all over, and are unpalatable. The flycatcher, *Bradyornis ater*, and the cuckoo-shrike,

245

FIGURE 95. MIMICRY OF THE MONARCH BUTTERFLY, *Danaus plexippus,* BELOW, BY THE VICEROY, *Basilarchus archippus,* ABOVE. (From Storer and Usinger, "General Zoology," 3rd Ed., McGraw-Hill Book Co., Inc., 1957.)

Campephaga nigra, mimic the drongos, but they are themselves edible. The tit, *Parus niger,* resembles the above birds ventrally, but has conspicuous white markings dorsally. Swynnerton offered a cat specimens of all five species, turned ventral surface up. All were refused. But when the birds were turned dorsal surface up, the cat quickly took the tit, while still refusing all of the others.

The Reaction against Protective Coloration. During the Romantic Period, almost every dull color pattern was interpreted as cryptic, almost every brilliant pattern as aposematic, and almost every resemblance between pairs of species as mimicry. This involved many gross errors, much uncritical acceptance of scanty and ill-founded data. In subsequent years, the selective value of color was commonly denied altogether, and was quite generally treated as one of the most doubtful aspects of evolutionary theory. The most important bases of this reaction, in addition to the purely psychological one, were three. First, the vision of predators might be sufficiently different (based upon ultraviolet light, for example) from that of man to invalidate the judgments of human observers. Second, the distastefulness of insects which serve as models for mimics had not been generally proved, and many biologists contended that the distastefulness was not genuine. Finally, McAtee published a study, based upon the examination of the contents of the stomachs of 80,000 birds, in which he reported that both protectively and nonprotectively colored insects were eaten by North American birds in numbers proportional to their respective populations. Because of its very broad experimental basis, this study has had great influence in discrediting the whole idea of adaptive coloration. Thus, in 1936, Shull * was able to write that "if the doctrine (of natural

---

* By permission from Shull, "Evolution," McGraw-Hill Book Co., Inc., 1936.

selection) can emerge minus its . . . warning colors, its mimicry and its signal colors, the reaction over the end of the century will have been a distinct advantage."

Resolution of Difficulties. Meanwhile, data have accumulated which answer satisfactorily the main objections to the theory of adaptive coloration. The first objection above was that predators might depend upon utraviolet or infrared vision, and hence that coloration as seen by man is irrelevant. This has been investigated extensively, both by testing the vision of predators in different portions of the spectrum, and by photographing presumed protected animals in ultraviolet and infrared rays, as well as with visible light. The results have shown that any part of the spectrum is used by some predator. Some animals which appear protectively colored to the human eye are not protectively colored when photographed by the type of light used by their usual enemies, but many are. And it is entirely possible, indeed probable, that many organisms which we do not suspect of being protectively colored would appear so if photographed at the appropriate wave length. This argument is plainly a two-edged sword.

The second objection was based upon the lack of proof that presumed distasteful aposematic animals, particularly insects, actually are distasteful to predators. In many specific instances, it has been proved that the animals in question actually are distasteful to some, at least, of their potential enemies. For example, when a large variety of freshly killed insects were placed on a feeding tray at the edge of a natural woods, birds were observed to come and feed. They took most of the cryptically colored insects, but very few of the aposematically colored ones. The recent experiments of Brower prove that some aposematic butterflies are, in fact, distasteful, for jays which had been conditioned by feeding on models rejected both models and mimics, while unconditioned birds took the mimics readily. Again, the tectibranch mollusc *Oscanius membranaceus,* which is aposematic, secretes dilute sulfuric acid. It is not taken by most fishes, and otherwise edible materials are refused by most fishes if treated with dilute sulfuric acid. When larvae of the magpie moth, *Abraxas grossulariata,* were offered to lizards and frogs, they were at once seized, then dropped, and refused on all future occasions. The predators then sat with mouths agape, rolling their tongues as though they were trying to get rid of an obnoxious taste. Similar experiments have demonstrated that mimics actually do, in some cases at least, share the protection of their models.

Finally, there is the important objection, based primarily upon McAtee's study, that protected as well as unprotected organisms are taken by predators in numbers proportional to their populations, and hence that there is no protective value demonstrable. McAtee's study has been seriously criticized because of the misleading method of presenting his data. The actual numbers of protected and unprotected insects taken were not recorded. Rather, the numbers of stomachs in which each was found was recorded. Thus, if 1000 specimens of a particular species of insect were found in a total of 80 stomachs, and if 100 specimens of another species were also found in a total of 80 stomachs, the published figures would be identical for each. Also, no distinction was made between different species of birds. Now it is hardly to be supposed that any animal would be perfectly protected from all of its potential enemies; but if an animal is even *somewhat* protected from *some* of its potential enemies, it ought to have an improved chance for survival in a highly competitive animal society.

There is some positive evidence that presumed protected animals are not taken by predators as often as are unprotected species. Examples have been given above for aposematic animals. Again, the protection need not be absolute. Thus, the porcupine is not preyed upon by the vast majority of carnivores, but the fisher (*Martes*) successfully takes porcupines by turning them over on their backs, then ripping open the unprotected ventral surface. But one would hardly wish to claim that, because a single carnivore takes porcupines, their protection is not effective. For there are many carnivores which would otherwise kill porcupines.

Experimental evidence is also available in support of the above observations. Sumner's experiments on predation on the mosquito fish (*Gambusia*) by penguins are most illustrative. These fishes slowly change color to match their background. If one group of fish is kept in a black tank and another in a light tank, each group will become adapted to its own background. If fishes from one of the tanks are then transferred to the other, they will contrast with their background until they can again become adapted. Sumner exposed such mixed groups of adapted and unadapted fishes to predation by penguins. Always, both types of fish were taken, but the unadapted fishes were taken in much greater proportion than were the adapted ones (Table 3). The differences are highly significant statistically in both types of tank, and they thus support the protective value of the color changes.

TABLE 3. PREDATION BY PENGUINS ON ADAPTED AND NONADAPTED MOSQUITO FISH [*]

COLORATION	DISCRIMINATION		
	CASUALTIES	SURVIVORS	TOTAL
Adapted "whites" in pale tank	103 (36%)	183 (64%)	286
Adapted "blacks" in black tank	73 (31%)	162 (69%)	235
Total	176 (34%)	345 (66%)	521
Nonadapted "blacks" in pale tank	165 (58%)	121 (42%)	286
Nonadapted "whites" in black tank	201 (86%)	34 (14%)	235
Total	366 (70%)	155 (30%)	521

[*] By permission from Cott, "Adaptive Coloration in Animals," Methuen & Co., Ltd., London, 1940.

Dice exposed variously colored races of the white-footed mouse, *Peromyscus,* to predation by owls. When several color races were exposed on a background nearly matching one of them, that one was taken less frequently than were the others.

The extensive literature dealing with all phases of adaptive coloration has been summarized and exhaustively studied by Cott. He has concluded that, although the concepts involved have been much damaged by too gullible acceptance during the Romantic Period, nonetheless those phenomena are real. He backs up his opinion with a very large number of examples, all of which are well thought out, but many of which could be explained away on the basis of coincidence. He believes, however, that their effect is cumulative, and that, taken together, they establish adaptive coloration as one of the main achievements of natural selection. It would be very difficult to read his book without inclining to his viewpoint.

SEXUAL SELECTION

A second special feature of Darwin's theory, that of sexual selection, has fared even less enviably than has that of adaptive resemblances. Darwin believed that the general theory of natural selection could not account for the color differences which so commonly characterize the sexes, nor for other types of ornamentation differentiating the sexes. He therefore proposed the theory of sexual selection to account for such differences. In brief, the theory is that the female selects her mate, and that therefore any male which is especially attractive will have improved chances of obtaining a mate and leaving descendants. This leads to development, in the male, of brilliant colors, elaborate combs, ornamentative hooks, in fact, of any secondary sex character which might be regarded as potentially attractive. Darwin included also antlers, tusks, and spurs, which, however, could also be accounted for by the general theory of natural selection.

The idea of sexual selection requires either that males be more numerous than females, or that polygamy be the rule; that the ornamentation of the males be attractive to the females; and that the females select their mates. Some positive evidence in favor of the theory has been obtained, but it is very scant. Generally speaking, the numbers of males and females seem to be about equal. There are some outstanding examples of polygamous societies among animals. One of the best is the fur seal, which breeds on the Pribilof Islands near Alaska. The males arrive at the breeding grounds before the females, and then engage in mortal combat, in which most of the males are either killed or driven away from the breeding grounds. Upon arrival of the females, each surviving male has an extensive "harem." Another fact which has been interpreted as supporting the theory concerns those instances in which there is a reversal of the usual pattern of sex behavior. For example, phalaropes (a kind of sandpiper) are characterized by brilliant plumage in the female and dull plumage in the male. But the female does the courting, while the male builds the nest and incubates the eggs.

There is much doubt how much influence ornamentation may have upon actual selection of mates. The male seals referred to above have powerful tusks, but the females do not appear to have any choice of mates: they simply accept the male which is at hand when they arrive on the breeding grounds. In some cases, secondary sex characters appear to excite sexual activity without actually influencing choice of mates. In *Drosophila*, for example, mating is preceded by courtship behavior which includes wing movements by the male. Wingless males can mate with normal females, but it takes much longer. Now Sturtevant has shown that, if a normal pair of flies is put into a bottle along with a wingless male, the female will mate after a normal time lapse, *but she will mate with the wingless male as readily as with the normal one.* It is evident that the courtship of the normal male has hastened receptiveness of the female without influencing her choice of mates.

Finally, most attempts to determine factors in choice of mates have been inconclusive. It may well be that courtship behavior in general has more to do with sexual excitement than with choice of mates. And it may well be that simple proximity has more influence on choice of mates than does any other factor for most animals. As the problem of sexual selection now stands, then, probably few biologists would care to state categorically that it plays no role in evolution, but equally few would care to ascribe to it a really important role.

SELECTION AND NONADAPTIVE CHARACTERS

One of the objections often raised against the theory of natural selection is that it cannot account for the many cases in which differences between related organisms do not have any evident adaptive value. There are several ways in which such cases may be harmonized with the theory. At the outset, it should be pointed out that it would be very difficult to prove in any particular case that there is no adaptive value. The endocrine glands of vertebrates were thought to be without function only a short time ago. The demonstration of their manifold functions awaited the development of suitable techniques. The same thing is at least potentially possible with respect to any character which appears to have no adaptive significance when studied by present methods.

A specific possibility is that an apparently nonadaptive character may have adaptive value during a limited phase of the life cycle. For example, many terrestrial animals return to water for breeding purposes. If studied only during the terrestrial phase, their aquatic adaptations would be rather puzzling. Again, a character may have no selective value under ordinary circumstances, yet be highly valuable in the extremes to which the organism may be occasionally subjected. Thus the plants of the San Francisco Bay region may for many generations never be subjected to any extremes of temperature. Because of this mild climate, a wide variety of introduced plants has been very successfully cultivated. But occasionally, freezing weather does strike, and it wreaks much havoc among the plants, with much greater damage among the introduced plants than among the native plants. As studied during "typical" years, the characteristics which adapt these plants to severe weather are difficult to understand. But the native plants are there not only because they can exploit the usual benign weather, but also because they are capable of withstanding the extremes. The occasional extremes may well constitute the most severe selective force to which these plants must become adapted.

Lastly, there is the possibility that a character may actually have no

PYTHON.

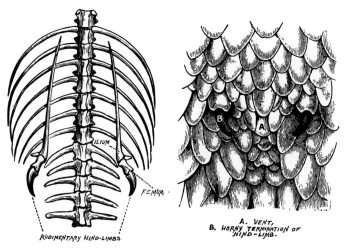

FIGURE 96. VESTIGIAL HIND LIMBS OF THE PYTHON. (From Romanes, "Darwin and After Darwin," Open Court Publishing Co., 1902.)

adaptive value. Such characters can be understood on several bases. They may be incidental effects of pleotropic genes—that is, genes with more than one phenotypic effect—so that, while one effect of such a gene is favored by selection, other effects are accidentally preserved along with it. Or the gene for a character of no selective value might be closely linked to one of definite selective value. Again, a gene may have become fixed in a species as a result of genetic drift (see Chapter 16). Finally, a character might be present in vestigial form because it had been valuable in the past, and has simply not been completely eliminated yet. Examples of such vestigial characters are numerous. Snakes, which have no use for limbs, are descended from typical reptiles with well developed limbs. Some snakes still have traces of the hind limbs present (Figure 96). A few of the many vestigial structures of man are illustrated in Figure 97.

Schmalhausen, a distinguished Russian student of evolution, has emphasized the conservative role which natural selection may play in evolution. He points out that, as the majority of mutants which may arise are disadvantageous, selection will tend to eliminate them and thus to conserve the status quo. This is undoubtedly true. But the relationships between natural species and their environments (i.e., the sum of the selective agencies) are continually in a state of flux, and so the creative effect of natural selection, at least in the long run, may be expected to outweigh the conserving effect.

Cooperation and Selection. Part of the early opposition to Darwinism was based upon Darwin's failure to discuss cooperative aspects of animal behavior. All social organisms cooperate in some degree. In social insects, like ants and termites, intracolony cooperation is the most striking aspect

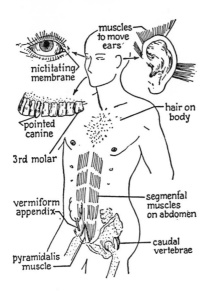

muscles to move ears'

nictitating membrane

pointed canine

3rd molar

vermiform appendix

pyramidalis muscle

hair on body

segmental muscles on abdomen

caudal vertebrae

FIGURE 97. SOME VESTIGIAL CHARACTERS OF MAN. (After Kahn, from Storer and Usinger, "General Zoology," 3rd Ed., McGraw-Hill Book Co., Inc., 1957.)

of the biology of the species. In man, cooperative and competitive aspects exist side by side. Some nineteenth century critics believed that this invalidated natural selection. This, however, was an error, for cooperation itself may have selective value, increasing probability of survival. The elaborate societies of ants have certainly aided these species in competition with others. The looser societies of grazing animals, like deer or caribou, provide mutual protection against predators. And surely it is obvious that human cooperation in many spheres is partially responsible for the great success of man. Thus animal cooperation, superficially contradictory to natural selection, is actually one of its products.

REFERENCES

BROWER, JANE V. Z., 1958. "Experimental Studies of Mimicry in Some North American Butterflies," *Evolution,* 12, 32–47; 123–136; and 273–285. The technical report on an excellent series of experiments. (Bates, Ford, Müller.)

COTT, H. B., 1940. "Adaptive Coloration in Animals," Methuen & Co., Ltd., London. An extraordinarily comprehensive review. (Swinnerton, Sumner.)

DOBZHANSKY, TH., 1951. "Genetics and the Origin of Species," 3rd Ed., Columbia University Press, New York, N.Y. (Dice, Quayle, Stakman, Sukatchew, Timofeeff-Ressovsky.)

FISHER, R. A., 1930. "The Genetical Theory of Natural Selection," Oxford University Press. A classic.

MCATEE, W. L., 1932. "Effectiveness in Nature of the So-Called Protective Adaptations in the Animal Kingdom, Chiefly as Illustrated by the Food Habits of Nearctic Birds," *Smithsonian Misc. Collections,* 85, 1–201. An excellent example of the evolutionary biology of the Agnostic Period.

MONTAGU, M. F. ASHLEY, 1952. "Darwin, Competition and Cooperation," H. Schumann, New York, N.Y. A systematic study of the role of cooperation in evolution.

SCHMALHAUSEN, I., 1949. "Factors in Evolution," McGraw-Hill Book Co., Inc., New York, N.Y. Develops the role of selection as a conservative factor.

CHAPTER SIXTEEN

Some Quantitative Aspects of Evolution

THE REALIZATION OF THE FACT that selection need not be an all-or-none force, but can operate through gradual changes in the frequency of certain characters or character combinations in a species, required that an attempt be made to analyze statistically the changes in the genetic composition of a species which might be expected under various conditions of mutation, selection, and population structure. This problem has been attacked mathematically by Fisher, Haldane, Wright, and others. The results of their calculations are one of the major achievements of the neo-Darwinian school. In the first edition of his brilliant book, Dobzhansky * prefaces the discussion of the statistical analysis of variation in populations with the statement that "... Only in recent years, a number of investigators ... have undertaken a mathematical analysis of these processes, deducing their regularities from the known properties of the Mendelian mechanism of inheritance. The experimental work that should test these mathematical deductions is still in the future, and the data that are necessary for the determination of even the most important constants in this field are wholly lacking." In the intervening years, many investigators in the new science of *population genetics* have ameliorated this picture, yet it is still true that the theoretical mathematics of evolution is more highly developed than is its experimental application to evolution in nature. It is still possible that the mathematical analysis of evolutionary phenomena, as it is now generally accepted, violates the dictum of Johannsen that "Biology must be handled with mathematics, but not as mathematics." Only some of the simplest and most important of the mathematical work will be presented below.

EVOLUTIONARY MATHEMATICS

The Hardy-Weinberg Law. Evolutionary mathematics begins with the Hardy-Weinberg Law, which states that, if alternative forms of a gene

---

* Dobzhansky, Th., "Genetics and the Origin of Species," 1st Ed., Columbia University Press, 1937.

are present in a population in a definite proportion, and if random mating and equal viability of all genotypes obtain, then the original proportions will be maintained in all subsequent generations, unless it is upset by some other factor, such as mutation or selection. In mathematical terms, the proportion of one allele, A, may be taken as q, the other, a, as $1 - q$, thus making the sum of their proportions 1. Then, in the F_2 and in all subsequent generations, the proportions of the possible genotypes will be $q^2AA:2q(1 - q)$ $Aa:(1 - q)^2aa$, and the proportions of the genes will be q in the case of A and $1 -q$ in the case of a. Let us substitute figures now for a monohybrid cross. In the cross $AA \times aa$, q will equal 0.5, and $1 - q$ will also equal 0.5, the meaning being simply that the alleles are present in equal numbers in the experimental population. The expansion of the binomial $[q + (1 - q)]^2$ gives the F_2 coordinates, $q^2 + 2q(1 - q) +$ $(1 - q)^2$, as stated above. Substituting numbers and genotypes, this would be $(0.5A + 0.5a)^2 = 0.25AA + 0.50Aa + 0.25aa$. As this is the familiar 1:2:1 genotypic ratio of the F_2 of a monohybrid Mendelian cross, it is clear that the elementary Mendelian ratios comprise special applications of the Hardy-Weinberg Law.

The Hardy-Weinberg Law is just as applicable to problems involving initial gene ratios not bearing any special relationship to standard Mendelian ratios. If the frequency of A is 0.8 and that of a is 0.2, the expanded formula would read $(0.8A + 0.2a)^2 = 0.64AA + 0.32Aa + 0.04aa$. The sum of the frequencies of the several genotypes still equals 1, which verifies the calculation. Now if this represented 50 organisms, it would represent 100 genes. Of these, 80 (all 64 genes of the 32 AA organisms, plus 16 of the 32 genes of the 16 Aa organisms) would be A, while 20 (the other 16 genes of the Aa organisms plus all 4 genes of the 2 aa organisms) would be a. Thus it may be seen that the values $q = 0.8$ and $1 - q = 0.2$ are maintained.

If the assumptions are changed now only by adding nonrandom breeding, such as self-fertilization, or preferential breeding so that organisms of similar phenotype tend to mate, the result will be an increase of the genotypes AA and aa at the expense of Aa. But the relative proportions of the *genes* will remain unchanged.

One can use this formula to determine *gene* proportions if character proportions are known. Thus, about 16 per cent of native white New Yorkers are Rh negative. As an Rh negative person is homozygous recessive ($rhrh$), $(1 - q)^2 = 0.16$, and $(1 - q) = 0.4$. The proportion of the recessive gene is therefore 0.4 and that of the dominant alleles, 0.6. Substitution of these figures in the expression $2q(1 - q)$ easily gives the frequency of heterozygotes as 0.48. Hence, the frequency of homozygous dominant persons must be 0.36. Thus the Hardy-Weinberg formula is a most effective tool for analyzing the genetic composition of populations. Again, the sickle cell trait (see Chapter 15), which depends upon a heterozygous genotype, Ss, is as frequent as 40 per cent in some equatorial African tribes, while 60 per cent are normal (ss). Substitution in the formula reveals similarity to the hypothetical case with $A = 0.8$ and $a = 0.2$.

But there is an additional factor here, to be discussed below: there is severe selection against genotype SS (sickle cell anemia).

The Hardy-Weinberg Law, then, operates to maintain the status quo. It is a conservative factor in evolution. In order to apply the formula to evolutionary problems, it is necessary to take into account the factors which might upset the equilibrium and cause a change in the relative frequencies of the alleles. The principal calculable factors are mutation and selection. As the mathematics of mutation and selection is rather complicated, the details will not be presented here. References to the papers of Fisher, Haldane, and Wright may be found in Dobzhansky's book.

Selection Pressure and Rates of Evolution. Calculations of the effect of selection are most easily made where selection favors a gene with complete dominance. If 1000 AA or Aa survive for every 999 aa that survive, the dominant form may be said to be favored by a positive selection pressure of 0.001 (or conversely, aa may be said to be subject to a negative selection pressure of the same magnitude). Thus, selection will modify the Hardy-Weinberg equilibrium by this small but calculable factor in each generation. Haldane has calculated the results of such a selection rate upon populations with varying initial proportions of the favored dominant gene. He found that the rate of increase in the proportion of the gene, in large populations, would be extremely slow when the initial proportion of the favored dominant is either very low or very high. But, when the initial proportion of the gene is moderate, the increase may be quite rapid. Thus, he found that it would require 11,739 generations to increase the proportion of a dominant gene from 0.000,001 to 0.000,002 (1 in a million to 2 in a million) with a selection pressure of 0.001. But a change of gene proportion from 0.00001 to 0.01 would require only 6920 generations; and the change from 0.01 to 0.50 would require only 4819 generations; but the change from 0.990 to 0.99999 would require 309,780 generations. It appears then that it is extremely difficult for mild selection pressure, unaided by any other factor, to establish a new dominant gene in a species, or to bring a well established dominant to 100 per cent incidence ("fix" it) in a species. But such pressures may readily result in a great increase in the relative proportion of an already well established gene. If selection favors a recessive gene, then the process is similar, but much slower. The initial step from 0.000,001 to 0.000,002 would now require 321,444 generations.

Purely mathematical studies of this type, and studies on experimental populations, are not uncommon, but their application to natural populations is more difficult. Kurtén has, however, reviewed the history of a pair of alleles over the past million years. In bears, the growth of the first upper molar is *allometric* (see below), that is, growth in height is more rapid than growth in length. The degree of allometry is genetically controlled, being quite marked in some bears, rather moderate in others. In any, the larger the bear, the higher the crown of the molar in relation to its length (Figure 98). The more extreme allometry is found in modern bears, *Ursus arctos,* while a more moderate degree characterized the late Pleistocene cave bear, *U. spelaeus,* now extinct. Both types of allometry

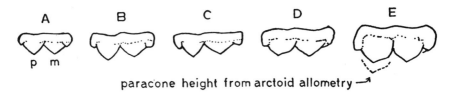

paracone height from arctoid allometry →

FIGURE 98. FIRST UPPER MOLARS OF BEARS; p, paracone, and m, metacone, the prominent cusps of the teeth. A and B, small and large *Ursus arctos;* D and E, small and large *U. spelaeus;* and C, a tooth from *U. arctos* with spelioid allometry. The dashed line in E indicates what the height of the paracone would have been had arctoid allometry prevailed. (From Kurtén, *Evolution*, V. 9, 1955.)

are found in *U. etruscus,* an early Pleistocene bear which was ancestral to both species. Kurtén has made the reasonable, but unproved, assumption that genes governing the two kinds of allometry are alleles, A_a for the arctoid, more extreme type, and A_s for the spelioid, more moderate type. Both alleles, then, were present in the ancestral *U. etruscus.* A sample from the mid-Pleistocene, which is probably ancestral to *spelaeus* and possibly to *arctos,* shows about 67 per cent A_a and 33 per cent A_s. The Hardy-Weinberg formula leads to an expectation of $4\,A_aA_a : 4\,A_aA_s : 1\,A_sA_s$. In this sample, many molars are intermediate, and hence probably based upon heterozygotes. The actual sample count was 42 arctoid, 50 intermediate, and 8 spelioid, not a significant deviation from the mathematical expectation of 44.45 : 44.45 : 11.1 ($= 100$, the sample size). In *U. spelaeus,* which appeared later, the gene A_a disappeared altogether, while both genes have been retained in *U. arctos,* but A_s is less abundant, making up only 23 per cent of the gene pool of a recent Finnish population of bears. Thus Kurtén has supplied an outline history of a pair of alleles over the past million years.

Some probable selection forces can be inferred. *U. spelaeus* was a large bear, and arctoid allometry would have produced a very high tooth, jutting out of the tooth row and not working harmoniously with the others. The spelioid pattern in *U. arctos,* however, leads to a low-crowned tooth which must wear down more quickly than its neighbors.

Mutation Pressure and Genetic Equilibrium. The above discussion has assumed that no mutations occur, but this assumption is, of course, not valid. In any particular case, it is possible that mutation might proceed only in one direction, as $A \rightarrow a$, or it might proceed in both directions, $A \rightleftharpoons a$. In the former case, even a slight mutation pressure would eventually lead to a species completely homozygous for the mutant gene, unless a selective disadvantage of the mutant were to prevent this. If the latter were the case, and if all of the genotypes had equal selective value, then the frequencies of the two alleles would reach an equilibrium, the numerical value of which would depend upon the actual magnitude of the mutation rates in the two directions.

The equilibrium point is related to the two mutation rates in a simple way. If rate $A \rightarrow a = u$ and $a \rightarrow A = v$, the frequency of a at equilibrium $=$ $q = \dfrac{u}{u+v}$. For example, if the rates are equal, then $q = \dfrac{1}{1+1} = \dfrac{1}{2}$, or equilibrium will occur when the frequencies of A and a are equal. If $u = 2\,v$, then a is being formed twice as rapidly as A, and $q = \dfrac{2}{2+1} = \dfrac{2}{3}$, or equilibrium will be reached when two-thirds of the genes are a. If $u = 4\,v$, then 80 per cent of the genes will be a at equilibrium.

The sickle cell trait is also apropos here. The Hardy-Weinberg formula gave an expectation of 0.64 ss : 0.32 Ss : 0.04 SS in the African tribes studied. Random breeding should give normals and heterozygotes in a ratio of 2 : 1, or 0.67 to 0.33, with SS being subject to 100 per cent negative selection. This should cause a drop of 7 per cent in the frequency of S (and a similar rise in s) in one generation. Thus S should be rapidly eliminated, yet the population appears to be in equilibrium. The reason is that selective elimination of ss individuals by malaria gives a selective advantage to the genotype Ss, which confers resistance to malaria, in spite of the severe selection against homozygotes for S.

In nature, neither mutation nor selection will ordinarily occur alone, and so the two will act simultaneously, perhaps in the same direction, perhaps in opposite directions, to upset the Hardy-Weinberg equilibrium. Most frequently, selection will work against mutation, as the majority of possible mutations are deleterious. This will result in very slow change, if any. But if a particular mutation is favored by selection, and if its mutation rate is appreciable, the combined action of mutation and selection might well cause a rather rapid change.

Haldane's calculations showing the extremely slow rate at which small selection pressures could establish a new mutant or fix one which is already well established have been referred to above. This type of calculation is more than an exercise in statistics, because of the importance of mutations of quantitative genes (see Chapter 13). And very minor mutations are unlikely to be subject to selection pressures of a much greater order than that used in the calculations. But the differentiation of species must involve the accumulation of a great many such differences, partly simultaneously, partly in sequence, if the neo-Darwinian theory be correct. Dobzhansky [*] has pointed out that "... The number of generations ... needed for the change may, however, be so tremendous that the efficiency of selection alone as an evolutionary agent may be open to doubt, and this even if time on a geological scale is provided." It is partly for this reason that Goldschmidt believed that the neo-Darwinian theory places too great a burden upon natural selection, and hence that the work of selection must be shortened by some other process, namely, systemic mutation.

---

[*] Dobzhansky, Th., "Genetics and the Origin of Species," 1st Ed., Columbia University Press, 1937.

Population Size and the Effectiveness of Selection. All of the above is based upon the assumption that mutation and selection are proceeding in a population of indefinitely large size, but Sewall Wright showed that the effectiveness of selection is greatly affected by the size of the population concerned. The mathematical basis for this proposition is complex, but the results are simple. It appears that mild selection pressures are relatively ineffective both in very small and in very large populations. Selection has its maximum effectiveness in moderate sized populations. What these relative population sizes mean in numerical terms is much less clear than might be desired. It should be pointed out, however, that it is the actual breeding population rather than the total population which is important here. Opponents of this idea point out that populations of natural species are, in most cases, immense, though good estimates are available only for a few endemic species. Its advocates, however, point out that such immense natural species are divided into more or less isolated subspecies; that these are further subdivided into local populations; and that these local populations are the significant breeding units. Migration from one unit to another will tend to blur the lines which separate them, but, generally speaking, the factors which tend to differentiate local populations and subspecies will be stronger than the migration pressure. This type of breeding population structure seems to be rather well established: the basic question which remains is, how large is large, and how small is small, from the viewpoint of population dynamics?

Wright believes that evolutionary changes might occur with "explosive" rapidity if favored by a combination of mutation, selection for a character (or combination of characters), and optimum population structure, as described above. This may be the answer to the question posed above, whether some new principle, in addition to mutation and selection, is necessary to account for the observed results of evolution.

Genetic Drift, the Sewall Wright Effect. Yet another factor which tends to upset the Hardy-Weinberg equilibrium is one which has been referred to under several names, including genetic drift (perhaps the best name), scattering of variability, and the Sewall Wright effect. The phenomenon was originally described by Fisher, but he has since disavowed its value completely, and it is Wright who gets (and deserves) the major credit for the development of this concept. Genetic drift refers to the accidental fluctuations in the proportion of a particular allele which depend upon the fact that the assortment of genes into gametes and the combination of gametes to form zygotes are random processes. It is well known that such accidental deviations from the theoretically expected assortment of alleles, and from their expected recombinations at fertilization, are responsible for the fact that Mendelian ratios are rarely obtained exactly. But because the deviations in each experiment are random, they tend to cancel out, and hence the validity of Mendelian principles is demonstrable by adding the results of many experiments, or by performing large-scale experiments.

In nature, the large-scale experiment is already provided. Such accidents of sampling do not have an important effect on large breeding populations, because an accidental increase of A in one part of a population will generally be counterbalanced by an accidental increase of a in another part of the same population. Or the genetic drift of one season will be reversed in the next season. But in small populations, the situation is very different. If there are only 100 individuals in a particular population (and smaller breeding populations do exist, for example, the whooping crane), and if a particular allele is present only once, then an accident of sampling might easily in a single generation remove it irrevocably from the population, or increase it many fold, say to 10 per cent. The result is that, in small, isolated populations, genes may be completely lost or completely fixed by genetic drift, without reference to selective value. Genetic drift is thus a force working in opposition to selection, for it tends to preserve or to destroy genes without distinction, whether favorable, neutral, or unfavorable. But selection tends to preserve those which confer some adaptive value and to destroy those which impair the adaptive value of a species. Severe selective forces will, of course, destroy disadvantageous genes irrespective of the population size.

Wright's viewpoint on the significance of genetic drift has been much

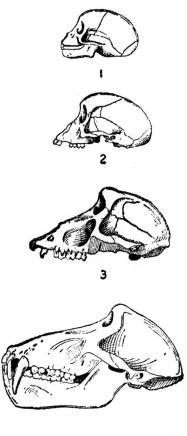

FIGURE 99. ALLOMETRY IN BABOON SKULLS, showing the great increase of facial length with respect to cranial length with increasing total size. Actually, $k = 4.25$, a very high figure. (From Huxley, "Problems of Relative Growth," Methuen & Co., Ltd., London, 1932.)

misunderstood by his proponents as well as by his opponents. It has often been said that he regards small, isolated populations as optimum for rapid evolution. But this is not the case. Because genetic drift predominates over selection in such populations, he believes that they will show a higher degree of homozygosity than will more typical populations, and that they will, on the whole, tend to be rather poorly adapted. As a result, they may well become evolutionary blind alleys.

A corollary to the Sewall Wright effect is what Stebbins has called the "bottleneck" phenomenon. It is often said that the numbers of a species tend to remain approximately constant in any locality. But every field biologist knows that a species which is very abundant in one year may be very difficult to find in another year, only to be followed by another increase. In years of scarcity, small populations assume an especial importance, for they are the only source from which the species can again be built up. Hence the term "bottleneck." Accidental changes in the genetic make-up of such bottleneck populations will therefore determine changes in the larger populations to be derived from them, and these changes will generally be nonadaptive in character. For example, the arctic hare periodically reaches a peak of population at which disease decimates the population. There follows a "no rabbit year" and then several years of recovery. Accidental changes in gene frequencies during the years of scarcity must have a profound effect upon the populations of years of abundance. The same thing must be true of the lemming, a small rodent of the Scandinavian mountains. Every few years its population reaches a prodigious level, and epidemics do great damage. The lemmings then

FIGURE 100. ALLOMETRY OF THE "TAIL" OF THE SWALLOWTAIL BUTTERFLY, *Papilio dardanus*. This moderately positive allometry is shown only by the males. (From Huxley, "Problems of Relative Growth," Methuen & Co., Ltd., London, 1932.)

migrate to the coast, plunge into the sea, and swim until they die. The population is re-established by the few which remain in the mountains. Similarly, those many species which have small over-wintering populations must show bottleneck effects.

Allometry. As an organism grows, the various parts grow at different rates, so that its proportions change. Thus a baby's head is relatively large, but its growth does not keep pace with that of the rest of the body, and hence its proportionate size in the adult is more moderate. One says that the head shows negative *allometry* (Greek, differential measurement) or *heterogony* (Greek, unlike development). On the other hand, the adult teeth are larger in proportion to the head than are the milk teeth, and so tooth development is positively allometric. D'Arcy Thompson first showed that trends of allometric growth can be analyzed mathematically. They correspond to the equation $y = bx^k$, where y is the size of the organ studied, x is the size of the animal as a whole (or of some part used for comparison), b is a constant determined by the value of y when x is 1, and k is the exponent of allometric growth. If k is less than 1, allometry will be negative, while if k is more than 1, allometry will be positive.

The most obvious application of this is to ontogeny. Thus young baboons are only moderately prognathous, but the jaws protrude ever more

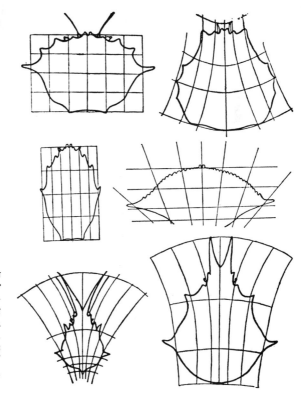

FIGURE 101. DERIVATION OF CRAB CARAPACES OF VERY DIFFERENT APPEARANCES BY CARTESIAN TRANSFORMATIONS OF A SINGLE ORIGINAL TYPE. (From Huxley, "Problems of Relative Growth," Methuen & Co., Ltd., London, 1932.)

strongly as they approach full size (Figure 99). In this series, $k = 4.25$, a very high value. In deer, antlers show positive allometry. In the red deer, *Cervus elaphus, k* is about 3 in young bucks, but it declines to about 1.6 as they grow fatter. Similarly, the "tails" of swallowtail butterflies show positive allometry (Figure 100).

Thompson and Huxley showed that phylogenetic changes in relative size can be analyzed in the same way as are ontological changes. That is, if selection favors increase in total body size, as long as the genes for allometry remain unchanged, this will require further increase in the proportionate size of specific parts. Thus the Irish Elk, *Cervus antiquus*, shared with other deer the positive allometry of the antlers. The species grew steadily larger throughout the late Pleistocene, and the antlers, continuing their allometric growth, became truly immense. It is often said that the Irish Elk became extinct because its antlers grew so large that it could not hold up its head. There is no evidence for this improbable assertion. Had selection against increased antler size been so severe, it would no doubt have favored genes for smaller body size. The cause of the extinction of this giant deer is unknown.

General body outline can be analyzed by Cartesian transformation, a special aspect of allometry, particularly in highly variable groups. The outline of a primitive member of the group is drawn upon a rectangular grid, then the grid is distorted by stretching particular parts in one direction or another. The results simulate the outlines of related species with different factors for allometric growth. Figure 101 shows this for a series of crabs. This is a strong indication that such evolutionary changes are based upon mutations of the genes influencing allometry. This conclusion is also supported by Kurtén's study on bear teeth.

REFERENCES

Huxley, J. S., 1932. "Problems of Relative Growth," Methuen & Co., Ltd., London. A thorough study of problems of allometry.

Kurtén, B., 1955. "Contribution to the History of a Mutation During 1,000,000 Years," *Evolution,* 9, 107–118. An interesting study, reported without difficult mathematics.

Li, C. C., 1955. "Population Genetics," University of Chicago Press. An authoritative treatment which, however, requires a good mathematical background. (Hardy, Weinberg.)

Simpson, George Gaylord, Anna Roe, and Richard C. Lewontin, 1960. "Quantitative Zoology," Revised Ed., Harcourt, Brace, & Co., Inc., New York, N.Y. An unusually lucid introduction to biometry, based upon zoological examples.

Thompson, D'Arcy, 1952. "On Growth and Form," 2nd Ed., Cambridge University Press. This book is a beautifully written classic, and the foundation of the allometry concept.

Wright, S., 1940. "The Statistical Consequences of Mendelian Heredity in Relation to Speciation," in Huxley, "The New Systematics," Oxford University Press. A succinct statement by one of the major architects of evolutionary mathematics.

CHAPTER SEVENTEEN

Continuous versus Discontinuous Variation

CONSIDERATION OF THE LIVING WORLD makes nothing more obvious than that variation is everywhere present. On the grandest scale, the Plant and Animal Kingdoms are differentiated. Almost as obviously, the major types, phyla, and classes, are differentiated within each kingdom. As one descends the taxonomic hierarchy, variation is demonstrable within each category, with the qualification that more refined methods of observation may be needed to establish the fact at the lowest levels. Equally important is the fact that the varying organisms are arranged into discontinuous clusters, more specifically, into a hierarchy of discontinuous clusters. This is the basic fact of taxonomy.

THE BASES OF DISCONTINUOUS VARIABILITY

Nonetheless, it is often maintained that the observed discontinuity in our present-day flora and fauna is illusory because it depends upon the extinction of intermediates which have actually existed in the past. According to this viewpoint, if representatives of every species that has ever existed could be arranged in order from the most primitive to the most specialized, an almost imperceptible transition would be observed, with abrupt discontinuities corresponding only to single gene differences. In other words, the entire living world would show no discontinuities greater than those which actually characterize single species.

Admittedly, no such assemblage could ever be possible because of the immense amount of extinction which has marked the history of life upon this earth. Nor is it possible to assemble a sufficiently complete fossil series to simulate the series suggested, for the fossil record is very incomplete. But the neo-Darwinian viewpoint is favorable to the possibility that an approximately continuous series may have existed, though never at a single time level. The alternative is, of course, that the larger steps in evolution may have been achieved through systemic mutations, or through some other type of macroevolutionary change. If this is correct, then at least part of the deficiency of the fossil record depends upon the supposed intermediates never having existed.

Permutations of Genes. Wright has approached the problem of continuous and discontinuous variation from the viewpoint of statistical analysis of permutations of gene combinations. He begins with the very modest assumption that every organism might have 1000 pairs of genes, and that each of these might form a series of ten multiple alleles. The number of possible recombinations of these alleles would then be 10^{1000}. If this entire array of genotypes could be formed, and if the resulting organisms were so arranged that each differed from its neighbors only by a single gene, they would undoubtedly form a smooth, continuous series from one end to the other. But it would be patently impossible to form this whole series, because the estimated number of electrons in the visible universe is only 10^{79}. Most of the potential genotypes, if formed, would be monstrosities which would be destroyed by natural selection. Of those genotypes which are actually formed, those which survive must cluster around "adaptive peaks," that is, character combinations which are physiologically harmonious, and ecologically sufficiently adapted to the demands of the environment in which the organism must face the test of natural selection. Separating these adaptive peaks are "adaptive valleys," which represent disharmonious, or unworkable, character combinations.

In such a system, a group of closely associated peaks represents a species, with each peak corresponding to a subspecies. Small ranges represent genera, while families and higher groups are represented by larger ranges. Discontinuities depend partly upon the impossibility of forming the entire series of genotypes, and partly upon the elimination by natural selection of many of those which are formed. It must be realized that this very ingenious explanation is an hypothetical model, which should not be expected to correspond completely to the facts of nature, for this would require that man and amoeba be conditioned by the same series of allelic genes. How far this model is applicable to nature is debatable: some believe that it is very generally applicable, while Goldschmidt believed that it is strictly applicable only to the differentiation of subspecies.

THE SPECIES CONCEPT

In any case, classification is possible because discontinuities do exist between varying series of organisms. The limits of the various discontinuous groups on the higher levels are rather easily ascertained, although the rank and relationships of a particular group may be disputed. Thus the Onychophora are treated by some systematists as a class of Annelida, by others as a class of Arthropoda, and by still others as an independent phylum. Some even unite all three groups in a single Phylum Articulata. But advocates of all of these ideas agree as to which animals are onychophorans and which are not. Again, a group treated as an order with several families by a "splitter" may be treated as a single family by a "lumper," while both men agree on what genera comprise the group. To the extent that described groups correspond to such actually discontinuous groups in nature, and that is generally the case, the conventional classifications may be said to be natural. But, to the extent that the rank accorded the various groups is arbitrary, the system of classification itself is arbitrary rather than natural. Taxonomists generally are agreed that, while the discontinuities of the higher groups are real, their assignment to systematic categories is primarily a matter of convenience. Linnaeus himself said this of the higher categories, but the species he regarded as a real thing, each one a specially created unit. The replacement of the archetypal concept with the evolutionary concept has not been accompanied by an abandonment of Linnaeus' idea of the definiteness of species. As Bateson has said, "Though we cannot strictly define species, they yet have properties which varieties have not, and . . . the distinction is not merely a matter of degree."

It becomes obligatory then to discuss the nature of this unique taxonomic unit, the species. In Chapter 3, species were defined as kinds of plant or animal the individuals of which differ "from each other only in minor traits, except sex; sharply separated in some traits from all other species; and mutually fertile, but at least partially sterile when crossed to other species."

Magnitude of Difference. Many other definitions of species have been published, and all of them, including the above, are unsatisfactory in the sense that they do not provide a basis upon which a practical taxonomist can decide whether two similar groups are distinct species or only subspecies. Some have tried to specify the degree of difference which is necessary to separate good species, but this is not practical not only because of the difficulty of formulating a quantitative expression, but because the degree of difference between species in some groups seems to be very much greater than in other groups; and because some undoubtedly good species, as *Drosophila pseudoobscura* and *D. persimilis*, show very little morphological difference, while different races of other undoubtedly single species, like man, show very pronounced differences. It appears that the degree of difference is much less important than the constancy of difference, that is, the discontinuity between the groups.

Discontinuity and Interspecific Sterility. But even strictly discontinuous differences need not indicate specific boundaries. Differences conditioned by a single gene will show complete discontinuity if dominance is complete (for either allele), and these differences may be of considerable magnitude. For example, the mutant *tetraptera* in *Drosophila* is characterized by the development of two pairs of wings, much as in the dragonflies and many other orders of insects. The type of discontinuity of greatest interest is that which results from the inability of related species to interbreed, the phenomena of interspecific sterility and hybrid sterility, concerning which there is an immense literature. Many definitions of species have emphasized this. It has the merit of being very generally true. But it also has some faults. At the outset, taxonomists must of necessity do most of their work with preserved specimens, and hence a sterility barrier would be difficult to use even if the validity of the concept were completely agreed upon. But there are some facts which cast doubt upon the validity of the concept. First, it is difficult to distinguish between cases in which organisms cannot interbreed, and those in which they can but do not for other reasons. Thus, the closely similar species *Drosophila pseudoobscura* and *D. persimilis* do not interbreed in nature because of a genuine sterility barrier. But the pariah and the Hindu do not interbreed, although there is undoubtedly no sterility barrier to prevent it. Sterility barriers may exist within an undoubtedly single species, as in the case of *Lymantria dispar* in which racial crosses may result in hybrids which are sterile because of intersexuality. Finally, there are many cases, especially among plants, in which crosses between well recognized species can be made easily. Frequently, the fertility of such crosses, or of the F_1 from them, is impaired partially, even greatly. But such cases make it difficult to maintain that sterility is an absolute criterion of species differentiation.

Nonetheless, interspecific sterility remains the common element of most definitions of species. Dobzhansky has defined species as the stage in the evolutionary process "at which the once actually or potentially interbreeding array of forms becomes segregated in two or more separate arrays which are physiologically incapable of interbreeding." Goldschmidt stated that groups which can be successfully crossed should be treated as a single species for evolutionary studies, while acknowledging that taxonomists may be justified on other grounds in separating them into distinct species for purposes of formal classification. Mayr[*] has recently defined species "as a group of actually or potentially interbreeding populations that is reproductively isolated from other such groups," and again, "in the strict sense of the word, speciation means the origin of reproductive isolating mechanisms." This concept of species based upon gene flow between related populations is often called the biological species concept. It has strong logical appeal, especially now that the successes of genetics in analyzing the problems of evolution are so striking. Sonneborn, however, has recently protested against it on two grounds. First, it is difficult to apply it to more than a few of the more thoroughly studied organisms because of the sheer volume of experimental study required for its adequate application. And second, it is not applicable to those many species of Protozoa, lower Metazoa, and plants which reproduce asexually or parthenogenetically. While some biologists have held that species in these organisms are not comparable to those of sexual organisms, Sonneborn makes a good case for the proposition that species in these groups are also based upon accumulation of genetic differences under the control of natural selection, and hence that a species concept which excludes them is unsound.

Because of the above described situation, including the difficulties of delimitation of species among asexually reproducing groups, many biologists have come to the conclusion that the species is an arbitrary unit in the same way that the higher categories are. Yet many biologists feel that

[*] Mayr, E., "Taxonomic Categories in Fossil Hominids," *Cold Spring Harbor Symposia on Quant. Biol.*, **15**, 112 and 115 (1950).

this is an error, and that the species is a valid, natural unit, even though we cannot define it adequately. The quotation from Bateson, with which this discussion was introduced, expresses this viewpoint nicely. Dobzhansky [*] has said that "Some biologists, lacking familiarity with the subject, have, in fact, fallen into this error (that species are arbitrarily determined units). In reality, no category is arbitrary so long as its limits are made to coincide with those of the discontinuously varying arrays of living forms. Furthermore, the category of species has certain attributes peculiar to itself that restrict the freedom of its usage, and consequently make it methodologically more valuable than the rest." Goldschmidt found that species which he studied have been so sharply, if not necessarily widely, differentiated, that he felt justified in speaking of a "bridgeless gap" separating species from one another.

Rassenkreise and Speciation. Dobzhansky believes that this confusion results from the mode of origin of new species. According to the neo-Darwinian conception, a widely distributed species should break up into partially isolated subspecies, which become differentiated by selection of different characters and by genetic drift. Thus a Rassenkreis, or circle of races, is formed, the terminal members of which should eventually become sufficiently differentiated that the sheer weight of difference would raise a sterility barrier. These would now be good species. With such a gradual origin, there would be no sharp break in morphological characters, hence taxonomists, who must depend mainly upon morphological characters, would have difficulty in determining whether the specific level of differentiation had been attained. All of this is based upon the assumption that the processes of evolution are the same both on the microevolutionary (changes within a species) and macroevolutionary (origin of new species or higher groups) levels, a plausible but unproven assumption which has been the center of much controversy.

Perhaps no more strongly suggestive study of a Rassenkreis has been published than that of *Ensatina eschscholtzii*, a Pacific coast salamander, by R. C. Stebbins. This species ranges very widely in the mountains of California and the Pacific Northwest (Figure 102). It is absent, however, from the great central valley of California, a low, hot, arid or semiarid region. Thus the distribution of this species forms an oval, with one segment corresponding to the coastal mountains, the other to the Sierra Nevada's. The two are connected at either end. As might be expected, this is not a homogeneous population, but it is broken up into seven subspecies, with intergrading populations usually being found wherever the ranges of two subspecies meet. It appears probable that the species originated in the northwest, then spread southward along the coastal and Sierra Nevada ranges, breaking up into subspecies along the way. The four coastal subspecies are uniformly dark brown to reddish brown dorsally, while the three interior subspecies show an increase in orange or yellow spotting from north to south. In southern California, the ends of

---

[*] Dobzhansky, Th., "Genetics and the Origin of Species," 1st Ed., Columbia University Press, 1937.

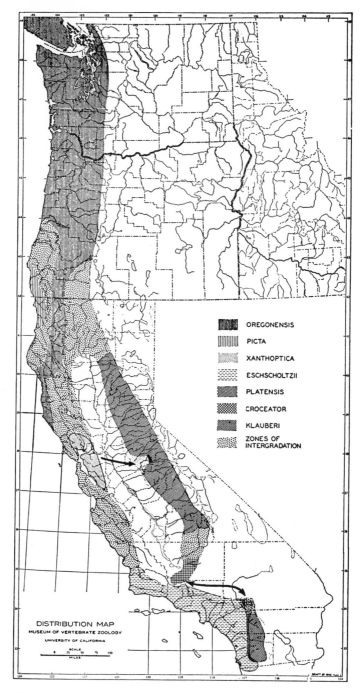

FIGURE 102. DISTRIBUTION OF *Ensatina.* (From Stebbins, *Univ. California Publ. Zool.,* V. 48, 1949.)

the chain are brought together, with *Ensatina e. eschscholtzii* representing the coastal division, while *E. e. croceator* and *E. e. klauberi* represent the interior division. The differences between them are very marked. Stebbins' first paper was based upon 203 specimens of *eschscholtzii*, 15 of *croceator*, and 48 of *klauberi*. Inland and coastal forms were not collected in the same locality, and no intergrades were found. Subsequent collections have enlarged all groups, and several *croceator-klauberi* intergrades were taken within a few hundred yards of *eschscholtzii* in the same canyon and the same ecological zone, yet no intergrades with *eschscholtzii* were found. Evidence was found for neither ecological isolation nor seasonal differences in breeding. Hence Stebbins believes that these terminal members of the *Ensatina* Rassenkreis must meet without breeding, and hence that they would be distinct species if the connecting members of the Rassenkreis were to become extinct. Scant data are available on their breeding behavior, and it is possible that intergrades may be absent for other reasons than intersterility, such as preferential mating, yet there is no evidence of this. Genetic studies upon this Rassenkreis, especially upon its terminal members, would be of great interest, although very difficult, because they do not breed readily in captivity. This is an exceptionally illustrative Rassenkreis, the facts of which comprise one of the strongest arguments for the neo-Darwinian theory.

The species question is, then, at once one of the most basic problems of biology and of evolution, and one for which no satisfactory answer is available. It is complicated by the difficulty of comparing sexual and asexual species, by cases in which subspecific or specific status is disputed, and by cryptic or sibling species which show only trivial phenotypic differences even though they are reproductively isolated. Darwin said that a species in any group is whatever a competent specialist on that group says a species is, and it may well be that species are not the same in different major groups. Discontinuity rather than degree of difference is likely to play the larger role in the achievement of an answer to the species question. Some aspects of the development of discontinuity will therefore be considered in the next chapter.

REFERENCES

Huxley, J. S. (Ed.), 1940. "The New Systematics," Oxford University Press. A valuable collection of essays on many aspects of taxonomy.

Mayr, E. (Ed.), 1957. "The Species Problem," American Association for the Advancement of Science, Washington, D.C. Another valuable collection of essays by many specialists. (Sonneborn.)

Stebbins, R. C., 1949. "Speciation in Salamanders of the Plethodontid Genus *Ensatina*," *Univ. Calif. Publ. Zoöl.*, 48, 377–526.

Stebbins, R. C., 1957. "Intraspecific Sympatry in the Lungless Salamander *Ensatina eschscholtzii*," *Evolution*, 11, 265–270. These two papers by Stebbins form the basis of the above discussion of the *Ensatina* Rassenkreis.

CHAPTER EIGHTEEN

Isolating Mechanisms and Species Formation

DARWIN AND HIS CONTEMPORARIES gave much thought to the possibility that the variations which form the raw materials of evolution might be "swamped" by crossing to the original type, with the result that no actual change in the species could occur unless such hybridizing were prevented by isolation of the new variant from the parent stock. This was based upon a pre-Mendelian conception of heredity, the *blending* theory of inheritance, according to which offspring should always be intermediate between the parents. If this theory were correct, then repeated backcrosses of a hybrid stock to the original type should result in an ever closer approach to the original type. In the hands of Moritz Wagner (1868 ff.) *isolation* of variant races became the necessary prerequisite and the inevitable cause of speciation. (This term, generally used as a synonym for species formation, is, unfortunately, etymologically incorrect. But it seems certain to remain a permanent part of our evolutionary vocabulary.)

MENDELIAN GENETICS, ISOLATION, AND SUBSPECIATION

With the rise of Mendelian genetics, the original idea of Darwin and Wagner became untenable. A gene-determined character could never be destroyed by crossing to the original type: it could only become heterozygous, with the possibility always present that homozygosity would be re-established. The gene might thus become evenly spread through a species, so that no tendency toward formation of a subspecies could be observed, but this would not entail loss of the gene, nor of the phenotype for which it was responsible. While breeding to the original type could never destroy a Mendelian gene, it could break up combinations of such genes. Selection can only operate on whole organisms, and so particular combinations of genes may have a value which would not be possessed by the separate genes of the genotype. Thus, mutants for a keen sense of smell and for rapid running (such as longer leg bones) would both be of selective value to a chase predator like the wolf. But the two in combination would be of much greater value than either one alone. While neither mutant could be destroyed by crossing to the original type, the combination would almost certainly be broken up either by independent assortment or by crossing over.

As subspeciation is the ordinary prerequisite to speciation in the neo-Darwinian scheme, the formation of breeding populations which have distinct complexes of genes (subspecies) is of great interest. Completely random breeding within a species would result in even distribution of all of its genetic variability and so no subspecies could be formed, although this would not prevent the mass transformation of a species into a new one if an appropriate selective force were operating, as Wright has pointed out. But complete random breeding probably never occurs, except in very small, endemic species. Typically, local populations breed largely among themselves, with relatively little outbreeding. The result is that different populations of single species can build up genotypes which differ consistently in some or many loci. On the basis of the resulting phenotypes, these may be classified as subspecies. Wherever two such subspecies meet, they ordinarily interbreed, and the expected Mendelian recombinations occur, with the result that a single, generally intermediate, and highly variable population is formed. Thus any start toward specific status which the subspecies may have made is lost. If, however, a subspecies is sufficiently isolated over a long period of time that interbreeding with its relatives is prevented, it may continue to accumulate differences until a physiological barrier to interbreeding is acquired. The subspecies may now be regarded as a new species. As long as the related groups remain geographically separated from each other they are referred to as *allopatric* species. There is frequently considerable doubt as to whether two geographically separated groups are in fact allopatric species or simply subspecies. But if they now move into the same territory and fail to interbreed, with the formation of intermediates, they are regarded as *sympatric* species, and there is much less doubt about their status.

Hence, the study of the means by which subspecies, and indeed species themselves, may be isolated from one another has played a major role in modern evolutionary studies. Very few would now care to accept the letter of the dictum of Romanes that "without isolation or the prevention of interbreeding, organic evolution is in no case possible." But, with some tempering, this has been the spirit of many recent studies.

ISOLATING MECHANISMS

The mechanisms by which subspecies and closely related species can be isolated from one another are classified by Mayr to include three main types of isolating mechanisms. First, there may be a *restriction of random dispersal* so that potential mates cannot meet. This is largely equivalent to the geographical mechanism of Dobzhansky and others. Second, there may be a *restriction of random mating* so that potential mates *do not* cross even though they have the opportunity. Finally, there may be a *reduction of fertility* so that a cross results in few offspring or none.

Restriction of Random Dispersal. Distribution maps generally indicate that a particular species is found continuously over broad areas, but, in fact, all species select in such large areas those restricted portions which present suitable ecological features. A checker board is thus a better model of species distribution than is the typical distribution map. For example, the distribution map for the American sycamore or plane tree, *Platanus occidentalis*, indicates a continuous distribution over more than half of the United States, from Texas in the South and Iowa in the North to the Atlantic coast. But if one wishes to find natural groves of sycamores in this vast area, it is necessary to look in rich bottom lands and along the banks of streams. Again, distribution maps show that the eastern meadowlark is found over most of the United States east of the Rocky Mountains, but it will be found only in open grasslands. The various local populations of most species are similarly separated by barriers of greater or lesser extent of territory which they cannot utilize for one reason or another. Within the broad areas of mapped distribution, there is almost no geographical feature which may not prove to be a barrier to the dispersal of some species.

Even a small amount of salt water is a nearly absolute barrier to amphibians. For this reason, the Pacific Islands are usually uninhabited by amphibians, except when they have been introduced by man, as in Hawaii. Salt water also separates many fresh-water fishes. For example, on the Pacific coast many fresh-water streams follow more or less parallel courses to the ocean. Typically, each stream will have its own subspecies or even species. Although the expanse of salt water separating the mouths of neighboring streams may be small, the fish do not cross it. But if the flood waters of the streams join together during the rainy season, then the streams share their fishes. This shows that it is actually the salt-water barrier, and not a homing instinct or other factor which typically keeps the fish faunas of neighboring streams separate.

Large bodies of water are among the most effective barriers to land birds. The Amazon River seems to be an absolute barrier to many birds, for in species after species the subspecies on opposite banks are different. Mayr has pointed out many instances in which neighboring tropical islands are inhabited by different subspecies even though the distances between them are rather short. When Darwin studied the birds of the Galapagos Islands, he found 26 species of land birds and 11 species of marine birds. But of these, 21 of the species of land birds were endemic, while only 2 of the species of marine birds were endemic. Islands which are separated only by short distances may have distinct subspecies or species, but, generally, those islands which are most isolated have the highest proportion of endemics.

Many mammals are also stopped by water barriers. Thus, subspecies of mice on opposite sides of major rivers are very likely to be different. The zoogeographical realms have been defined primarily on the basis of their mammalian fauna, and it is noteworthy that water barriers, the oceans, separate many, though not all, of these realms.

Mountains have sometimes been described as islands in a sea of lowlands. The flora and fauna of mountains are similarly restricted in their dispersal by their inability to cross the lowlands intervening between neighboring mountains or mountain ranges. Mayr studied the mountain birds of New Guinea, and found that almost all of them have broken up into subspecies in much the same fashion as have the birds of archipelagos. In some cases, there are even series of distinct altitudinal races, including lowland, mid-mountain, and alpine races. Many mammals such as the mountain goat and bighorn sheep are limited to the highest mountain habitats. Turel found that a very high proportion of alpine plants belonged to endemic races.

Similarly mountains will serve as barriers to lowland organisms. The American opossum, *Didelphis virginiana,* ranges widely in the eastern United States. The low eastern mountains do not comprise a barrier to it. It was introduced into California nearly 90 years ago for sporting purposes. It has thrived in the low coastal range, but has been unable to invade the Sierras. The thirteen-lined ground squirrel, *Citellus tridecemlineatus,* is widely distributed in the prairies of north central United States, but it stops short of the Rockies. The cottontail rabbit, *Sylvilagus floridanus* ranges over most of the United States east of the Rockies, but has been replaced in the mountains by its cousin, the jack rabbit. The white-footed mouse, *Peromyscus leucopus,* is similarly distributed, while its near relative, *P. maniculatus,* has successfully invaded the mountains.

Similarly, extensive forests will serve as barriers to dispersal of grassland organisms, prairies will serve as barriers to forest organisms, and even finer classifications are practical. Thus, the red tree mouse, *Phenacomys longicaudus,* lives on a diet of fir needles. It also nests in fir trees, and spends most of its life in them. Thus, not only will prairies serve as a barrier to its distribution, but non-fir forests will be very effective barriers. Even simple distance may serve as a barrier, for Dice and Blossom found that seven species of small mammals were subspecifically distinct at Tucson and at Yuma, yet there seemed to be no barrier to their free dispersal other than distance. As stated at the beginning of this discussion, there is scarcely any natural feature which may not be a barrier to some plant or animal. And the barrier of one is the highway of dispersal of another.

Thus far, only geographic factors in the restriction of random dispersal have been mentioned. But Mayr has included several characteristics of animals which tend to limit their dispersal even in the absence of geographic barriers. One of these is the sedentary character of many animals. Surprisingly enough, it is not sessile animals, like the sea anemones, to which this applies with greatest force, for the small, free-swimming larvae of these marine invertebrates may be widely distributed by ocean currents. Similarly, most plants, although strictly sessile, have means of dispersal of seeds which are at least as efficient as typical means of dispersal for animals, and commonly much more so. But the larger animals, particularly the vertebrates, whose locomotor organs are among their most obvious characteristics, seem to use their powers of locomotion to maintain their home range far more than for range expansion. The very extensive experiments in bird banding comprise the best evidence for this, as well as the best known example, but similar evidence is accumulating for all groups of vertebrates.

The homing instinct of many migratory birds is well known, and it is clear that this would also serve to limit random dispersal. The homing instinct has been reported for other vertebrates, and even for some invertebrates. The social structure of some species may have a similar effect. Thus, in the geese, family groups do not break up at the end of the nesting season, but rather they migrate together, and separate only after returning to the nesting ground in the following year. The result is a high degree of inbreeding.

Restriction of Random Mating. The principal factors which limit random mating, in the absence of geographical separation, are ecological differences, behavior differences, and mechanical differences of the copulatory organs. The first two are well established and will be considered in some detail, but the third has limited validity.

An important type of ecological difference between subspecies (or species) which may tend to prevent interbreeding is the selection of different habitats, so that potential mates from different populations do not meet even though they exist in the same general area. For example, Dice found that the ranges of two subspecies of *Peromyscus maniculatus* overlap in northern Michigan, but there is no evidence of their interbreeding in nature, although they will interbreed in the laboratory. One of these races is found principally in forests, the other on sandy beaches. A similar case in the same species has been studied by Murie in Glacier National Park. Here, one of the races is confined to forests, the other to open prairies. The common water snake, *Natrix sipedon*, presents a comparable situation in Florida, where fresh-water and salt-water races may come very close together, but they are kept separate by their habitat preferences. Pictet has described a curious case in Swiss moths. *Nemeophila plantaginis* occurs in different altitudinal races, with a race above 2700 meters and one below 1700 meters differing in a single gene. At 2200 meters, there is a hybrid population, in which *all* of the moths are heterozygous. When the two pure races are crossed in the laboratory, the offspring include all three types in the F_2 so it appears that the homozygous types must be subject to a very severe selective elimination at the 2200 meter level, if these data are correct. Data similar to the above are being gathered for many different groups, both plant and animal, and it seems clear that such differences in habitat requirements are of general importance as barriers to random mating.

Another ecological factor is seasonal isolation, that is, breeding seasons are sufficiently different to make interbreeding improbable. Thus, in California there are five described species of cypress trees (*Cupressus*). These are subdivided into ten distinct entities which could be called subspecies, all of which have quite limited distribution, some being represented only by a single grove. These may grow side by side with only a very few hybrid trees formed, or none at all. The factor which prevents interbreeding appears to be simply this, that the several races pollinate at somewhat different times, and so do not have the opportunity to cross. But the occasional hybrid trees can be explained easily by the fact that some trees of an early variety may pollinate later than usual, and some of a late variety may pollinate earlier than usual. Anderson has shown that such seasonal differences are important in isolating some species of the *Iris*. The same has been demonstrated for some other genera, and it seems probable that seasonal isolation is common among plants.

Seasonal isolation is by no means unknown among animals. Generally speaking, the breeding season of warm-blooded vertebrates is rather long, and the seasons of many members of the same group coincide or overlap broadly. Nonetheless, Mayr cites several instances in which seasonal isolation appears to be effective among birds. But among the cold-blooded vertebrates and among the invertebrates, breeding seasons may be quite restricted, and so seasonal isolation may be highly effective. In northeastern United States *Rana clamitans*, *R. pipiens*, and *R. sylvatica* may all breed in the same pond. But generally *sylvatica* will begin breeding before the others appear in the ponds. And *clamitans* may not begin breeding until the others are through. Water temperature is the decisive factor in determining when each species begins breeding. *R. sylvatica* begins to croak at water temperatures as low as 44° F in southern Michigan, and probably breeds at temperatures only slightly higher. *R. pipiens* begins to breed when the water temperature reaches 55° F, or perhaps somewhat higher. *R. clamitans* does not appear in the ponds until the water temperature exceeds 60°, and the minimum temperature for breeding has not been determined for this species. Actually, crosses between these three species fail in the early embryonic stages, but the seasonal separation may have preceded the sterility barrier and made possible its development. A similar situation applies to the salamanders, *Ambystoma tigrinum* and *A. maculatum* in the same area. *A. tigrinum* begins breeding very soon after the spring thaws, and is usually through breeding by the end of March. But *A. maculatum* does not begin breeding until late in March or early in April. Blair has published comparable data for the toads of the genus *Bufo*.

There are many instances known in which behavior differences, particularly in regard to courtship, restrict random mating between members of different species. Among plants, mating is usually based upon more or less mechanical (from the viewpoint of the plant) situations, such as wind pollination or insect pollination. But among animals, some preparation usually precedes mating. This may be very slight for some marine animals like the sea urchin in which discharge of gametes, together with some sex stimulating substance, by a single individual may induce all or most of a large colony to shed their sex cells. Gametes are then mixed by the water currents, and fertilization follows. At the other extreme, elaborate courtship, sometimes lasting for many days, may precede mating. Mayr has pointed out that in many birds pair formation may precede copulation by periods up to several weeks. Among such species, he finds that interspecific hybrids are very rare, while they may be rather common between closely related species which do not have such an "engagement" period.

275

He attributes the absence of hybrids among the former to the effect of differences in behavior pattern which may *break* the "engagement" of pairs which are not of the same species.

Spieth has made a detailed study of the sexual behavior of the six species of *Drosophila*. He found that courtship and mating can be divided into six phases, at any one of which incompatible behavior of the potential mates may break off the courtship. Although several hundred attempts were made to obtain interspecific crosses, only once did actual copulation result. In the majority of cases, courtship was stopped during the first stage. Yet the observable differences in the courtship patterns of the several species are rather minor, being concerned with tapping of the female by the male, postural reflexes, and similar movements. In other species, however, differences in courtship patterns may be more pronounced. In crabs of the genus *Uca*, species can be recognized at a distance by their motions during courtship. Birds which show only minor morphological differences may be easily recognized by their songs. The mating "dances" of salamanders and turtles may be very striking.

Scents, songs, and recognition marks perhaps all belong here. It is well-established that scent plays a major role in mating reactions of the Lepidoptera. If a female of a rare species is placed in a screen cage, many males will gather around it soon after the cage is placed outdoors. But if the female is instead exposed in a glass container, the males do not assemble. That these scents are highly specific can be shown by the selective response of males when two closely related species of female are put out in the same area. "Wrong" associations rarely occur. Petersen studied a case in which no less than 37 species of a single genus of moths live in a single valley without interbreeding. Visible differences between these species are minor, and Petersen believes that conspecific matings are guaranteed by the scents of the moths.

Songs of birds are well known for their role in mating, and many instances are known in which the sounds produced by insects also play the same role. In addition to the recognition of conspecific mates, these may serve to stimulate sexual activity. As pointed out above, the wing movements of the male of *Drosophila* serve to hasten the receptiveness of the female, but once excited, she will accept a *wingless* male. Mayr believes that the many phenomena formerly described as sexual selection may be properly understood simply as sex-stimulating mechanisms. This is very plausible, and if correct would immediately bring these phenomena back into accord with the general theory of natural selection.

Mechanical factors were once regarded as important isolating mechanisms, but it now appears that their importance was overrated, if not completely in error. Among insects, the morphology of the genitalia may be very complicated, and it frequently presents the best available taxonomic characters. On this basis, Dufour long ago proposed the "lock and key" theory, that there must be a very exact correspondence between the morphology of male and female parts to permit copulation. The female genitalia are thus compared to a lock which can be opened only by one key, namely the male genitalia of the same species. This theory is very suggestive, but unfortunately there is much more evidence against it than there is for it. On the positive side, a few interspecific crosses in moths have been described in which death resulted from the inability of the copulating pair to separate. Two snails of different species were observed trying to copulate over a period of several hours, but they failed, presumably because of mechanical difficulties. On the negative side, there is a great array of evidence. In some cases, extreme differences in the genitalia do not prevent successful copulation. Copulation between insects of strikingly different morphology has been observed. In many genera, the female genitalia may be identical throughout, while the males show differences of taxonomic importance. Even extreme size differences may be no bar to copulation: the Dachshund and the St. Bernard have been successfully crossed.

Character Displacement. An isolating mechanism which might be discussed as restriction of random mating or as reduction of fertility is *character displacement,* a concept recently developed by Brown and Wilson. They observed many cases in which closely related species are most strongly differentiated where they are sympatric, yet convergent where they are allopatric. Thus the ant *Lasius flavus* has a wide holarctic distribution, while *L. nearcticus* is confined to northeastern U.S. In this common area, the two species are quite different both morphologically and ecologically, yet elsewhere in its range, *flavus* is rather similar to *nearcticus.* Their interpretation is that, when recently separated species become sympatric, selection favors strong differentiation which will reduce the probability of wastage of gametes by matings of low fertility. Also, ecological differentiation reduces direct competition between the species, thus permitting a greater total population in a given region. Character displacement may influence any aspect of the biology of species, including fertility. Brown and Wilson have cited many examples from groups as diverse as birds and ants, fishes and crabs.

Reduction of Fertility. The final category of isolating mechanisms comprises those which act through a reduction of fertility. This is ordinarily subdivided into interspecific sterility, in which there is a failure to produce an F_1, and hybrid sterility, in which a good F_1 is produced, but this hybrid is sterile, and so no F_2 results. Yet the case is already overstated, for the phenomenon is not absolute: fertility may be reduced without producing absolute sterility, and this is very commonly the case, particularly among plants. The literature on interspecific sterility and hybrid sterility is immense, going clear back to Aristotle. While it cannot be summarized here, some of the more salient facts will be discussed below.

Interspecific sterility may be based upon the failure of the pollen or sperm to reach the ovule or egg, or upon the production of an inviable zygote. The first type is particularly well known in plants. It frequently happens that the growth of pollen tubes is slowed down in interspecific crosses, or that the pollen tube bursts, so that no fertilization is possible. If the species crossed are only remotely related, the pollen tubes may not grow at all. If they are more closely related, they are likely to grow more slowly than normal. As a result, if species A is pollinated both by A and by B pollen, practically all of the fertilizations will be conspecific ($A \times A$) rather than heterospecific ($A \times B$), because A pollen on A styles grow at a normal rate, while B pollen on A styles grow at a reduced rate. While this phenomenon is well known in interspecific crosses, it is not unknown in intraspecific crosses. Thus, *sugary* and *starchy* maize differ only by a single gene, yet the growth rate of *sugary* pollen tubes on *starchy* silks is slower than that of *starchy* pollen tubes. The two types of pollen tubes have equal growth rates on *sugary* silks.

The bursting of pollen tubes is particularly likely to occur in crosses in which the male parent has a chromosome number greater than that of the female parent. Wherever the chromosome numbers of the species are equal, the ratio of the chromosome number in the style to that in the pollen tube will be 2 : 1 (diploid to haploid). But when the ratio comes closer to 1 : 1, then bursting of pollen tubes commonly results. This is particularly likely to occur in crosses between polyploid species (see Chapter 19) and their diploid ancestral species. For example, it appears that the commercial tobacco, *Nicotiana tabacum*, with a chromosome number of 48, is derived from *N. sylvestris* and *N. tomentosa*, both of which have 24 chromosomes. The cross of either of the latter species to *tabacum* can be made easily, provided that *tabacum* is used as the female parent. This will give a ratio of stylar to pollen tube chromosomes of 4 : 1. But if the same cross is attempted using *tabacum* as the pollen parent, the ratio is 1 : 1, and the cross usually fails because of bursting of pollen tubes. It has been shown that, in some cases, the bursting of pollen tubes results from a greater osmotic pressure in the pollen tube than in the style. While this is correlated with chromosome number, it is clear that the higher osmotic pressure of the pollen tubes is a physiological effect of the genotype rather than a direct effect of the number of chromosomes, for chromosomes never exist in sufficient numbers to have important effects on osmotic pressure. Nothing strictly comparable is known in animals. However Serebrovsky has shown that pH and osmotic pressures of vaginal secretions of various domestic mammals have small but consistent differences. Interspecific inseminations generally result in death of the sperm, and this may be a result of osmotic incompatibility.

Once a hybrid zygote is formed, there is no assurance that it will reach maturity, for death may occur at any stage during development. Thus in interfamilial crosses of sea urchins, the paternal chromosomes are extruded, and the resulting haploid larva dies at an early stage. In the cross between *Datura stramonium* and *D. metel* (Jimson weeds), development proceeds to the eight-cell stage, then stops. In many plant hybrids, lethality appears to result from an inadequate nutritive relationship between the endosperm and the embryo, for embryos which would otherwise die may be brought to maturity if they are dissected out of the seed and raised on an artificial nutritive medium. The adult plants so obtained may be fully as vigorous as the parent species.

Little progress has been made in analyzing the genetic basis of interspecific sterility, and for this reason, the two cases discussed below have especial interest. Hollingshead, using the extensive materials of Babcock's laboratory, has made many interspecific crosses in the genus *Crepis*, a weed of cosmopolitan distribution. When *C. capillaris* is crossed to *C. tectorum*, the outcome depends upon the strain of the latter which is used. The F_1 from some strains is fully viable; from other strains, the F_1 includes fully viable plants and plants which die in the cotyledon stage in a ratio of 1 : 1; while from some strains all of the progeny die in the cotyledon stage. It appears, then, that *tectorum* has a gene which, in crosses to *capillaris*, behaves as a dominant lethal. Every effort to find a phenotypic effect of this gene in pure *tectorum* has failed, but the gene has also been found to behave as a lethal in crosses with several other species of *Crepis*. This gene evidently functions harmoniously with the *tectorum* genome, although in an unknown way. But, in combination with the *capillaris* genome, it is so disharmonious that lethality results.

A similar, and perhaps more thoroughly understood, situation has been discovered in tropical fishes by Bellamy. In the moonfish, *Xiphophorus maculatus*, there is a dominant, sex-linked gene N which causes an increase in the amount of black pigment, with NN being darker than Nn fishes. Only the recessive allele of this gene is found in the closely related swordtail, *X. helleri*. The two species can, however, be crossed, and the F_1 is also fertile. If NN moonfish are used for this cross, then the resulting hybrids will all be Nn, and will in all other respects combine one genome from each of the parent species. In these Nn hybrids, the amount of black pigment exceeds that of the NN moonfish parent. If the F_1 hybrids are now backcrossed to pure nn swordtails, the backcross generation should have approximately three-fourths of its chromosomes derived from swordtail, and some of these will also be Nn. In these, the pigmented tissue becomes tumorous, and so the gene may be regarded as a lethal. The gene N, then, produces a normal color variant in pure moonfish; in the hybrids, it produces a more extreme expression of the same character; but in a genotype closer to pure swordtail, this same gene N produces a more extreme effect, melanotic tumors, and has thus become an interspecific lethal. In other words, a gene which is beneficial, or at least harmless, in its proper genetic background may become lethal when placed in a different genetic situation. It is not known whether these data have any relationship to the fact that pigmented growths of mammals are sometimes precursors of cancer.

The problems of hybrid sterility are not fundamentally different from those of interspecific sterility. In fact, hybrid sterility could be regarded as a special form of interspecific sterility in which the defect is simply delayed for one generation. But this delay makes it possible to get somewhat more insight into the genetic and cytological mechanisms which are operative. When sterile hybrids are studied cytologically, much the most common anomaly found is that the chromosomes fail to synapse. The unsynapsed chromosomes will typically be distributed without division to the two daughter cells at one of the maturation divisions, with the result that the distribution of the chromosomes is random. All of the chromosomes might go to one pole and none to the other, or equal numbers might go to each pole, or any intermediate result might occur. The other division is usually equational. Most of the gametes so produced are inviable, but there is a possibility that the genomes from the parent species might be accurately separated by random division, with fertile gametes resulting. There are authenticated records of several mule mares which have produced offspring. And in plants, the inclusion of all of the chromosomes of the hybrid within a single gamete may also result in viable offspring, which, however, will be polyploid (see Chapter 19).

Yet the failure of the chromosomes to synapse cannot be the whole explanation of hybrid sterility, for the reproductive systems of species hybrids are often grossly abnormal. In hybrids between *Drosophila melanogaster* and *D. simulans*, the gonads are rudimentary, and meiotic divisions never begin. In other cases, the chromosomes actually do synapse, but the hybrid is sterile anyway. Dobzhansky has described a particularly instructive case in hybrids between *Drosophila pseudoobscura* and *D. persimilis*. If the proper strains of these two species are selected for hybridization, synapsis in the hybrid will be complete; but other strains give partial synapsis, or none at all. If failure of synapsis were the essential factor in hybrid sterility, then the first group of hybrids should be fertile, the second partially fertile, and the last highly sterile. But in point of fact the outcome is the same in every case: the first maturation division proceeds normally up to the metaphase, but the anaphase is grossly abnormal and results in a single, binucleate cell. The second division does not occur at all, and the giant spermatids degenerate. It is evident that hybrid sterility may result from a derangement at any point in the long and complicated series of processes which extend from the zygote to the mature gametes for the production of the next generation.

Whenever synapsis of the chromosomes fails completely, no insight into its causes can be obtained. But *partial* synapsis frequently occurs, and this may permit a study of the factors which prevent its completion. It frequently turns out that chromosomal rearrangements are present, and that these place purely mechanical obstacles in the way of completion of synapsis. Horton's study of synapsis in hybrids between *Drosophila melanogaster* and *D. simulans* is most illustrative. The meiotic chromosomes of this hybrid are not available for study because the gonads are rudimentary, as was mentioned above. However, there is every reason to believe that the salivary gland chromosomes, which are well developed in the hybrid, give an accurate picture of the synaptic behavior of the chromosomes of the gonads. Any rearrangements which the chromosomes of the two species show with respect to one another may be assumed to have arisen since their separation but there is no way of judging which species has the more primitive arrangement.

In general, the salivary gland chromosomes of this hybrid are well synapsed, but there are a considerable number of regions in which synapsis is irregular or lacking (Figure 103). In these regions, ten rearrangements have been identified with certainty. Six of these are inversions, five of them being quite small (two to twelve bands in length) and one rather large (involving nine sections of the third chromosome, and including

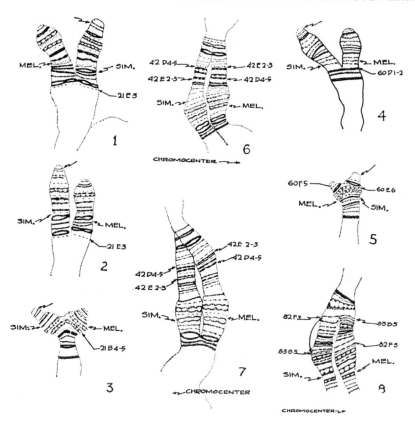

FIGURE 103. FAILURE OF SYNAPSIS BECAUSE OF CHROMOSOMAL REARRANGEMENTS IN A HYBRID BETWEEN *Drosophila melanogaster* AND *D. simulans*. In each figure, the species from which each strand is derived is marked. Numbers identify the bands with respect to the standard map of the salivary gland chromosomes of *D. melanogaster*. (From Horton, *Genetics*, V. 24, 1939.)

something on the order of 250 bands). The small inversions were too short to permit the formation of typical inversion loops, but they did suppress synapsis not only within their own length but also for a variable distance beyond the ends of the inverted sections. Occasionally, a pair of homologous bands did synapse in very short inversions, showing that identical orientation of the bands is not essential for synapsis. As Horton pointed out, this raises the possibility that single band inversions may be undetectable by any means now available, yet they could have a mutational effect. The other four identified rearrangements were at the ends of the chromosomes, and were more difficult to analyze; however, the evidence indicates that they were based upon a series of small translocations. In addition, there were fourteen regions in which suppression of synapsis was not associated with definitely identifiable rearrangements, but Horton believes that small, cytologically undetectable rearrangements are most probably responsible for the inhibition of synapsis in these regions. He concludes that the differentiation of *Drosophila melanogaster* and *D. simulans* from their common ancestor has involved as many as twenty-four chromosomal rearrangements.

CHROMOSOMAL REARRANGEMENTS AS ISOLATING MECHANISMS

This raises a very important question as to the extent to which such rearrangements, interfering as they do with the normal course of meiosis in the hybrids, may constitute genetic barriers between related species. It is evident, of course, that failure of synapsis cannot be imputed to extensive chromosomal rearrangements in every case, for simple lack of homology will also prevent synapsis in a hybrid. But closely related species must have a considerable amount of homology between their chromosomes if there be any validity to the modern genetic approach to problems in evolution. And there are many specific cases in which, like that described above, chromosomal rearrangements produce considerable disturbance of the mechanism of meiosis.

Examples from Drosophila. The fundamental requirement for genetic interspecific isolating mechanisms is that both of the homozygous types (pure species) should be fully fertile, while the hybrid is largely or entirely sterile. Generally speaking, chromosomal rearrangements studied in the laboratory do not meet this requirement: while the heterozygous types do show a reduced fertility, the homozygous rearrangements are commonly inviable in *Drosophila* (but often not in plants). Yet that this need not be so is proven by the fact that fully fertile species are known in nature, the chromosomes of which can be shown to be rearranged with respect to one another. In the example above, if *D. melanogaster* be regarded as retaining the ancestral chromosomal pattern (an arbitrary assumption), then *D. simulans* must be homozygous for more than twenty rearrangements, including certainly both inversions and translocations, and possibly duplications and deletions as well. It is evident then, that while the majority of rearrangements are deleterious, like the majority of gene mutations, nonetheless there is always a possibility that a particular rearrangement or *combination of rearrangements* may establish a new harmonious genetic system which can be maintained in the homozygous state and which is isolated from the parent type by the sterility of the hybrids. Species pairs such as *D. melanogaster* and *D. simulans* prove that this has actually happened in the formation of species now existing.

Perhaps the most thoroughly studied case in which inversions differentiate the chromosomes of species is that of the *Drosophila pseudoobscura* group, studied by Dobzhansky and his collaborators, as reported in Chapter 14. For the present, the major fact may simply be restated, that a series of overlapping inversions has made it possible to reconstruct the exact phylogeny of three species, *pseudoobscura*, *persimilis*, and *miranda*, together with the details of subspeciation in the first two species.

Oenothera and Translocation Complexes. An unusual situation is found in the genus *Oenothera,* the evening primroses. This plant came into prominence in the early experiments in genetics, for this was the subject of De Vries' studies during which he rediscovered Mendel's laws. Yet it soon became apparent that some aspects of the genetic behavior of *Oenothera* were very anomalous. A single species, *O. hookeri,* which is native to the Pacific coast of North America, behaves like a typical plant genetically. It has large flowers and is normally cross-fertilized. The remainder are found east of the Rocky Mountains. They are difficult to treat taxonomically, and there is no agreement on the number of species, lumpers giving as few as one at one extreme, and splitters as high as one hundred at the other extreme. All are characterized by small flowers and self-fertilization, and all show unusual genetic behavior.

The distinctive genetic features of most *Oenothera* species are the following four. First, they show only 50 per cent fertility as compared to *O. hookeri.* This is a result of the formation of defective seed rather than of a small seed set. Second, when crossed to *O. hookeri,* "twin" hybrids are formed; that is, there are two classes of F_1 plants which differ in a considerable number of traits. Third, in spite of the high degree of heterozygosity which is demonstrated by the formation of the twin hybrids, the plants breed true when selfed (their normal method of reproduction). Finally, crossing over rarely enters into the results of *Oenothera* crosses, but, when it does, it always involves large blocks of characters. The results of crossing over in *Oenothera* are so striking that cross-over products were originally interpreted as large mutations.

All of these characteristics are understandable in terms of a series of translocations in the various species of *Oenothera.* But first, a more thorough discussion of the behavior of heterozygous translocations may be helpful. Let us assume that two pairs of chromosomes are named, respectively, a·a' and b·b', each letter identifying a chromosome end (Figure 104). At synapsis, then, normal pair formation occurs, and each pair behaves independently of the other. But now let us suppose that a translocation occurs, so that only one normal chromosome of each pair remains, and the other two now have the constitutions a·b' and b·a'. Homologous point still synapses with homologous point, so that the translocated chromosomes will bind the two tetrads together, forming a cross in the pachytene (a particularly clear phase immediately after completion of synapsis), and a ring of four chromosomes on the metaplase plate. In the division of such a ring, alternate members typically go to the same pole in the anaphasic movement, with the result that half of the gametes formed get both of the normal chromosomes and half get both of the translocated chromosomes.

But there is no reason why only two pairs of chromosomes should be involved in a translocation complex. If three chromosomes, designated as a·a', b·b', and c·c' are translocated so as to yield, in addition to the normal chromosomes, strands of the constitution b·a', c·b', and a·c', then a ring of six would be formed, as illustrated in Figure 104. Again, alternate disjunction occurs at anaphase, so that half of the gametes have only un-

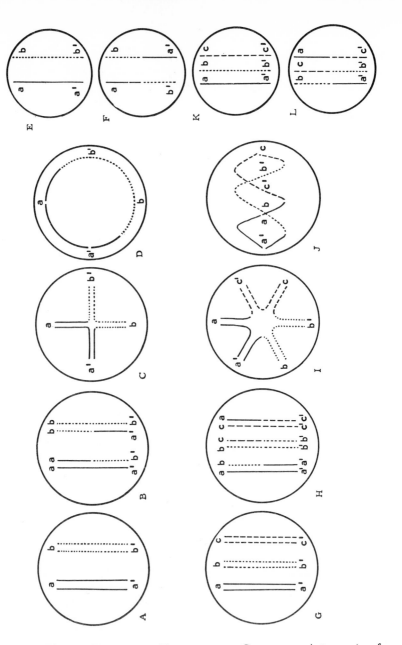

FIGURE 104. MEIOTIC BEHAVIOR OF TRANSLOCATION COMPLEXES. *A,* two pairs of untranslocated chromosomes. *B,* a heterozygous translocation between the two pairs of chromosomes. *C,* Pachytene configuration of the translocation heterozygote. *D,* Metaphase ring, with translocated chromosomes alternating with untranslocated chromosomes. *E* and *F,* the two types of gametes formed by alternate disjunction. *G,* three pairs of untranslocated chromosomes. *H,* a triple translocation. *I,* pachytene in the complex translocation heterozygote. Note that the central segment of each chromosome is not synapsed. *J,* early anaphase, showing alternate disjunction of the chromosomes. *K* and *L,* the two types of gametes formed by alternate disjunction.

translocated chromosomes, while the other half have only translocated chromosomes. Theoretically, the limit to the number of chromosomes which may be joined together in such a translocation complex is set only by the total number of chromosomes which a plant possesses, and this limit is actually reached in some species of *Oenothera*. *O. hookeri* has seven pairs of chromosomes, all of which behave independently. The other species all show translocation ring formation of some degree, and in some, all fourteen chromosomes form a single large ring at the metaphase of the first meiotic division.

The genetic peculiarities referred to above all follow from the behavior of translocation heterozygotes if one more special feature be added. Because of the alternate distribution of chromosomes of these rings, only two types of gamete are formed. This should permit the formation of three types of zygote, the two homozygous types, and the heterozygous type. But only the heterozygous type ever appears in the progeny of selfed plants. This is because each chromosome complex (a complex being a group of chromosomes inherited as a unit because of alternate disjunction) contains a recessive lethal gene, but it is a different gene in each complex. The result is that neither complex can become homozygous.

Thus *O. lamarckiana*, a species studied by De Vries, is a heterozygous, self-fertilized species. But self-fertilization yields only parental type plants, and only 50 per cent of the seed is viable. When crossed to *O. hookeri*, twin hybrids are formed, and this makes possible the demonstration of genetic differences between the two chromosome complexes of *lamarckiana*. The two complexes present in this species are called *gaudens* and *velans*. *Gaudens* carries the lethal gene l_1 while *velans* carries l_2. Each carries the normal allele of the lethal carried by the other. As a result, $g \cdot g$ and $v \cdot v$ zygotes die, while $g \cdot v$ zygotes survive. At synapsis, the chromosomes of *O. lamarckiana* form a ring of twelve chromosomes and a pair. This does not mean, however, that all of the chromosomes of one set are normal, and six of the chromosomes of the other as translocated. If *O. hookeri* be regarded as the standard, then translocated chromosomes are present in both the *gaudens* and *velans* complexes. If the *hookeri* chromosomes be named a·a′, b·b′, c·c′, d·d′, e·e′, f·f′, and g·g′, then the chromosomes of other *Oenothera* species can be described in terms of translocations of this standard set. The *velans* complex has the constitution a·a′, b·b′, c·d′, d·c′, e·e′, f·f′, and g·g′. Available data will permit two possible interpretations of the structure of the *gaudens* complex, as follows:

a·a′, b·f, c·c′, d·f′, e·b′, d′·g′, g·e′   or
a·a′, b·f, c·c′, d·g, e·g′, e′·b′, f′·d′.

Further data are necessary in order to determine which of these structures is correct.

The 50 per cent fertility which characterizes most *Oenothera* species is, then, a result of the balanced lethals which the several chromosome complexes carry. The twin hybrids are formed when any species is crossed to *O. hookeri* because each complex of the heterozygous parent (*hookeri* is largely homozygous) is radically different in genetic content from the

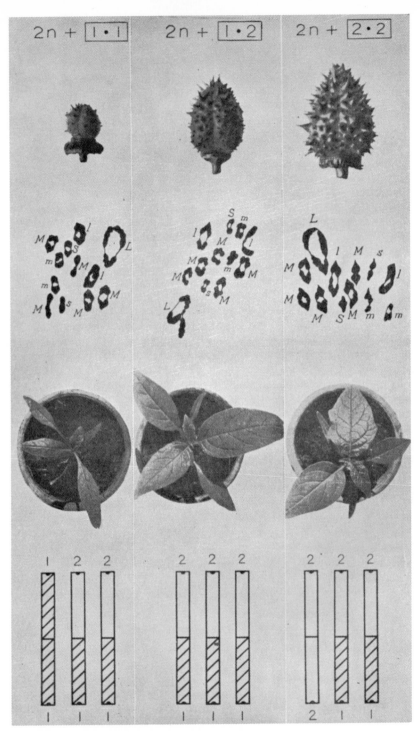

FIGURE 105. See opposite page for legend.

other. In other words, most species of *Oenothera* are highly heterozygous. Hence two types of contrasting offspring are formed when any other species of the genus is crossed to *hookeri*. Yet, in spite of their heterozygosity, the plants breed true when selfed. Independent assortment is prevented by the alternate distribution of the chromosomes of the rings, which results in all of the translocated chromosomes being segregated into one class of gametes, while all of the untranslocated chromosomes are segregated into another class of gametes. Only parental type zygotes form the progeny, because the homozygous ones are eliminated by the balanced lethals.

The problem of the absence of cross-over effects in *Oenothera* crosses is somewhat more complex. An inspection of the diagram of the pachytene chromosomes of translocation complexes involving several pairs of chromosomes will reveal that each chromosome consists of two ends which are perfectly paired and a central segment which is mechanically held out of contact with its homologue. It is evident that crossing over should occur normally in the pairing arms of such chromosomes. But the pairing arms appear to be largely homozygous throughout the genus, with the result that crossing over does not have genetic effects. The central segments, however, are protected from crossing over because homologous parts are prevented from establishing synaptic contact. The genetic differences between the various chromosome complexes in the genus are concentrated in these segments, and so the sets of genes characterizing the various complexes are highly permanent. Rarely, crossing over does occur in the central segments, and when it does, radically different character combinations occur. These are the "half mutants" (apparently because they affect half of the progeny) of early genetic literature. Thus the distinctive genetic features of the genus *Oenothera* are simple results of the fact that these plants are permanent translocation heterozygotes.

Chromosomal Mutants of Datura. In the genus *Datura*, the Jimson weeds, Blakeslee and his collaborators have found translocation phenomena which are, in some respects, even more complicated than those found in *Oenothera*. The normal chromosome number is 24, so that there should be 12 pairs of chromosomes on the metaphase plate of the first maturation division. This is generally the case, but in interracial and interspecific crosses, rings of four and of six commonly are found. This shows that the various species, and the races within a single species, differ by a few translocations for which the several races and species are themselves homozygous. Each translocation complex appears to be associated with a different phenotype.

Trisomics are also common in this genus, that is, one chromosome of the set may be present in triplicate, making a total of 25 chromosomes

FIGURE 105. TRISOMICS IN *Datura*. The center column shows seed capsule, meiotic chromosomes, and plant for the primary trisomic 2n + 1·2, called "Rolled." The left hand column shows one of its secondary trisomics, 2n + 1·1, or "Polycarpic." The right hand column shows its other secondary, 2n + 2·2, or "Sugarloaf." (From Blakeslee, A. F., *J. Hered.*, V. 25, 1934.)

(2n + 1) rather than the usual 24 in the zygote. Any one of the twelve chromosome pairs may be so affected, and the phenotype of the plant depends upon the particular chromosome which is present in triplicate. Each is called a "prime type." Although any trisomic causes phenotypic changes throughout the plant, they have been named on the basis of changes in morphology of the seed capsules. At meiosis, the extra chromosome should be distributed to half of the gametes, but it tends to lag behind the other chromosomes and be destroyed in the cytoplasm, and so less than half of the gametes formed by a trisomic plant will transmit the trisomic condition. As the n + 1 condition is a pollen lethal, it can be transmitted only by the ovule parent. Such trisomics, in which a completely normal chromosome is present in triplicate, are called *primary* trisomics. All of the twelve possible primary trisomics in *Datura* are known.

But *secondary* trisomics are also known, in which the extra chromosome represents only one half of a normal chromosome, the half, however, being duplicated (Figure 105). These again are associated each with a distinctive phenotype. Twenty-four secondary trisomics are possible, but not all of these have been found. Finally, there are *tertiary* trisomics, in which the extra chromosome is a translocation product, being made up of halves of two different chromosomes. These again are characterized each by a distinctive phenotype. At meiosis, the extra chromosome binds together the two tetrads with which it shares homology, thus making a ring of *five* on the metaphase plate.

Evaluation of the Data. It is thus clear that in some well demonstrated instances homozygous inversions and translocations are among the characters which differentiate races and species. Less detailed evidence is available with respect to duplications and deficiencies (Chapter 14), yet these rearrangements are known to be involved in the differentiation of *Sciara* species, and differences in the metaphase chromosomes of many insects for which the salivary gland chromosome technique is not available are most easily understood in terms of duplications and deficiencies.

This brings us back to the question with which this discussion began: to what extent do chromosomal rearrangements function as genetic isolating mechanisms? And are they also responsible for the phenotype differentiation of related species, or is this a matter of independently accumulated gene mutations?

With respect to the latter question, all geneticists agree that position effects are a common result of chromosomal rearrangements. But many rearrangements have been investigated without any evidence of corresponding position effects coming to light, particularly in maize, perhaps the most thoroughly known of all plants genetically and cytologically. Hence the majority of geneticists doubt that chromosomal rearrangements play a major role in the phenotypic differentiation of species. Singleton has found it useful to disregard the distinction between chromosomal and gene mutations, while Goldschmidt regarded the former as the basis for systemic mutations of fundamental importance for speciation.

With regard to the first question, there is less difference of opinion. There seems to be no room for doubt that chromosomal rearrangements

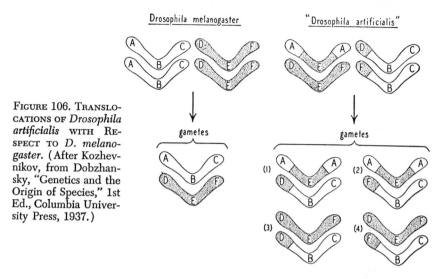

FIGURE 106. TRANSLOCATIONS OF *Drosophila artificialis* WITH RESPECT TO *D. melanogaster.* (After Kozhevnikov, from Dobzhansky, "Genetics and the Origin of Species," 1st Ed., Columbia University Press, 1937.)

do serve an important function in isolating related populations one from another. Kozhevnikov performed a particularly interesting series of experiments on this problem. By combining in a single stock of *Drosophila melanogaster* two. translocations between the second and third chromosome, he has obtained a strain which he regards as a synthetic species, *Drosophila artificialis* (Figure 106). *D. artificialis* forms four types of gametes, which combine to form sixteen types of zygotes. However, only four of these survive, as the other twelve contain large deficiencies and duplications. It is completely isolated from the parent species, *D. melanogaster*, for all zygotes formed by the "interspecific" cross are inviable, because of the large deficiencies and duplications. But a "species" with only 25 per cent viability would not be likely to fare well in nature. If, however, an additional rearrangement should stabilize this genotype, so that it would become viable in the homozygous condition, a good species would for all practical purposes have been synthesized in the laboratory, by means of a fortunate combination of a few chromosomal rearrangements.

While much progress has been made in identifying and classifying isolating mechanisms, and while they have been successfully fitted into the neo-Darwinian theory as a fundamental part of the mechanism of evolution, yet very little progress has been made in analyzing the genetic basis of isolating mechanisms. The chromosomal rearrangements are an outstanding exception to this statement even though their exact role in evolution is not clear. Yet everyone can agree with Dobzhansky that "... there can be little doubt that chromosomal changes are one of the mainsprings of evolution." [*] There is, however, a very general feeling among geneticists that isolating mechanisms and phenotypic differentia-

---

[*] *Op. cit.*

tion are both generally based upon quantitative characters which are influenced in a small degree by each of many different pairs of genes.

It is interesting to note that, if Goldschmidt was right in his opinion that systemic mutation, in contrast to gradual accumulation of small mutations, is the basis for speciation, then only genetic isolation is important for speciation. For other types of isolating mechanism, while they would then be important for the biology of a species, could not affect its evolution above the subspecific level. The decisive chromosomal changes, which at once would be the basis both for genetic isolation and for phenotypic differentiation, could arise as well in a continuous population as in a population which was already partially isolated. And even with very long isolation by other means, the decisive chromosomal isolation might fail to occur.

FAILURE OF ISOLATING MECHANISMS

Regardless of the manner in which one evaluates the role of isolating mechanisms, one of their important properties is their occasional failure. Subspecies and species which are ordinarily separated by one or several isolating mechanisms may occasionally produce hybrids. And these may be most instructive. If the barrier is geographical, it is essential that this be spanned before hybridization can occur. This may happen because of natural events or through the agency of man. In the first class are the many cases in which populations were separated during the glacial ages, during which time they diverged, perhaps to a specific degree, perhaps only to a subspecific degree. With the recession of the glaciers, the diverging populations have again migrated into their former range, and now hybridization may occur. Mayr has assembled many such cases from the fauna—particularly the avifauna—of central Europe. During the Pleistocene glaciation, Scandinavian and Alpine ice caps approached within about 300 miles of each other in central Europe, with the result that the temperate flora and fauna which had formerly inhabited this zone were forced to take refuge either in southern France and Spain, or in the Balkans. Thus segments of formerly continuous populations were isolated from each other at opposite ends of the Mediterranean Sea. While thus isolated, the two populations of the various species diverged. But, with the recession of the ice, the eastern and western populations again moved into their original territory.

The behavior of these once isolated but now sympatric populations is very different in different cases. In some, no interbreeding occurs, indicating that the divergent populations have reached the status of distinct species. In others, such as the hedgehogs, *Erinaceus europaeus* (western) and *E. roumanicus*, hybrids are rare, but they do occur. These are treated as good species, but their status may be debatable. In still other cases, for example the crows *Corvus corone* and *C. cornix*, there is a rather stable hybrid population where the eastern and western groups meet. These are now treated as a Rassenkreis, under the former name.

In the case of the several species of plane trees (*Platanus*) the great geographic barriers have been spanned by the intervention of man. This genus was once widely distributed throughout the Holarctic Region. Sometime during the Tertiary period, however, its distribution became quite discontinuous. The submersion of the Bering Straits area separated the Old World and New World populations. In the Old World, the elevation of the great mountain ranges restricted its range to Asia Minor and the eastern Mediterranean region, where it has been described under the name *P. orientalis*. Meanwhile, the elevation of the western mountain ranges and formation of deserts caused the plane trees in the United States to break up into three discontinuous populations. One of these, which inhabits most of the United States east of the Rocky Mountains, is called *P. occidentalis*. The southwestern representative is called *P. Wrightii*, while the California representative is called *P. racemosa*. There is also a Mexican species. In western Europe, the London plane tree, *P. acerifolia*, is a common cultivated shade tree. It is generally regarded as hybrid between *P. occidentalis* and *P. orientalis*, although these species cannot survive in western Europe today. All of the *Platanus* species are fully interfertile when crossed artificially. Whether ecological or other barriers would limit the crossability of these "species" if the hybrids were not cultivated is problematical. The possibility that hybridization might lead to the formation of new species has often been discussed. For those who are able to accept *P. acerifolia* as a good species, this is a clear-cut example. But the least that can be said is that the *Platanus* species are less distinct than are typical good species, and very little support for the idea of speciation via hybridization is available elsewhere. An important exception, allopolyploid plants, will be discussed in the next chapter.

Introgressive Hybridization. A more important result of the hybridization of species is what Anderson has called "introgressive hybridization." This is a rather imposing name for a very simple phenomenon. If a natural hybrid is formed, it is very probable that it will be mated not to another hybrid but to one of the pure parental species. As a result of this backcross, some of the genes of each parental species will "introgress" into the genotype of the other. Heiser's study of the sunflowers, *Helianthus annuus* and *H. Bolanderi*, exemplifies this phenomenon well. *H. Bolanderi* is restricted to the west coast of the United States. *H. annuus* appears to be originally an eastern species, but it has been introduced into the coastal states by man, and has become well established and widely distributed. There is some ecological separation of the two species, but they are found together in areas disturbed by man, so that habitats intermediate between those usually occupied by these species result. In such areas, several natural hybrids were found, and one large "hybrid swarm" was found along a roadside. A hybrid swarm is a population in which F_1, and F_2, and later generations of hybrid segregation are intermingled with backcross progeny of various degrees. Naturally, such a population shows extreme variability. The hybrids of recognizable degree all showed some reduction in fertility, sometimes a very drastic reduction, with only about 3 per cent of the gametes being viable. On backcrossing to either pure species, the fertility increases. Thus genes from each of these species can be transferred to the other. Heiser has found that each of these species of sunflower varies *in the direction of the other*. While this could be accounted for by parallel mutation in the two species, he believes that introgression is a much more likely explanation.

It was noted above that hybridization of these species is particularly likely to occur in areas disturbed by man. Anderson, who has made extensive studies on introgression in plants, believes that this may be a necessary requirement for the survival of hybrid swarms. The reason is that the hybrids are likely to require habitats intermediate between those to which the parent species are adapted. He therefore speaks of the "hybridization of the habitat." As man disturbs nature with his manifold activities, such "hybrid habitats" are likely to be formed, with the result that he facilitates the exchange of genetic variability between related species, and hence increases the range of types from which selection can choose the most advantageous.

Introgressive hybridization has not been widely investigated among animals, yet Hubbs has stated that his studies on fishes indicate that this process is also active in the evolution of the fishes, and examples have also been published for birds and other groups.

REFERENCES

ANDERSON, E., 1949. "Introgressive Hybridization," John Wiley & Sons, Inc., New York, N.Y. Still the major work on this subject.

DOPZHANSKY, T., 1951. "Genetics and the Origin of Species," 3rd Ed., Columbia University Press, New York, N.Y. (Dice and Blossom, Spieth.)

GOLDSCHMIDT, R. B., 1940. "The Material Basis of Evolution," Yale University Press, New Haven, Conn. (Turel.)

MAYR, E., 1942. "Systematics and the Origin of Species," Columbia University Press, New York, N.Y.

(The above three books, all of which have been introduced in connection with preceding chapters, all make important contributions to the subject of isolation.)

CHAPTER NINETEEN

Polyploidy

ABOUT FORTY YEARS AGO, Winge published a study of the numbers of chromosomes in plants. Although diploid numbers ranging from 4 to well over 200 were noted, the frequencies of these numbers were by no means random. Twelve was the most frequent number, while 8 was the second most frequent. About 50 per cent of all plants have numbers below 12. Among those plants with higher chromosome numbers, the most frequent numbers are multiples of the lower ones. Within a single genus, it commonly happens that there is a series of species in which the chromosome numbers of some are multiples of that of another species. For example, there are species of wheat with 14, 28, and 42 chromosomes. Seven chromosomes appears to be the basic haploid number in this genus. If the haploid numbers of many plants be plotted on a frequency histogram, it turns out that none of the maxima fall on prime numbers.

POLYPLOIDY, A MAJOR PHENOMENON IN PLANT EVOLUTION

Winge concluded that the most probable explanation of these facts lay in the assumption that over half of the higher plants were *polyploids,* that is, their haploid chromosome set consisted of two or more basic sets of chromosomes existing side by side in the same nucleus. This could happen in either of two ways. Either a single haploid set of chromosomes might be present more than twice (autopolyploid), or two different sets of chromosomes might be present, making a total of more than two genomes (allopolyploidy). Winge ventured the guess that the latter type would prove to be the more frequent, basing his opinion on the assumptions that lack of homology would prevent pairing of the chromosomes in species hybrids, and that the necessity of pairing would therefore stimulate doubling of the whole chromosome complement in such hybrids. While his reason may be doubted, this conclusion, and the rest of Winge's conclusions, have been substantiated since his original publication. The development of polyploid series seems to have been one of the major phenomena of plant evolution, and one of the most thoroughly understood.

Colchicine Induction of Polyploids. The study of polyploidy has been greatly facilitated by development of experimental techniques for its artificial production. Many methods of moderate efficiency have been introduced. These include selection of bud variants with typical polyploid characters; treatment of seed with temperature shocks; radiation of seed; and the cutting off the stem of a plant, then selecting tetraploids from among the callus shoots (shoots which develop just below the wound). This method gives a yield of about 15 per cent tetraploids in the tomato. In other plants, treatment of the wound with heteroauxin may be necessary in order to get good results. But all of these methods have been made obsolete by the colchicine method. Colchicine is an alkaloid drug derived from the root of *Colchicum autumnale,* the autumn crocus. It has been known for many years that this drug interferes with the metabolism of nucleic acid (a major constituent of the chromosomes, as well as of some other parts of the cell), but it was only in 1937 that it became known that this drug is a mitotic poison. The prophase of a colchicine-influenced mitosis is apparently normal, and the doubling of the chromosomes proceeds as usual. However, the spindle is either very defective or absent entirely, and so the chromosomes, which have already been duplicated, are all included in a single restitution nucleus. Thus a tetraploid condition is established. Colchicine is usually used either in aqueous solution or in an ointment of lanolin. In either case, a concentration of about 0.4 per cent is adequate. The solution may be sprayed on flowers, or seed may be soaked in it. Seedlings may be soaked, or a cut stem may be smeared with the ointment, so that callus shoots will be tetraploid. Yields of 50 to 100 per cent tetraploid plants are not uncommon when colchicine is used.

The Gigas Habitus. Much the most common polyploids are tetraploids, with four haploid genomes (4n) in the somatic chromosome complement. The gametes are therefore 2n. Autotetraploids are known both in nature and in experimental materials. As a matter of fact, one of the bases of the mutation theory of De Vries was a mutant strain of *Oenothera lamarckiana* (Figure 107) which has turned out to be a spontaneously produced tetraploid, as it has 28 chromosomes instead of the usual 14. This plant illustrates well a complex of characteristics which are generally, though by no means universally, found in tetraploid plants. At the outset, it is considerably larger than diploid *lamarckiana.* Because of this, De Vries named it *O. gigas,* regarding it as a new species. The stems are thicker, and the leaves are shorter, broader, and thicker than those of the diploid plants. The most obvious physiological difference is a slower growth rate, but increased vitamin content has been reported for tetraploid tomatoes, and it seems probable that the physiology of the tetraploid plant is modified as much as its morphology. These traits collectively are referred to as the "gigas habitus," because they so commonly characterize tetraploids. Yet there are exceptions to all of them. There are tetraploids, for example, which are dwarfed as compared to the diploid.

It may be noted that the tetraploid *Oenothera lamarckiana* could be regarded either as an autopolyploid or as an allopolyploid. Because the *gigas* variety is derived from a single diploid parent species, it could be regarded as an autotetraploid. But because the diploid *lamarckiana* itself is a permanent structural hybrid, the *gigas* variety could be regarded as an allo-

FIGURE 107. *Oenothera lamarckiana*, DWARF, NORMAL, AND GIGAS FORMS. (By permission, from "Principles of Genetics," 4th Ed., by Sinnott, Dunn, and Dobzhansky. Copyright, 1950. McGraw-Hill Book Co., Inc.)

tetraploid. Generally, no such confusion exists, yet the distinction between these two types of polyploidy is perhaps always one of degree rather than an absolute distinction. By definition, the several haploid sets which comprise the chromosome complement of an autopolyploid do not differ from one another in any greater degree than do the two haploid sets of the corresponding diploid. But the first requirement for the formation of an allopolyploid is that the two parent species must be able to form a viable (though not necessarily fertile) hybrid. As discussed in the preceding chapter, this becomes less and less probable as the relationship of the potential parent species becomes more remote. The most usual situation is that both parental species belong to the same genus, although many cases are known in which allopolyploids have been formed between different genera of a single family. Most usually, the relationship is close enough so there can be no doubt that there is a significant degree of homology between the chromosomes of the parental species. Frequently, this may be evident in terms of a limited amount of synapsis in the F_1 hybrid. Thus it seems probable that some degree of homology between the several genomes of a polyploid is necessary for its formation. In autopolyploids, this homology is substantially complete, while in allopolyploids it is markedly incomplete.

AUTOTETRAPLOIDY IN EVOLUTION

By the use of colchicine, a large number of experimental autotetraploids has been formed and investigated, but autotetraploids are also known from nature. Müntzing reviewed fifty-eight well authenticated examples in 1936 and expressed the opinion that a list of over one hundred cases could have been compiled. The study of polyploidy in all of its phases was greatly stimulated by the introduction of the colchicine technique, and there can be no doubt that the list would be much longer if a comprehensive review were to be published now. Autopolyploidy does not seem to lead to the formation of new species, but only to well-marked varieties. Yet there is considerable reproductive isolation between a diploid and its autotetraploid, because the hybrid between them is a triploid (three haploid genomes in each somatic cell). Triploids are highly sterile because the distribution of the chromosomes at meiosis is irregular. A few of Müntzing's cases may be discussed below.

Phleum alpinum (a near relative of Timothy) is a grass which has a diploid race with 14 chromosomes and a tetraploid race with 28 chromosomes. The two are morphologically distinct, and are partially separated geographically. The diploid race is found only in Scotland, while the tetraploid race is found both in Scotland and in northern Scandinavia. It is quite characteristic that tetraploids range more widely than their diploid ancestors, frequently invading territories which are geologically more recent. Crosses between the two races are successful, but the F_1 is sterile, being triploid. The case of *Nasturtium officinale* is similar. Races with 32, 48, and 64 chromosomes are known, and are respectively diploid, triploid, and tetraploid. As in the case of *Phleum*, the polyploid races have a more northerly distribution than does the diploid race. Also, the polyploids are perennial, while the diploid is annual. This is also a very common, though by no means universal, difference between the chromosomal races. Crossing of the tetraploid and diploid races is more difficult than in the case of *Phleum*. Almost all of the seed obtained from this cross was shrunken and inviable.

A particularly instructive example is that of the diploid and tetraploid races of *Tradescantia caniculata,* the spiderwort, in southern United States. The chromosome numbers are 12 and 24. This species is widely distributed over the great plains from the Rocky Mountains eastward to the Mississippi River. Over most of this area the plants are tetraploid. But in a small area in northern Texas, the diploid race is found. This area geologically is the oldest part of the total range. It appears that only the tetraploid race has been able to invade those territories more recently opened up to floral colonization. This greater aggressiveness of polyploid plants is quite characteristic. Another species of the same genus, *T. occidentalis*, overlaps the eastern part of the range of *T. caniculata*. Interestingly enough, this species, which extends eastward, is also tetraploid over most of its range, but the diploid form is found in the same refuge in northern Texas. In both of these species, the tetraploid races are much more vigorous than are the diploid races. Crosses between the races can be made, but the triploid offspring are sterile.

In most of the plants studied by Müntzing, polyploidy resulted in well-marked races, but not in speciation. However, he studied sixteen pairs of chromosomal races in which taxonomists had either assigned the different chromosomal races to different species, or had debated the advisability of doing so. These occur in such well known genera as the grasses *Phleum* and *Festuca,* and the garden flowers *Viola, Dianthus,* and *Chrysanthemum.* The differences between these pairs of chromosomal races or species are qualitatively similar to those discussed above, but quantitatively they are greater. Structural differences include some or all of the characters of the *gigas* complex. They are usually separated geographically and ecologically. For example, *Chrysanthemum shimotomaii* is a seashore plant, while *C. indicum* inhabits mountains and inland fields. In crosses, they are highly sterile.

Experimental autotetraploids have been thoroughly studied genetically and cytologically. Although the plants are frequently more hardy than their diploid relatives, they have not been outstandingly successful, because they show markedly reduced fertility, and because they tend to revert to diploidy. The cause of these characteristics is revealed by a study of meiosis in the autotetraploids. For normal meiosis, the chromosomes should form only pair-wise associations. But *four* homologues of each kind are present. Frequently, only pairs are formed, and normal gametes result. But sometimes three chromosomes of a kind synapse, with the fourth one behaving independently. In such a case, two of the three synapsed chromosomes pass to one pole and one to the opposite pole. The independent chromosome may now balance this situation, or it may make it even more unbalanced. Again, all four chromosomes might synapse to form a tetravalent. Ordinarily, a tetravalent divides so that two members go to each pole, yielding normal gametes. But tetravalents also give a 3 and 1 distribution. Now if only one or a few chromosomes are missing or are in excess in an otherwise tetraploid zygote, it may be successful. Such trisomic and monosomic strains are well known to plant breeders. But if more than a small portion of the chromosome pairs are so unbalanced, lethality results. This is believed to be the cause of the reduced fertility of the autotetraploids. Reversion to the diploid condition is probably based upon development of unfertilized ovules.

Now the study of meiosis in naturally occurring autotetraploids reveals abnormalities comparable in kind and in degree with those of the experimental autotetraploids. In the face of such a disadvantage, one may well ask how it is that these have ever become established in nature, let alone become more widespread than the diploid parent, as is so commonly the case. However, the tetraploids are commonly more vigorous, and adapted to more severe environments, and it seems probable that their selective value more than compensates for their reproductive liability. Yet most of the naturally occurring polyploids which have been analyzed have been of the allopolyploid type, and it may well be that it is this reproductive liability which has restricted the role of autopolyploidy.

ALLOTETRAPLOIDY IN EXPERIMENT AND IN NATURE

Allotetraploidy has also been produced experimentally. Many methods are available, of which much the simplest is the treatment of the F_1 hybrids between two species with colchicine. Other methods, however, give more insight into the means by which naturally occurring allotetraploids may be formed. One method is to cross two different autotetraploids. For example, if A and B each represent different haploid chromosome sets, then $AAAA$ and $BBBB$ would be the corresponding autotetraploids. With normal reduction, these will form gametes with the formulae AA and BB. Upon cross fertilization, the allotetraploid, $AABB$, will be formed. Because this actually comprises two different diploid groups existing side by side in the same nucleus, the terms *amphidiploid* and *double diploid* are often used as synonyms for allotetraploid. But, because allotetraploids are more common in nature than are autotetraploids, and because it is generally more difficult to cross two tetraploids than to cross the corresponding diploids, it seems unlikely that this method has had general importance in nature.

Two methods are based upon the occasional failure of reduction divisions, which is especially frequent in plants with chromosome complements which do not synapse readily. Thus, in the cross $AA \times BB$ (using the terminology introduced above), the F_1 should be AB. But if there is insufficient homology between the chromosomes of the A and B genomes to permit synapsis, the probability of nonreduction becomes considerable. Thus a significant percentage of AB gametes may be produced. In a self-fertilized plant, fertilization of some AB ovules by AB pollen would be quite probable, thus producing the allotetraploid, $AABB$, at once. Each chromosome now has a homologous mate, and, since there is little tendency of A chromosomes to synapse with B chromosomes, there is no tendency to form complex synaptic associations. The meiotic divisions therefore proceed perfectly normally, with gametes of constitution AB resulting, and with no impairment of fertility.

Another method of production of allotetraploids involves a two-step utilization of nonreduction. If the hybrid AB is cross-fertilized, it is especially likely to be backcrossed to one of the parental species, let us say AA. If AB has undergone nonreduction, the backcross progeny will then be AAB. These plants may again undergo nonreduction, producing gametes of formula AAB. If backcrossed to parent BB, the resulting progeny would again be the allotetraploid, AABB. The allotetraploid might also be formed by doubling of the chromosomes in the zygote of the original hybrid, AB, in a fashion comparable to that produced by treatment with colchicine. These allotetraploids show characters of both parental species, together with some typical tetraploid characters, in new and distinctive combinations. They breed true, and are effectively isolated from the parent species by the sterility of the hybrid between the allotetraploid and either of the parental species. This sterility results from the behavior of the chromosomes in meiosis. In the hybrid between AABB and AA, the chromosomal formula will be AAB. The chromosomes of the two A genomes will synapse normally and will be distributed to the daughter cells normally. But the chromosomes of the B genome have no synaptic mates, and so they are distributed at random. They might all go to one pole, with the result that gametes of formulas A and AB would be produced in equal numbers. Or they might pass in equal numbers (but not equal genetic endowment) to each pole, or any intermediate result might occur. All of these latter combinations would be inviable, and, since they comprise most of the gametes of an AAB plant, these triploids are highly sterile. Some viable gametes would also be produced by nonreduction. Thus allotetraploids may properly be regarded as good species, produced in one or a few steps. It has even been suggested that new genera, families, and higher categories might be produced in this way. But available cases indicate that, while species and genera are formed through allopolyploidy, higher groups are not. Stebbins believes that this is inherent in the fact that no really new genetic material (mutations) is involved, but only new combinations of old and related genomes.

The above discussion has been entirely in terms of tetraploidy. While this is both the simplest and the most common type of polyploidy, many higher degrees are known both in nature and in experimentally produced plants. But no new principles are involved in the higher polyploids, and so all degrees will be discussed without distinction. Any polyploid having an even number of genomes in the somatic cells (tetraploid, hexaploid, octaploid, decaploid, etc.) should be fully fertile, while those with an odd number of genomes (triploid, pentaploid, septaploid, etc.) should be highly sterile because of abnormal meiosis, and so these can survive in nature only if they have efficient means of asexual reproduction.

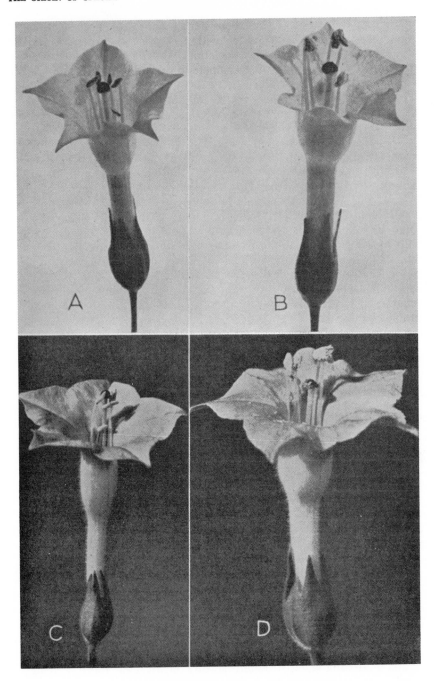

FIGURE 108. See opposite page for legend.

Species Synthesis. A very interesting allotetraploid was produced by Karpechenko as long ago as 1924 while experimenting on crosses between the radish, *Raphanus sativus*, and the cabbage, *Brassica oleracea*. In each of these species, the haploid chromosome number is 9. In the hybrid, there were 18 chromosomes ($9R + 9B$), but they behaved as 18 singletons at meiosis, there being little or no tendency to synapse. As a result, most gametes were inviable. But a few fertile hybrids were obtained. Cytological examination showed that these had 18 perfectly synapsed *pairs* of chromosomes at meiosis. Thus it was evident that allopolyploidy had arisen spontaneously in a few of the hybrid plants. These showed a combination of characters of the two genera which was completely different from anything previously encountered. The allotetraploid bred true, and was reproductively isolated from both parents. And so Karpechenko felt justified in describing this new plant as a new, synthetically produced genus, under the name *Raphanobrassica*.

A similar synthesis of a new species of tobacco was performed in 1925 by Clausen and Goodspeed. Commercial tobacco, *Nicotiana tabacum*, was crossed with a wild species, *N. glutinosa*. The hybrid was generally sterile, but a single plant was fertile. This bred true, had distinctive morphological traits, including the gigas habitus, and was reproductively isolated from the parent species. Hence the fertile hybrid was described as a new, synthetically produced species, *Nicotiana digluta* (Figure 108). Its synthesis was later repeated with much greater facility by the colchicine method. *Nicotiana tabacum* has 24 pairs of chromosomes, while *N. glutinosa* has only 12 pairs. As expected, the sterile hybrids had 36 chromosomes (not pairs), but the fertile hybrids had 36 pairs of chromosomes, and so were allopolyploids. Actually, these were allohexaploids, for *N. tabacum* itself is an allotetraploid species, as will be shown below.

Genome Analysis. There is much evidence that allopolyploids are widespread in nature. For the most part, the evidence for this consists of series of chromosome numbers in a single genus or family which are multiples of a low number found in the same group. In addition, the species with higher chromosome numbers frequently show combinations of the characters of more than one of the basic species. But the most convincing evidence is what Clausen has called *genome analysis*. This means that a series of experimental crosses is made in order to establish the actual source of the different genomes of a suspected allopolyploid. The analysis of *Nicotiana tabacum* is an illustrative example. The first step in genome analysis is to select probable parent species on the basis of morphological characters held in common by the allopolyploid and the suspected parents. Any species so chosen must of course have a chromosome number such that it

---

FIGURE 108. *Nicotiana* HYBRIDS. *A, N. tabacum* × *N. glutinosa* (36 chromosomes) is sterile, while *B*, its colchicine-induced tetraploid with 72 chromosomes (= *N. digluta*), is fertile. *C, N. glutinosa* × *N. sylvestris* has 24 chromosomes and is sterile, while its colchicine-induced tetraploid, *D*, with 48 chromosomes, is fertile. Note that all parts of the tetraploid flowers are larger than the corresponding parts of the diploid flowers. (From Warmke, H. E., and Blakeslee, A. F., *J. Hered.*, V. 30, 1939.)

could have contributed one or more genomes to the species being analyzed. On morphological grounds, Clausen and Goodspeed regarded *N. sylvestris* and some member of the *N. tomentosa* group as the most probable parent species of *N. tabacum*. Examination of the chromosomes, the second step in the analysis, showed that *N. tabacum* had 24 pairs, while *N. sylvestris* and *N. tomentosa* each had 12 pairs. Thus the numbers corresponded to the requirements of the problem. Further, the chromosomes of each of the basic species resembled some of those of the polyploid species in size and shape. Finally, a series of crosses was made to test the actual relationship of the chromosomes.

The genomes of *N. sylvestris* and *N. tomentosa* may be designated as S and T respectively, so that the normal diploid plants will have the formulas SS and TT. On the assumption that these genomes have been differentiated by mutation and by chromosomal rearrangements since the original formation of *N. tabacum*, this species is designated by the formula $S^1S^1\ T^1T^1$. In the hybrid between *sylvestris* and *tabacum*, $S^1\ ST^1$, 12 pairs and 12 singletons are formed at meiosis. This indicates that all of the chromosomes of *sylvestris* had a close enough homology with those of *tabacum* to synapse normally. This hybrid is, of course, sterile because of the irregular distribution of the unsynapsed T^1 chromosomes to the gametes. Similarly, if *tomentosa* is crossed to *tabacum*, the primary gametocytes show 12 pairs and 12 singletons. Thus the 12 chromosomes of *tomentosa* also have their homologues among the 24 chromosomes of *tabacum*. The next step was to cross *sylvestris* and *tomentosa*. This produced a viable but sterile plant ST, very similar to the haploid *tabacum* which can be produced experimentally. A small amount of pairing of chromosomes did occur in this hybrid—2.5 pairs per meiotic cell was the average. This demonstrated the fact that there is some homology between the chromosomes of *sylvestris* and *tomentosa*, but not so much but what pairing frequently was completely absent. This hybrid, ST, was then treated with colchicine to produce the allotetraploid, SSTT.

The resulting plant resembled *N. tabacum* much more closely than it did either of the parent species. While it does not morphologically duplicate any of the many known varieties of *tabacum*, it does not differ from them more than they do from one another. And the possibility remains that, if the right varieties of *sylvestris* and *tomentosa* were selected, a more exact duplication of naturally occurring *tabacum* might be achieved. And so this may be regarded as a fairly successful duplication of the natural origin of one species, *Nicotiana tabacum*. Yet there is one disappointing feature. While the artificially produced *tabacum*, SSTT, is fully fertile when used as a pollen parent, it is female sterile. For this reason, as well as others, it is suspected that another member of the *tomentosa* group may be the actual parent of *tabacum*, and *N. otophora* appears to be a probable selection.

A more frequently quoted example of genome analysis is Müntzing's study of *Galeopsis tetrahit*, the hemp nettle. This species has a haploid number of 16, whereas most species of this genus have haploid numbers of 8. This led Müntzing to suspect polyploidy. On the basis of comparison of the morphology of *tetrahit* with that of the 8-chromosome species, he selected G. *pubescens* and G. *speciosa* as the most probable parent species. The hybrid between *pubescens* and *speciosa* was highly sterile, but some good gametes were formed. A single F_2 plant was obtained, and this proved to be a triploid, apparently the product of an unreduced gamete and a gamete with the *speciosa* genome. This triploid was backcrossed to a pure *pubescens* plant, and again a single viable plant was obtained. As this plant was tetraploid, it must have been formed by union of an unreduced gamete from the triploid parent with a normal gamete from the *pubescens* parent. This "artificial *tetrahit*" was fully fertile, both to itself and to natural *tetrahit*, which it resembles. It breeds true, and is sterile when crosses are attempted with either of the parent species. Thus there can be no doubt that Müntzing has duplicated the natural "synthesis" of *Galeopsis tetrahit*.

The Wheats. A polyploid series of great economic importance is that of the genera *Triticum,* the wheats, and *Aegilops,* a related grass of no commercial value. Wheats with three different chromosome numbers are known. The Einkorn group comprises three species, of which *T. monococcum* is the most important, and has 7 pairs of chromosomes. These species are not commercially valuable. The Emmer group includes eight species, two of which, *T. durum* and *T. turgidum,* are important crop plants. These eight all have 14 pairs of chromosomes. Finally, the Vulgare group comprises four species, two of which, *T. vulgare* and *T. compactum,* are of great economic value. The wheats of the Vulgare group all have 21 pairs of chromosomes. This seriation—7—14—21—strongly suggests a polyploid series with diploid, tetraploid, and hexaploid members. Further, however, it is possible to produce haploid plants of the Emmer and Vulgare groups. In these haploids, there is no pairing of the chromosomes at meiosis, hence it is believed that the homologies between the different genomes of the polyploid species must be rather remote. And so these are allopolyploids. If *T. monococcum* is crossed to an Emmer wheat, seven pairs of chromosomes and seven singletons appear at meiosis, hence the seven chromosomes of *monococcum* do have homologues in the Emmer group. If *T. monococcum* is crossed to *T. vulgare,* seven pairs and fourteen singletons appear. Thus it appears that the Einkorn genome, A, is found throughout the genus. Similarly, an Emmer wheat crossed to a Vulgare wheat gives 14 pairs and 7 singletons, indicating that all of the Emmer chromosomes are also present in the Vulgare genome. The closely related genus *Aegilops* has species characterized by 7 and by 14 pairs of chromosomes. When *Aegilops* is crossed to Einkorn or Emmer wheats, no pairing results, indicating that no demonstrable homology exists. But when *Aegilops cylindrica,* with 14 pairs of chromosomes, is crossed to a Vulgare wheat, 7 pairs and 21 singletons (7 from *Aegilops* and 14 from the wheat) appear. This indicates that the third genome of the Vulgare wheat is homologous with one genome of *Aegilops,* and has evidently been obtained by an intergeneric cross, followed by doubling of the chromosomes to form the fertile allopolyploid species.

These data may be interpreted in the following way. The Einkorn wheat, being the only diploid species in the group, is no doubt the most primitive, and may be represented by the formula AA. By mutation and chromosomal rearrangements, this A genome has become sufficiently different, presumably in an extinct race, to be designated as a different but related genome, B. Hybridization between AA and BB races resulted in the formation of the allotetraploid species of the Emmer group, $AABB$. *Aegilops*, as it is known today, is also an allotetraploid genus, $CCDD$, but no doubt the genus has included in the past a diploid species, CC. Hybridization between this primitive *Aegilops* and the Emmer wheats resulted in the formation of the allohexaploid wheats of the Vulgare groups, $AABBCC$.

Polyploidy in Bromus: a Complex Case. An even more complicated case has been analyzed by Stebbins and his collaborators in the genus *Bromus*, a wide-spread complex of range grasses. Stebbins believes that seven is the basic number of chromosomes in this group, but most of the American species, of which *B. carinatus* is typical, have 28 pairs of chromosomes, or 56 chromosomes in the somatic tissues. Hence they are octoploids. These chromosomes include 21 medium-sized and 7 large pairs. A South American species, *B. catharticus*, however, has only 21 pairs of chromosomes, and these are all medium-sized. Hence this species is hexaploid. A single American species, *B. arizonicus* has been found to have 84 chromosomes, all medium-sized. Hence this species is a duodecaploid (12-ploid)! Crosses between these species have been made and the behavior of the chromosomes at meiosis in the hybrid studied. When *B. carinatus* and *B. catharticus* are crossed, 21 pairs are formed by the medium-sized chromosomes, while the 7 large chromosomes from *B. carinatus* behave as singletons. The three sets of medium-sized chromosomes are called A, B, and C, while the set of large chromosomes is called L. On this basis, the hexaploid species, *B. catharticus*, has the formula $AABBCC$, while the octoploid species, *B. carinatus*, has the formula $AABBCCLL$. When *B. carinatus* is crossed to *B. arizonicus*, a very complex meiotic pattern results. Such a hybrid will receive 42 medium-sized chromosomes from *B. arizonicus*, and 21 medium together with 7 large chromosomes from *B. carinatus*. At meiosis, the 7 L chromosomes of *carinatus* and 14 of the chromosomes of *arizonicus* behave as singletons. Fourteen bivalents (normal tetrads) are present, indicating that two sets of chromosomes, arbitrarily designated as the A and B sets, are held in common by the two species. But in addition to this, as many as seven trivalents (complex associations of three pairs of chromosomes) may be formed. Thus it appears that one of the *carinatus* sets of chromosomes has considerable homology with *two* sets in *arizonicus*. This set is arbitrarily designated C_1, while the second C set in *arizonicus* is called C_2. The two sets of chromosomes in *arizonicus* which are not represented at all in *carinatus* are called D and E. Thus the hexaploid species, *B. catharticus*, has the formula $AABBC_1C_1$; the octoploid species, *B. carinatus*, has the formula $AABBC_1C_1LL$; and the duodecaploid species, *B. arizonicus*, has the formula $AABBC_1C_1C_2C_2DDEE$.

SOME GENERALIZATIONS ON POLYPLOIDY

It appears, then, that allopolyploidy is a very general phenomenon. In 1942, Goodspeed and Bradley published a review in which they listed 124 well-authenticated cases of allopolyploidy, including both natural and experimental examples. There is no doubt that the list would be much longer now if it were brought up to date. Natural polyploids seem to be generally allopolyploids rather than autopolyploids, and so this type evidently has especial interest for evolution. As pointed out, inspection of tables of chromosome numbers makes it appear probable that over half of the higher plants are polyploids of one degree or another. While the formation of polyploid series does not entail any new genic material, it does produce new combinations upon which selection can act, combinations which may be very different from anything formed in any other way. It is not surprising, therefore, that polyploids often invade territories not occupied by their diploid parents. Polyploids seem to be much more aggressive invaders of new territory than are their diploid relatives. For example, Anderson has shown that *Iris versicolor* is almost certainly an allopolyploid derived from *I. virginica* and *I. setosa*. *I. virginica* is widely distributed in the southeastern portion of the United States. *I. setosa* is found as two widely separated races, one on the coast of Alaska, the other in Labrador, Nova Scotia, and Newfoundland. These species are regarded as remnants of the preglacial floras of their respective areas, and they have not extended their ranges into the glaciated parts of North America. But *I. versicolor,* their allopolyploid offspring, is distributed from Labrador and northeastern United States westward through the Great Lakes region to Wisconsin and Winnipeg. It is thus widely spread in the glaciated part of North America.

An interesting and largely unsolved problem is that of the effect of polyploidy on the occurrence of new genetic variability due to mutations. Some geneticists have expressed the opinion that polyploidy should increase the total genetic variability rapidly, because random mutations might occur in any of the genomes and become homozygous. Others have pointed out that a new recessive mutant occurring in one genome would now be covered by its dominant allele in *three* other genomes (or more in higher polyploids) so that its phenotypic expression would be much less probable than in a diploid. It may be that these viewpoints are not completely irreconcilable. From a short-range viewpoint, the latter is probably correct, but, when time is provided on a geological scale, differentiation of the genomes by random mutation should finally result in greater total variability than could be achieved in a diploid.

It appears to be generally true that diploids are found in the older part of the total range of a group, while polyploids invade the geologically more recent parts. Thus, in a group in which polyploidy is common, the diploids tend to become relics, while the more diversified polyploids fill the available niches. Stebbins pointed out that one consequence of this is that, when conditions become unfavorable for such a group, the diploids are the first to become extinct. And, when a mere relic remains of a once prominent group, it is likely to consist of one or a few polyploid species without near relatives. Thus the entire order Psilotales, once a dominant group, is now represented by only two genera, which are regarded by many botanists as monotypic. These have over a hundred pairs of chromosomes, and so it is almost certain that they are the last remnants of a once great polyploid complex.

POLYPLOIDY IN THE ANIMAL KINGDOM

Although polyploidy has been a major phenomenon in plant evolution, its role in animal evolution has not been adequately assessed, and it is generally considered to be of minor importance. The reason for this difference between the kingdoms is not known with certainty, but Muller has suggested that it may be based upon the fact that the sexes are usually separate in animals, while plants are usually hermaphroditic (monoecious). Random segregation of the several pairs of sex chromosomes in a polyploid organism would result in sterile combinations. This explanation has been widely accepted. Vandel has reviewed all of the known cases of polyploidy in animals, and the data which he has assembled lend support to Muller's theory. Thus he finds, among plants, only eleven cases in which polyploid plants are also dioecious (sexes separate). These include *Fragaria elatior,* a hexaploid species of strawberry. It has been proven that there is only one pair of sex chromosomes present in this species. Whether the other two pairs have lost their original sex-differentiating function, or whether polyploidy first developed in a monoecious ancestor, with the separation of the sexes occurring later, cannot be ascertained. Vandel favors the former hypothesis. In any event, it appears that polyploidy, so common among plants in general, is rare among those plants which have acquired the dioecious habit.

The majority of the animals which have been reported to be polyploids are parthenogenetic. Curiously, all of these parthogenetic polyploids are arthropods. The common waterflea, *Daphnia pulex,* occurs in a diploid, sexually reproducing form, and in a hexaploid, parthenogenetically reproducing form. *Artemia salina,* the brine shrimp, occurs in tetraploid and octoploid parthenogenetic forms, but there is also a tetraploid race which reproduces sexually. The European sow bug, *Trichoniscus elisabethae,* is a triploid parthenogenetic species, while it is uncertain whether the parthenogenetic ostracod, *Cypris fuscata,* is triploid or tetraploid. The "walking stick" insects, *Carausius morosus* and *C. furcillatus* are, respectively, triploid and tetraploid, as well as parthenogenetic. The psychid moths *Solenobia triquetrella* and *S. lichenella* are also parthenogenetic tetraploids, but a bisexual, diploid race of the former is known. Finally, the parthenogenetic beetle, *Trachyphlaeus,* also appears to be tetraploid.

A very few tetraploid animals reproduce bisexually. *Artemia* has already been mentioned. *Parascaris equorum* (= *Ascaris megalocephala* of older literature), an important nematode parasite of horses, is known in diploid, tetraploid, and hexaploid forms, all of which reproduce bisexually. The starfishes *Asterias forbesii* and *A. glacialis* and the sea urchin *Echinus microtuberculatus bivalens* all are tetraploid forms which reproduce bi-sexually. The golden hamster, *Cricetus auratus,* may belong here, for it has 22 pairs of chromosomes, while its nearest relatives have 11 pairs. It has been claimed, but not proven, that this species is an allotetraploid of *Cricetus cricetus* and *Cricetulus griseus.* Thus it appears that any sexual unbalance which may be caused by random segregation of the sex chro-mosomes in polyploids can be overcome. But that it is not frequently overcome is indicated by the rarity of polyploids among bisexually repro-ducing plants and animals, and by their over-all rarity in the Animal King-dom, where separate sexes are the rule.

A related phenomenon is fragmentation of the chromosomes, with the result that multiples of a basic number appear in a series of chromosome counts. Vandel lists examples in most of the major groups of animals in which species show chromosome numbers which are twice the number of some more primitive species. Like polyploidy among plants, increase in the number of chromosomes by fragmentation among animals appears to go hand in hand with increasing specialization. Thus, among the lower mammals, the diploid number is generally 24; among the Eutheria, it is generally 48; among the highly specialized eutherians, such as the ungu-lates, it is often 60; and among such highly specialized mammals as the rodents, it may be as high as 84. Why fragmentation of the chromosomes should be related to evolutionary specialization is an obscure point, but it may be that this is based upon position effects, or comparable phe-nomena.

REFERENCES

Goodspeed, T. H., and M. V. Bradley, 1942. "Amphidiploidy," *Botan. Rev.,* **8,** 271–316.

Müntzing, A., 1936. "The Evolutionary Significance of Autopolyploidy," *Hereditas,* **21,** 263–378.

Stebbins, G. L., 1950. "Variation and Evolution in Plants," Columbia University Press. (Karpechenko, Winge.)

Vandel, A., 1937. "Chromosome Number, Polyploidy, and Sex in the Animal King-dom," *Proc. Roy. Zool. Soc.,* London, *Series A,* **107,** 519–541.

(Three of the above papers are thorough reviews of the subjects stated. Unfortu-nately, none is up to date. Stebbins' book is a comprehensive review of evolution in plants, and it includes a more timely study of polyploidy.)

CHAPTER TWENTY

Distribution of Species

FROM THE TIME OF LINNAEUS until well into the post-Darwinian era, biogeography was a very active field, in which many of the most distinguished biologists worked. New discoveries of major importance could be expected as the reward of competent work in this field. New phyla were described fairly frequently, new classes and orders with regularity. But, by the end of the nineteenth century, all of the major biogeographical realms were fairly well known, their floras and faunas were catalogued, and only problems of detail and revision confronted young biogeographers. Meanwhile, the rise of experimental biology made biogeography and taxonomy rather passé, and their practitioners were often regarded as hack workers. The modern revival of evolutionary studies has brought with it a renewed interest in taxonomy, and more recently biogeography has shared in this renaissance. There are problems in evolution which can be profitably attacked only if detailed biogeographical information, both present and past, be studied. Some of these problems will be considered below.

In one sense, problems of distribution overlap those of isolation. The minimum biogeographical requirement for relationship between two species is that their ancestors must at some time have lived in the same area. In other words, permanent isolation and relationship are mutually exclusive. Thus, if the magnolias of southeast China are related to those of southeastern United States, their present isolation must not have characterized all past ages. As a matter of fact, it is well established that their distribution was continuous in the Tertiary. On the other hand, the development of geographic isolation between related populations is certainly an important aspect of biogeography.

ISLAND LIFE

It was the study of the flora and fauna of the Galapagos Islands that first caused Darwin to consider the possibility that species might be mutable; and the island life of the Malay Archipelago played a major role in bringing Wallace to the same conclusion. Since their time, the study of oceanic island life has always been an important aspect of the study of evolution. Many features inherent in the island locale combine to cause this. The fact that the inhabitants of any island generally resemble those of the nearest mainland leaves little doubt that the island dwellers have migrated from the mainland. However, unless an island is very close to the mainland, the difficulties of crossing the water barrier will keep out a large portion of the potential migrants. Thus competition is less rigorous, and the selection pressure lower than on the mainland. In the absence of the check of normal interspecific competition, and often in the absence of predators, the populations may reach prodigious numbers relative to the space available. Yet the total population is small as compared to continental populations, and it is likely to be broken up by ecological factors into much more restricted breeding colonies. In such a situation, genetic drift may be stronger than selection, with the result that island forms are likely to be less well adapted than their mainland relatives. Apparently (but not actually) contradictory to this is the fact that closely related island dwellers may become adapted to situations so different that only remotely related organisms of the mainland could be compared to them. An example of each situation will be discussed below.

Genetic Drift on Oceanic Islands. A celebrated example of the former type is that of the snails of the genus *Partula* on the islands of Tahiti and Moorea, neighbors in the Society Islands, about 2400 miles south of Hawaii. These are typical volcanic islands, characterized by a central volcano from which deep valleys and narrow separating ridges radiate to the sea. The snails feed on the plants of the valleys, and the intervening ridges are almost impassable barriers to them. Several species of *Partula* are represented on the islands, some species being found in many of the valleys, others in few or only a single valley. In each case, the inhabitants of every valley comprise a distinct race, characterized by such things as size, direction of coiling, details of shape, and color. Only those species which are represented in only one valley, such as *P. tohiveana*, are monotypic. The variation of the several races of a species definitely is not clinal. Races of neighboring valleys may be strongly divergent, while races at opposite ends of an island may be quite similar. Different species in a single valley may have identical ecological requirements, living on the same food plants, and yet they may show no tendency toward parallel variation. All attempts to interpret this situation in terms of different selective forces in the different valleys have failed. It appears probable, then, that all of the races of *Partula* are subject to substantially identical selective forces, and the differences between them result from genetic drift. But genetic drift on so grand a scale is possible only because the geographic features of the islands enforce almost complete isolation upon all of the local breeding populations. These snails have been studied intensively three times, at wide intervals, in the past eighty years, and it appears that significant changes in many of the races have occurred even in that short time.

Rapid Selective Differentiation on Oceanic Islands. One of Darwin's studies in the Galapagos Islands was made upon a family of finches, the Geospizidae, which had become adapted to an amazing range of ecological niches in the islands, so that their superficial differentiation was much greater than usual within the confines of a single family. Another family of birds, the Drepaniidae or Hawaiian honey-creepers, has undergone a similar adaptive radiation within a geologically brief time. The Hawaiian Islands appear to have originated no earlier than the Pliocene. As the honey-creepers are forest-dwellers, their original ancestor could not have reached Hawaii (by migration from Central America) before the forests themselves were established. This might have occurred by mid-Pliocene or later, thus giving a maximum of five million years for the differentiation of the family in Hawaii. Lest one be tempted to think of this as a long time, let us recall that many instances are known in which only subspecific differentiation has been achieved after periods as long as 50,000,000 years.

When the remote ancestor of the Drepaniidae first migrated from the mainland to the Hawaiian forests, it found a rich field for which it had no competitors. The result was a rapidly expanding population which soon provided its own competition. That is, the population began to outstrip the food supply available *by the original method of feeding.* This drepaniid progenitor may well have been similar to the living *Loxops virens chloris,* which has a moderate-sized, slightly curved bill, adapted to feeding upon insects in foliage. But the birds occasionally dig for insects in loose bark, or probe flowers for nectar and insects. Now the development of races or species with different feeding habits would permit the survival of a much larger total drepaniid population. How this might have occurred is perhaps indicated by living members of the genus *Loxops.* *L. v. chloris* is widely distributed through the islands, but on Kauai, one of the most isolated islands, this species is represented by another subspecies, *L. v. stejnegeri.* Another species, *L. parva,* is found only on Kauai, and the characteristics of the bills of these two *Loxops* representatives are most suggestive. The bill of *L. v. stejnegeri* is somewhat larger, heavier, and more strongly recurved than that of *L. v. chloris.* While it still has feeding habits similar to those of *L. v. chloris,* it depends more upon digging for insects in loose bark. It also visits flowers for nectar and small insects. But the bill of *L. parva* has deviated from that of *L. v. chloris* in just the opposite way. It has become shorter and straighter. *L. parva* depends primarily on insects on the surfaces of branches and leaves. Its bill is not well adapted to digging, and it rarely attempts this. While it is not well adapted to visiting flowers, it does frequently visit Acacia flowers. It seems probable that, when these two species of *Loxops* were brought into competition, any variations which tended to adapt them to different sources of food were strongly favored by natural selection. The result is that these two species, so strongly divergent in bill structure, differ very little in other respects.

But the genus *Loxops* only begins to indicate the range of variations of bills in the Drepaniidae (Figure 109). The genus *Hemignathus,* most members of which have become extinct in recent times, showed a much greater range of variation than usually characterizes whole families. *Hemignathus obscurus,* which was probably closely related to *Loxops virens,* had a very long, slender, and strongly curved bill. It was adapted for probing fine crevices in the bark of trees, and for visiting flowers. In *H. lucidus,* the upper mandible is much like that of *obscurus,* and it was

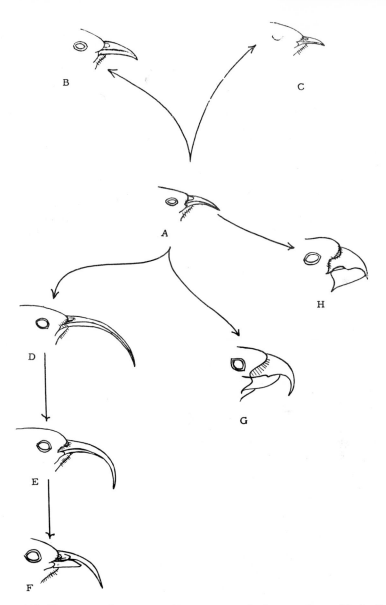

FIGURE 109. EVOLUTION AMONG THE DREPANIIDAE. A, *Loxops virens chloris;* B, *L. v. stejnegeri;* C, *L. parva;* D, *Hemignathus obscurus;* E, *H. lucidus;* F, *H. wilsoni;* G, *Pseudonestor xanthophrys;* H, *Psittarostra kona.* (Redrawn from Amadon.)

similarly used for sweeping insects out of crevices in the bark of trees. But the lower mandible was much shorter and thicker. This was used in wood-pecker-fashion to pry or chisel bits of bark to expose its prey. This species rarely visited flowers because its bill was not adapted to sucking the nec-tar. *H. wilsoni,* which is still extant, differs from the others in that its lower mandible is very heavy and chisel-like. It uses its lower mandible to break away wood and bark, then sweeps its upper mandible through the crevices so exposed. It never visits flowers. A closely related genus, *Pseudonestor,* has a heavy bill which is adapted for crushing dead twigs to expose beetle larvae. In a fourth genus, *Psittarostra,* which appears to be closely related to *Loxops,* the bills are finch-like, that is, they are short, heavy structures for cracking hard seeds. Thus there is within this family a very wide range of bill structure and feeding habits, and this has been achieved within a short span of time. Differentiation of other structures has been much slower. Competition for a limited food supply by closely related birds has given a strong adaptive value to variations which tend to open up new food sources to them. Thus there has resulted a rapid adaptive radiation. This is an excellent example of character displacement (see Chapter 18).

ENDEMISM

Closely related to the phenomena of island life is that of endemism. *En-demic* is defined as restricted to or prevalent in a particular district. Thus defined, all species are endemic, for all are confined to a definite area, even though that area may be very great. For practical purposes, a species is regarded as endemic if its distribution is very much more restricted than that of typical species. Thus, the northern white pine, *Pinus strobus,* is widely distributed over northeastern United States and Canada, and is not regarded as endemic; but the redwood, *Sequoia sempervirens,* which is confined to the coastal valleys of California, is an endemic. A still more restricted endemic is the recently discovered living member of the genus *Metasequoia,* most of the members of which are extinct. This species is confined to a single valley in central China. Two different types of en-demic species are recognized. A species may have a very restricted distri-bution because it is a *young* species, and has not had time to expand its range. Or a species may have a restricted range because it is the last remnant of an old group nearing extinction. Some biologists reserve the term *endemics* for the first type, while calling the second *epibiotics.* Yet most species of restricted distribution fall into the latter class. They have few living close relatives, and they are often well represented in paleon-tologic series, as for example, *Sequoia.*

Island life, particularly that of oceanic islands, is replete with endemics. It has been estimated that more than 90 per cent of the flora of the Hawai-ian Islands is endemic, and it is probable that the figure for animals would be comparable if calculated. As already mentioned, the high proportion of endemics in the Galapagos Islands was among the factors which di-rected Darwin's thoughts toward the possibility of the transformation of

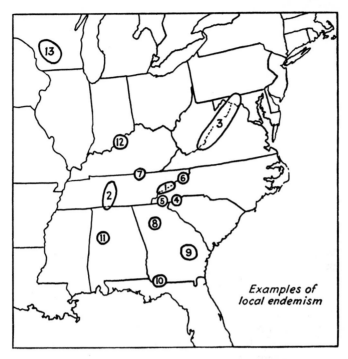

FIGURE 110. EXAMPLES OF ENDEMIC PLANTS IN EASTERN UNITED STATES. Common names are given for the better known genera. 1, *Calamagrostis Cainii, Senecio Rugelia, Rubus carolinianus* (raspberry); 2, *Lesquerella Lescurii, Petalostemon Gattingeri* (prairie clover), *Lobelia Gattingeri, Psoralea subacaulis;* 3, *Allium oxyphilum* (wild onion), *Eriogonum Alleni, Oenothera argillicola* (evening primrose), *Pseudotaenidia montana, Solidago Harrisii* (goldenrod), *Aster schustosus, Convolvulus Purshianus* (bindweed); 4, *Shortia galacifolia;* 5, *Lindernia saxicola;* 6, *Buckleya distichophylla;* 7, *Conradina verticillata;* 8, *Amphianthus pusillus;* 9, *Penstemon dissectus* (beard-tongue); 10, *Torreya taxifolia;* 11, *Neviusia alabamensis;* 12, *Penstemon Deamii;* 13, *Penstemon wisconsinensis.* (From Cain, "Foundations of Plant Geography," Harper & Brothers, 1944.)

species. Mayr has pointed out that many of the birds of south Pacific islands which he has studied are endemics. It is easy to understand why the flora and fauna of an oceanic island should be endemic. The great water barrier prevents all but pelagic species from extending their ranges. But endemism is by no means confined to islands. The problem of continental endemism in animals has not been well studied, but plant geographers have given much attention to this problem. The cases of *Sequoia sempervirens* and *Metasequoia* were cited above. *Sequoia gigantea* is similarly restricted to the Sierra Nevada Mountains. Cain lists thirteen localities in which no less than twenty-five species of plants are endemic (Figure 110). Examples of endemic plants could be multiplied indefinitely, but these are sufficient to indicate the prevalence of the phenomenon, and we may now ask why continental endemics are restricted to so

313

limited an area. In the case of young endemics, the answer is obvious: they simply have not had time to achieve the range extension which may be expected in the future. Whether this is a sufficiently frequent occurrence to have general importance will be discussed below in connection with the Age and Area theory of Willis. The more usual situation appears to be that endemics are relics of ancient groups, like the Sequoias and the Cypress. Such endemics generally appear to have very little genetic variability, with the result that they are adaptable only to a narrow range of environments. This must be based in part upon a very low mutation rate. Also, it may be that much variability which such species possessed in the past may have been lost by genetic drift as the population contracted.

CONTINENTAL DISTRIBUTION

But island life and endemism are specialized phenomena. The continental distributions of most organisms must be based upon other principles. Simpson has shown that the history of the mastodonts, an extinct family closely related to elephants, illustrates clearly the more general principles of distribution. Mastodonts first appear in the fossil record of north Africa in the Oligocene. As other Oligocene faunas from all parts of the world are well known, it is quite certain that north Africa was the place of origin of the mastodonts, and that they did not occur elsewhere at that time. But they immediately began to expand their range by active migration, and by the beginning of the Miocene period they had reached mid-continental Africa to the south, the Baltic area to the north, and India to the east. By mid-Miocene, they occupied about half of Africa and most of Europe and Asia except the most northerly parts. Late in the Miocene, they crossed the Bering Strait to North America, over which they spread during the Pliocene. Only toward the end of the Pliocene did the mastodonts reach South America, over much of which they spread during the Pleistocene. About this time, the mastodonts became extinct in the Old World, and by the late Pleistocene, the Americas were their last remaining refuge. Finally, the American mastodonts also disappeared.

This history may be restated in more general and more explanatory terms. A species (or larger group) originates in a definite, more or less restricted area (north Africa in the case of the mastodonts). It then tends to spread by active migration in all directions, occupying whatever suitable habitats it may find, until it reaches an impassable barrier (the limits of the continents of Africa, Europe, and Asia in the present example). Further distribution may then stop, or the group may find a way to cross a barrier, as the mastodonts did cross the Bering Strait to North America. This may be accomplished by *removal of the barrier* by geologic or climatic changes. Once such a major barrier is crossed, the point of crossing becomes a new center of dispersal from which the group again expands its range until it meets an impassable barrier. After a time, climatic, topographical, or biotic conditions change sufficiently that the group can no longer compete adequately over all or part of its range. Local populations then become sparse, then altogether extinct. Finally, only an isolated population, which can be regarded as an endemic, remains. With the extinction of this remnant, the history of the group comes to an end.

DISCONTINUOUS DISTRIBUTION AND BRIDGES

It is in connection with this type of history that the problem of widely discontinuous distributions must be understood. In any particular case, discontinuity may be brought about by extinction in the intermediate parts of a wide distribution, or by the bridging of a barrier between two distinct areas, or by a combination of the two. This brings us to the problem of how barriers may be bridged. Simpson has analyzed this problem and has classified bridges into three types: corridors, filter bridges, and sweepstakes bridges. A corridor is a broadly continuous connection, existing over a long period of time, so that it permits an extensive interchange of the floras and faunas of the connected regions. Such a connection now exists between Europe and Asia, and these two accordingly comprise a single biogeographic region. A filter bridge is more temporary in duration, and more restricted in extent. Conditions are more uniform upon it, with the result that it "filters" the flora and fauna which might use it; only those with appropriate characteristics can pass. The Bering Strait acted as a filter bridge for mammals during the Pliocene period. Only those mammals could cross which were capable of making a rapid crossing and of withstanding cold weather. A sweepstakes bridge, on the other hand, does not involve migration across a land connection, but rather it depends upon accidental transportation in the absence of any real connection. Corridors and filter bridges should operate equally well in either direction, and should cause the exchange of numerous forms; but a sweepstakes bridge should operate only in one direction, and only a few forms should succeed in crossing it. It is a one chance-in-a-million phenomenon. But this may allow many chances indeed if time be available on a geological scale.

Corridors. Corridors are most striking after they no longer exist, that is, after geological events have separated land masses which once were continuous. Simpson has pointed out that New Mexico and Florida can be regarded as being connected by a corridor at present. As expected, they share most of their major groups and a large number of genera, but species are likely to be different because the climatic conditions of the two states require rather different adaptations. One gives little thought to such a situation as a "bridge," simply because the question never arises as to why the areas concerned should not have a substantially similar flora and fauna. But it is perfectly possible that they might be separated in future ages by, for example, an inland sea occupying the present Mississippi drainage basin. And then paleontologists might use the fossils of our time, or common genera of their own time, to prove that once a corridor existed between New Mexico and Florida.

Such is the situation with respect to eastern Asia and eastern North America. During much of the earth's history, Asia and North America were broadly connected across the north Pacific. Early in the Tertiary, much of this connection was submerged, leaving only the islands of the Bering Strait. In warmer times, during the Mesozoic Era and until sometime in the Eocene, a very extensive exchange of floras and faunas occurred across this north Pacific bridge. How complete this exchange was is indicated by the fact that no less than 156 living genera of plants are known to be common to the two regions at present. In some cases, such as *Symplocarpus foetidus* (skunk cabbage), the species appear to be identical and even the races are closely similar. The exchanged genera are by no means confined to such rapid migrants as herbs, for many genera of trees are common to the two areas, as *Acer* (maple), *Catalpa*, and *Magnolia*. Many known as fossils in both areas are now living in only one. Thus *Castanea*, the chestnut tree, still survives in eastern Northern America, while *Ginkgo*, the very primitive maiden hair tree, still survives in Asia. This problem has been studied less in animals, but parallel examples are known. Thus, the alligator is known only from the United States and China, and the salamanders *Triturus* and *Cryptobranchus* are also found in these remote places. The present isolation of these forms was accomplished by geologic and climatic changes. Geologically, the north Pacific corridor was largely submerged, and the mountains of western North America were elevated, thus making climate and topography unfavorable for the former inhabitants of this area. Further, the climate of the entire northern part of the world became colder, with the result that these temperate and subtropical organisms became extinct over much of their former range, leaving the distributions as they are found today.

Filter Bridges. The main characteristic of a filter bridge is that it does filter out many of the organisms of the connected regions, while permitting the passage of others. Also, a filter bridge is typically of brief duration, while a corridor lasts for periods which are long even on the geological time scale. But, while genuine corridors between continents appear to have been rare, perhaps only those described above having existed, filter bridges have been fairly common. The Bering Strait had this character repeatedly in the Pleistocene (Figure 111). Its filtering action has probably been due in part to the fact that the land connection may not have been completely continuous. Any organisms for which small expanses of salt water form an impassable barrier would have been unable to cross. Also, because of its short duration, it would be crossable only by plants and animals capable of migrating fairly rapidly. But a more important factor concerns the location of the bridge and the prevailing climatic conditions. In this example, the location was just below the Arctic Circle, and the climate was cold, for it was during the Pleistocene that the great glaciation occurred. Thus, only those Palearctic and Nearctic animals which were adapted to cold could make the crossing. Temperate and tropical animals were excluded simply because they were unable to approach the bridge. Mainly mammals seem to have made the crossing, perhaps because of their superior powers of locomotion. Animals such as bears, cats, bison, deer, and mammoths seem to have crossed from Asia to North America, while dogs, horses, and camels crossed in the other direction (Figure 111).

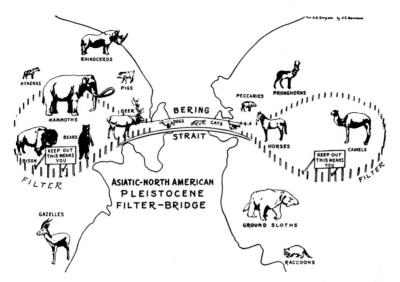

FIGURE 111. DIAGRAMMATIC REPRESENTATION OF THE PLEISTOCENE FILTER BRIDGE ACROSS THE BERING STRAIT. (From Simpson, *J. Washington Acad. Sci.*, V. 30, 1940.)

Australia was probably colonized from Asia via a filter bridge. At present, the Malay Archipelago extends toward Australia in a long arc from the Malay Peninsula. But during the Mesozoic Era, when the primitive mammals first arose, there was a continuous or near continuous land connection, and via this the monotremes and marsupials reached Australia. As this bridge was broken before the origin of more modern mammals (placentals), these did not reach Australia, with the exception of bats and small rodents which probably reached Australia by a sweepstakes route.

Another rather thoroughly known filter bridge is the Central American bridge between North and South America, which has functioned at least twice. In the Cretaceous and earliest Paleocene, the most ancient mammals of the South American fossil beds reached that continent from North America via this filter bridge. These included a wide variety of marsupials, many of which evolved in South America to form the various carnivorous types. The placental mammals which entered South America at this time were very primitive, including ferungulates which had not yet become clearly differentiated into herbivorous and carnivorous types, but were omnivores, and edentates, such as the armadillos, anteaters, ground sloths, and tree sloths. In South America, ferungulates evolved primarily along herbivorous lines, perhaps because the carnivorous niches were already filled by marsupials. Selection favors diversification. During most of the Paleocene, however, this intercontinental connection was completely submerged, and so evolution proceeded entirely independently on the two Americas. During the late Eocene and Oligocene, the connection again rose, but only enough to form a chain of more or less widely separated islands. Thus a sweepstakes bridge was formed, a bridge likely to be crossed only by animals which could be carried from island to island by such accidental means as floating on driftwood. Naturally, only a relatively few small mammals succeeded in crossing. Rodents of the guinea pig type (cavioids) and monkeys were the principal "island hoppers." But the islands were again submerged in the late Oligocene. During the Pliocene, the elevation of the land again established a chain of islands, but this time the elevation proceeded until a continuous land connection, which has persisted to the present, was established in the Pleistocene. Over this bridge, a wide variety of mammals have passed. From North America to South America went such mammals as deer, camels, tapirs, horses, mastodonts, cats, weasels, raccoons, bears, dogs, mice, squirrels, rabbits, and shrews. But other North American mammals were "filtered out," because of ecological or topographical barriers which generally prevented their gaining access to the bridge. These included such animals as pocket gophers, beavers, bobcats, bison, and sheep. But porcupines, armadillos, capybaras, and ground sloths, as well as other mammals, invaded North America from the southern continent. Representatives of the first two types have survived in North America, but the others have long since become extinct. Simpson has said that there were twenty-nine families of land mammals in South America and twenty-seven in North America before the Pleistocene connection was established, with only two of these families common to the two continents. This makes a total of fifty-six families in the Americas. Soon after the connection was established, there were no less than twenty-two families in common. Fourteen of these were originally North American, and seven were originally South American, while one was of uncertain origin. Nine families were still confined to North America, while seventeen were still confined to South America. Thus the total was only forty-eight families, as eight had already become extinct. At present, there are fourteen families in common, nine confined to North America, and fifteen confined to South America, for a total of only thirty-eight families in the Americas. Thus extinction has continued since the Pleistocene. This filter bridge, then, has resulted in an extensive exchange between the two faunas, but it has not in any way merged them into a single fauna. This example has been discussed in terms of mammals, because the mammalian paleontology for the Americas has been intensively worked out, but it is probable that comparable results would be obtained for any group if adequate data were studied.

Sweepstakes Routes. Sweepstakes routes are much less tangible because no actual land connection is present. It has often been stated that natural rafts—driftwood or uprooted trees, for example—might carry plants, particularly as seeds, and animals from one place to another. This is the type of transportation most commonly envisaged in connection with sweepstakes routes. But the concept need not be restricted to water barriers. It can apply to any type of barrier if the crossing is improbable but not impossible. Simpson has summarized the characteristics of a sweepstakes route as follows: Generally, only small animals, and particularly arboreal types, can cross. The chances are much greater for some of these than for others, but at any particular time, the probability of a successful crossing by any organism is small. But, chance is the major factor in determining that a crossing is made, and so less probable organisms may succeed while more probable ones fail. This is comparable to a lottery (whence the term sweepstakes) in which a person who holds only one ticket may win, while a person who holds many tickets may lose. Finally, a sweepstakes is likely to be a one-way route, in contrast to corridors and filter bridges. Island life is commonly established via the sweepstakes method, and because of this it is likely to be rather unbalanced.

Simpson has summarized the results of the sweepstakes colonization of Madagascar in Figure 112. Lions, elephants, apes, antelopes, and zebras are selected to typify animals which cannot cross the Mozambique Channel to Madagascar either because they are too large for the natural rafts, or because they do not approach the seashore for ecological reasons. They do not hold sweepstakes tickets. Many animals, however, hold tickets as there is no apparent reason why they could not cross the channel as easily as those that have done so, yet these "disappointed ticket-holders" have not crossed. Out of the multitudes of ticket-holders in the African fauna, only a few have "won," and these have been determined by chance, not by the characteristics of the animals. These are typified by some mice; by certain cat-like carnivores of the family Viverridae, including the fossa which is illustrated; by lemurs and some other primitive Primates; by the tenrec, a peculiar insectivore; and by a pigmy hippopotamus which may have crossed the channel by swimming. These are all types which are represented in Africa by widely diversified groups.

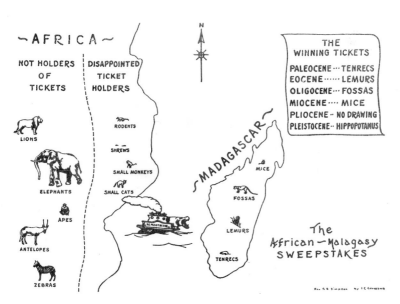

FIGURE 112. THE AFRICAN-MALAGASY SWEEPSTAKES. (From Simpson, *J. Washington Acad. Sci.*, V. 30, 1940.)

Had a filter bridge, let alone a corridor, existed, a much broader representation of each order should have entered. Thus the carnivores of Africa include a wide variety of cats, but only the related viverrids reached Madagascar. The Primates of Africa include many monkeys and apes, but none of these have reached Madagascar. Only two ungulates, out of a large African ungulate fauna, reached Madagascar. These are a bushpig and a pigmy hippopotamus which is now extinct.

PROBLEMATICAL DISTRIBUTIONS

Discontinuous distributions of many species may appear at first inspection difficult to understand. As geological data are assembled, these frequently become readily understandable on the basis of corridors or filter bridges which are no longer extant. Of course, if the bridge is still extant, there is no difficulty at all. But some distributions can be understood only on the assumption that the discontinuity was established via a sweepstakes route. These cases are likely to remain in dispute even after thorough study, because many biologists are unable to accept the reality of so indeterminate a route. They prefer to look for land connections, and then search for some factor that would prevent its use by a broad sampling of the floras and faunas. The great difficulties to which this has at times led may well indicate the artificiality of the goal.

Cain lists no less than seventeen pairs of areas between which major discontinuities exist. He does not regard his list as exhaustive, but suggests that other authors may wish to extend it. Only a few more examples will be discussed below.

The Eastern Asia-Eastern North America Case. When the biotic similarities of eastern North America and eastern Asia were first discovered, an explanation was by no means easily given. Here were two closely similar biotic regions separated by nearly half of the circumference of the globe and a great ocean, yet they had many genera and even some species in common! But geological and paleontological studies revealed that these regions were broadly continuous during the Mesozoic Era and during much of the Tertiary period. The climate was much warmer, and thus these temperate and subtropical plants and animals were able to inhabit the great space now separating them. Details of their subsequent separation have been discussed above.

Bipolar Mirrorism. Du Reitz has studied what he calls bipolar "mirrorism" of floras of the northern and southern hemispheres, in which the same or closely related species may be present in the temperate and boreal zones of both hemispheres. Thus he finds a botanical correspondence between the Mexican plateau and Peru; between Texas and Argentina; between Chile and California; and between the Straits of Magellan and Arctic North America. The plants which he studied are principally mosses, lichens, sedges, grasses, and other plants the species of which are not generally familiar to laymen, but some examples may be cited. The crowberry shrub, *Empetrum,* is found both in North America and in the Andes. Alder trees, *Alnus,* are found in the mountains of Mexico and

Central America, and also in the Andes of Peru. The beech tree, *Fagus*, is widely distributed in the northern hemisphere, and the closely related tree *Nothofagus* is found in both South America and Australia. Many mosses, grasses, and sedges show comparable distributions. This bipolar mirrorism cannot be the result of a corridor, because the regions concerned do not have enough in common, and those plants that they do have in common are mostly small. Very few trees are common to the areas under discussion. Yet a corridor should cause a very general exchange, resulting in floral unification of the areas connected. Du Reitz suggested a filter bridge via the Pacific Islands. But the chain of islands extending from Lower California toward South America ends in the tropics, and the mirrorism begins not in the tropics but in the temperate zones. Thus the facts of bipolar mirrorism are opposed both to a corridor and to a filter bridge, and so the sweepstakes route seems to be the most probable explanation for this type of distribution. Transport of seeds by migrating birds could be a factor.

The Australia-South America Case. Yet another difficult problem of distribution is that of the similarities of the inhabitants of South America and the Australian region, including New Zealand. The zoological evidence centers around the marsupials. Perhaps the fact that the mammalian fauna of Australia is predominantly marsupial is better known than any other fact of Australian natural history. In the Tertiary period, South America had a large marsupial fauna, predominantly carnivorous, and paralleling the Australian marsupial carnivores rather closely. Although most of these became extinct when the placental mammals from North America arrived in the Pleistocene, there are still more species of marsupials in South America than on any other continent except Australia. Many genera of plants are common to the two continents. *Nothofagus* has already been mentioned. Other examples include the sedge *Carex* and the moss *Sphagnum*. Because of these facts, some biogeographers have postulated a land connection in earlier ages between South America and Australia. But this is highly improbable, not only because no positive geological evidence favors such a corridor, but also because the biological evidence itself is not consistent with a corridor. There are common elements in the floras and faunas of these continents, but there is no semblance of unity between them. Further, a corridor should be freely used by mammals, yet the marsupial families of the two continents are all different, and none of the South American placentals reached Australia. Perhaps the most likely theory is that exchanges occurred by a sweepstakes route via Antarctica and intervening oceanic islands during warmer ages. This is supported by the fact that the only extant vascular plant of Antarctica, the grass *Deschampsia antarctica*, is represented by other species of the same genus in both South America and New Zealand. It should be added, however, that there is every reason to believe that the marsupials reached the southern continents by invasion from the north.

THEORIES OF DISTRIBUTION

Several comprehensive theories of distribution have been proposed, frequently with the solution of these difficult cases as a primary objective. Theories to be discussed below include the continental drift hypothesis of Wegener, the age and area theory of Willis, and the climate and evolution theory of Matthew.

Continental Drift. The continental drift hypothesis is generally associated with the name of Wegener, although the idea was not strictly original with him. It has been modernized and corrected by Du Toit. In brief, the theory holds that from the Paleozoic Era until late in the Mesozoic Era there were only two major land masses, Gondwana and Laurasia, and these were in contact at times. Gondwana centered around the South Pole, while Laurasia overlapped the equator and extended well into the northern hemisphere. During the Cretaceous, these masses fragmented to form the present continents, and these have since drifted apart, very slowly, toward their present positions. Gondwana gave rise to the southern continents, Africa, South America, Australia, and Antarctica, as well as to the Arabian and Indian Peninsulas and the major Pacific islands, such as Madagascar and New Zealand. Laurasia broke up into North America and Eurasia. There has since been a general northward drift of all of the continents and an east or west drift of specific ones, so as to reach the present positions. As the drift is always assumed to have been very slow, the continents must not have been widely separated until well into the Tertiary period, if the hypothesis is correct.

This hypothesis was originally proposed on the basis of some facts of Paleozoic floras, such as the presence of fossils of tropical and semitropical plants in Alaska. A generally southern origin of the continents would readily explain this. However, there are some serious difficulties too. The geological evidence for continental drift is rather scant, consisting mainly of reciprocal curves of coast lines of lands which should have been rent apart. For example the outline of the east coast of South America can be fitted to the west coast of Africa in jigsaw puzzle fashion. However, the correspondence of geological strata is much less convincing, and so geologists are generally rather skeptical of the continental drift hypothesis. From a biological viewpoint, the major difficulty is not that it cannot explain specific distributions but that there is not enough to be explained. According to the theory, the continents were substantially continuous until late in the Mesozoic Era and could not have been widely separated until well into the Tertiary period. This would mean that corridor connections were generally present late enough that a very general exchange, such as actually occurred across the north Pacific corridor, should have been world-wide. As the various biogeographic realms are actually quite distinct, it is difficult to believe that they have been broadly connected so recently.

The Age and Area Theory of Willis. The Age and Area theory of Willis was proposed on the basis of a very extensive study of plant geography. Because a species must originate in a definite locality, and because plants tend to enlarge their range by slow migration in all directions,

Willis concluded that the age of a species (or higher group) must in general be proportional to the area now occupied. World-wide species are thus presumed to be very old, and endemics are presumed to be young species. To this he added what he called the principle of evolution by differentiation. This means that evolution proceeds from higher groups to lower, rather than from lower to higher as is most generally supposed. By this he means that a large mutation may produce at a single step a new class, order, or other higher group. The new order, let us say, is at first monotypic, but successive large mutations produce subdivisions, the families. These then break up into genera, and the genera into species. As the oldest section of any group will be that one which is at the place of origin of the group, this should also be the center of diversity for the group, for here there will have been a maximum amount of time for differentiation.

The major thesis of Age and Area is a truism, simply a restatement of the expanding phase of the history of any group, as exemplified above in the history of the mastodonts. But the major question is, is the expansion of a species or higher group likely to be slow enough that the present distributions of species and larger groups can give a good indication of their relative ages? Or are present distributions perhaps more generally maximal, and indicative of the limits imposed by physiographic or climatic barriers? Fernald has pointed out that many cases are known in which spread of plants has been very rapid, too rapid at least for Age and Area to have useful application when time is available on a geological scale. For example, it has been less than 25,000 years since the glaciers receded from northern United States and Canada, yet in that brief interval a large and not unbalanced flora has occupied the vast glaciated area, amounting to about one fourth of the continent. This includes such trees as the white spruce, *Picea canadensis;* the canoe birch, *Betula papyrifera;* the white cedar, *Thuja occidentalis;* and the mountain maple, *Acer spicatum.* Smaller plants of this association include burreeds of the genus *Sparganium,* pond weeds of the genus *Potamogeton, Iris, Viola,* and thousands of other plants of many types. All of these plants have, in less than 25,000 years, colonized a vast area, stopping only when stopped by barriers to their dispersal. As most species are much older than 25,000 years—a mere moment of geological time—it appears that age must generally play a minor role in determining the distribution of species.

Another serious defect of the Age and Area theory was Willis' emphasis upon the nature of endemics as young species. He recognized that endemism could be produced also by extinction of a species over all but a small part of its range. But he regarded this as an unusual situation. He seems to regard extinction (which is likely to be preceded by endemism) as a more or less unusual phenomenon. Yet it is clear from the paleontological record that extinction has been the fate of the overwhelming majority of species. As mentioned above in the discussion of endemics, most endemics are regarded as relics rather than young species by most biogeographers. His idea of evolution by differentiation is based entirely upon statistical evidence, and it has never been favorably regarded by most biologists.

Matthew's Theory on Climate and Evolution. Matthew based his ideas on climate and evolution upon his long experience as a geologist and mammalian paleontologist. He believed that the continents and the ocean basins were substantially permanent, and that theories involving continental drift, lost continents, or former land bridges across deep oceans were not in accord with known geological facts. In contrast, the climate of the world has alternated throughout the history of the world between warm, moist phases in which a mild climate prevailed throughout, and severe, arid phases in which climatic zones differed one from another. In the warmer phases, shallow seas covered much of the continental lowlands, and tropical organisms could inhabit northerly regions. But in times of arid, zonal climates, the polar regions were cold, the continents were elevated, glaciation might occur over large portions of the northern continents, and only the tropics remained mild. By the use of polar projection maps (Figure 113), he showed that the great land masses of the

FIGURE 113. POLAR PROJECTION MAP OF THE WORLD. (Redrawn from Matthew.)

world are mostly located in the northern hemisphere, with the three principal southern continents being more or less isolated, and centered in regions which should have mild climates even in the more severe times.

When severe climate begins in the polar regions and progresses southward on the Holarctic land mass, the organisms which are there must either adapt themselves to the new conditions, migrate southward, or become extinct. As a result, the more progressive organisms, the ones which succeed in adapting themselves to new conditions, should originate in Holarctic centers, while the southern continents should become refuges of primitive, less adaptable types. Matthew found that the paleontological record supported this sequence of events for the mammals. All of the major groups of mammals have arisen in the Holarctic, and the most primitive mammals are and have been centered in the southern continents, as the monotremes and marsupials in Australia. His study of the past and present distribution of the mammals convinced him that extensive geographical changes are not necessary to explain the distribution of mammals. That is, a sufficient explanation is contained in the permanence of the continents with minor changes in level, alternating uniform warm and zonal climates over the geological ages, and the use of the bridges described above. Thus no appeal to undemonstrated phenomena such as continental drift or former land bridges across the deepest of present oceans is necessary.

Matthew did not claim competence to judge the evidence from distribution of invertebrates or plants, but he did believe that his theory would have to be consistent with evidence from *all* groups in order to be valid. He believed that this would prove to be true.

Darlington's recent reappraisal of zoogeography reaffirms Matthew on permanence of continents, but he finds much evidence that the major groups of vertebrates arose in the Old World tropics, whence they invaded the Holarctic and the rest of the world. All classes of vertebrates are most abundant and varied in the tropics, and almost all of the dominant groups are represented there, while few dominant groups are confined to the north temperate zone. Much the same thing has been urged for plants for many years by Camp and more recently by Axelrod.

Related to this is Brown's idea of *centrifugal speciation*. As the largest populations occur in the large, favorable areas, probability favors these as places of origin of most of the progressive mutations. Migrations then occur because of population pressure, and so the variability of a species is spread over an increasing range. Should conditions become less favorable, the species will decline and retreat from much of its territory to the most favorable parts. This leads to isolation and possibly to speciation. Should some of the new species again be brought into contact, character displacement should accelerate their further differentiation. The result is a pattern of speciation from the center to the periphery of those large areas which are most favorable for any particular group.

REFERENCES

CAIN, A. S., 1944. "Foundations of Plant Geography," Harper & Brothers, New York, N.Y. Sound botanical geography. (DuReitz, Fernald.)

DARLINGTON, P. J., 1957. "Zoogeography," John Wiley & Sons, Inc., New York, N.Y. This very important book is the first thorough reassessment of its subject since Wallace. (Amadon, Brown.)

DU TOIT, A. L., 1937. "Our Wandering Continents," Oliver & Boyd, Edinburgh. The basic reference on continental drift.

HESSE, R., W. C. ALLEE, AND K. P. SCHMIDT, 1951. "Ecological Animal Geography," 2nd Ed., John Wiley & Sons, Inc., New York, N.Y. Primarily ecology, but very informative in relation to animal distribution.

MATTHEW, W. D., 1939. "Climate and Evolution," New York Academy of Sciences. A reprint of a classic.

SIMPSON, G. G., 1940. "Land Bridges and Mammals," *Proc. Wash. Acad. Sci.,* **30,** 137–163. A brilliant paper, and the basis for the discussion of bridges in this chapter.

WALLACE, A. R., 1876. "The Geographical Distribution of Animals," The Macmillan Co., London.

WALLACE, A. R., 1911. "Island Life," 3rd Ed., The Macmillan Co., London. These two books are still sound and deserve careful study by every serious student of biogeography.

WILLIS, J. C., 1949. "Age and Area," Rev. Ed., Cambridge University Press. A side issue in biogeography.

PART FIVE

Retrospect and Prospect

CHAPTER TWENTY-ONE

Retrospect and Prospect

IN RETROSPECT, we have reviewed the main outlines of the concept of organic evolution as modern scientists have developed it, as well as the reasons which have led them to this viewpoint. We have reviewed the fossil record, and seen that this is the most impressive evidence of evolution, in spite of its grave deficiencies. Finally, we have reviewed the modern attempts to analyze the genetic, cytological, ecological, geographical, and other causative factors in evolution, and we have seen that these complement one another to form a synthetic whole.

The dominant theme of recent work in evolution has been the neo-Darwinian theory, according to which the basic phenomenon of evolution is the slow accumulation of small mutations, the screening out of combinations of these by ecological factors which comprise natural selection to form subspecies, and finally the formation of good species by the same processes, aided by isolating mechanisms which prevent the subspecies (incipient species) from merging with the general population from which it came.

A second theme, urged by Goldschmidt and others, has emphasized the changes of chromosomal architecture which commonly distinguish related species, and which may be the basis for systemic mutations which profoundly change the organism. By this means, new species could be formed in one or a few steps, but they would then immediately be subject to the test of natural selection, and to subspeciation by neo-Darwinian evolution. Polyploidy, especially allopolyploidy, might also be properly included here, but both schools agree that this is a special, though important, phenomenon from which generalization is not justified.

The clash between these viewpoints has been acrimonious at times, yet they may prove to be supplementary rather than mutually exclusive. They were originally based upon different theories of the gene, morphological and atomistic for the neo-Darwinians, physiological and integrated for Goldschmidt. The physiological consequences of structure and the structural consequences of physiology were not clear. Today, much evidence indicates that patterns within nucleic acid macromolecules are critical for the gene. In terms of chromosomes, the gene may be a more or less broad field, with a point focus comparable to a physical center of gravity. These functional areas may vary in size, and they may overlap broadly, so long as their foci remain separate. The data of the corpuscular gene deal with these point foci, while position effect deals with the broader fields of genic action. Such a theory of the gene includes elements of both of the major theories of a few years ago, and it may subserve evolution along lines related to both theories, and perhaps to others not yet envisioned.

Another major source of disagreement centered around rates of evolution calculated for the mild selection pressures expected under neo-Darwinian theory. Even so convinced a neo-Darwinian as Dobzhansky expressed doubt that such a slow process could achieve the observed results, even in the great reaches of geological time, and Goldschmidt was certain that it could not. He proposed the systemic mutations in part to provide the needed acceleration. Many other accelerating mechanisms are also available. For example, character displacement, acting simultaneously upon an array of characters and under strong selective pressure, might simulate systemic mutation, as Brown has said. Again, selection forces may be very strong, and this should produce rapid changes. For example, both physical and biotic factors in most of North America have been profoundly changed during the past 400 years. Many species have become extinct as a result of changed selection pressures, and many others must have undergone profound changes adaptive to the new conditions. Such strong selective forces are probably always acting during major transitions, as from fish to amphibian, or from reptile to bird, for the intermediates are perhaps ill-adapted to both modes of life and under strong selection pressure to complete the transition. This may be why such transitions often seem abrupt in the fossil record. Again, it is improbable in any particular instance for predominant direction of mutation and direction of selection to coincide, but, if this should happen occasionally, evolutionary change might be very rapid indeed.

Each of these accelerating mechanisms has probably played a role in evolution, and it is perhaps premature to estimate their relative importance. Collectively, they provide an answer to one of the apparent contradictions of a few years ago, and they help to provide middle ground between evolutionary hypotheses which once seemed to be irreconcilable.

The most significant aspect of current research in evolution is the effort to synthesize data from all aspects of biology and from many of the physical sciences into a meaningful whole. Great advances have been made in all branches of biology under the stimulus of the Modern Synthesis, and its productivity may be expected to continue well into the future. This may result in profound modification of the major evolutionary theories of today, but these will have served science well by providing the basis for such fruitful investigations.

MAN AND THE FUTURE

In prospect, what does the future hold? And what influence will man have upon the future of evolution, and what will be the character of the future evolution of man himself? Obviously, it is impossible to answer these questions, but, as speculation upon them is always fascinating, some possibilities will be briefly outlined.

Plant and Animal Breeding. One of the major ways in which man has influenced evolution is through plant and animal breeding for agricultural and other purposes. The achievements in this field are great. There is probably not a single plant which now grows in the farms and gardens of the world which is the same as when man first cultivated it. Indeed, the very survival of those plants which man has domesticated may be due to the protection given them by man. For the plants which thrive best in the wild are often the most difficult to grow under cultivation, while those which thrive best under cultivation, such as corn, may quickly die out in competition with wild species. Because of these facts, Mangelsdorf has suggested that man has domesticated plants with character complexes already unsuited for competition in the wild, so that it may be said that man has rescued them from impending extinction, to the mutual advantage of plant and man. By artificial selection, agriculturally desirable characters, such as volume of seed production in grain plants, have been accentuated, while other characters, perhaps more important in a wild state, have dwindled. The bearing season of many plants has been much extended. Even the biochemical characters of plants have been altered by artificial selection, for protein and vitamin content of many plants has been significantly increased by selection of favorable breeding stocks. By the same method, the range over which particular kinds of plants can be successfully grown has been greatly extended. And the development of new, resistant strains of plants is one of the chief weapons in combating plant diseases, such as wheat rust. Although not generally so described, many of the agricultural productions of man may properly be described as good subspecies, not qualitatively different from naturally produced subspecies. But the new wheats and other plants which have been produced by the induction of allopolyploidy are best regarded as artificially produced species, or even genera, again not qualitatively different from those which occur in nature. Indeed, in some cases, such as *Galeopsis tetrahit,* the natural species itself has been resynthesized artificially.

The achievements in animal breeding have been more modest but still significant. From the wild horse, such divergent breeds as the Thoroughbred race horse and the Percheron draft horse have been formed by selection of breeding stock for the purposes desired. Different breeds of cattle have been perfected for high milk production or for high beef production. Each has been adapted to a wide variety of climates, from tropical to subarctic. Sheep have been specialized for production of wool or of mutton. Chickens with much increased egg production have been developed. The immense variety of dogs, from Pekinese to St. Bernard, have been produced by artificial selection under the hand of man. If these extremes were found in nature, no one would hesitate to call them distinct species. But because we know their history, we refer them all to a single species, *Canis familiaris.*

While the past achievements of plant and animal breeding are great, there is every reason to believe that far greater achievements lie ahead, for the greatest past achievements, such as the development of hybrid corn, have been accomplished during the past fifty, or even thirty, years, using the tools of genetics which were previously not available. It is reasonable to believe that the continued application of these principles will lead to even better agricultural products, both plant and animal. Thus man is truly achieving dominion over the world of life.

Distribution and Evolution of Wild Species. Man has also had a great influence on the distribution and evolution of wild species, and that influence may be expected to continue and to increase. So far as his effects on other organisms were concerned, primitive man probably was not much more important than were many of the larger wild animals. But, as his proficiency in the use of tools and the tilling of the soil increased, so also did his influence on other organisms. Soon the presence of man became a major selective force to which wild species, both plant and animal, had to adapt themselves in order to survive. With the development of the Industrial Age and the tilling of a very large portion of the arable soil in all civilized countries, man's selective influence on other organisms has reached a peak. In order to survive now, all living things must either become adapted to the presence and activities of man, or else they must be restricted to those dwindling refuges in which man's influence is least prominent.

Thus man has caused the extinction of many species and the extreme reduction of others within recent times. The case of the passenger pigeon, *Ectopistes migratorius*, is a well known example. Within the memory of some persons still living, the passenger pigeons were so numerous that flocks of them literally blackened the sky for hours at a time. The supply of pigeons was regarded as inexhaustible, and they were very intensively hunted and trapped for the market. But by 1880, passenger pigeons were noticeably less numerous. By 1890, their numbers were seriously depleted. By 1900, they were a rare bird, and movements were afoot to save the pigeons. But all efforts to save them failed, and the last one died in the Cincinnati Zoo in 1914. Extinction of the passenger pigeons is commonly attributed to intensive hunting, but the equally intensive destruction of the forests in which they found cover may have been quite as important a factor.

The buffalo were also prodigiously numerous when the white man invaded North America. They are now reduced to a few isolated and protected herds, and to exhibition specimens. Again, intensive hunting may have been an important factor in the decimation of the buffalo, but the fencing of the range was equally important. Comparable stories could be recited for many other species which inhabited North America before the white man came. But they all simply emphasize the statement with which this discussion began, that if organisms are to survive, they must either become adapted to the presence and activities of man, or they must be restricted to those dwindling refuges in which his influence is not great.

But man's influence is not always negative. Some organisms have profited greatly by the activities of man. This is obvious in the case of domesticated plants and animals, but it is also true of many wild species. Rats and mice, originally Palearctic animals, have been inadvertently carried to all parts of the world on cargo ships. They live in direct competition with man, invading his buildings for shelter, and robbing his larder for food. On the whole, they have been so successful that they must be very much more numerous than they were before the rise of civilization. The crow similarly has profited by farm lands, and small carnivores, such as foxes and weasels, have not only learned to live in close proximity to man, but they have learned to attack his small domestic animals, such as chickens and rabbits, and still escape his wrath sufficiently well to maintain themselves in competition. The success with which many weeds have invaded civilized habitats is too familiar to require discussion.

Radiation and Evolution. Another activity of man has great potential importance for evolution. This is the induction of mutations by radiation, including atomic radiation. Since the discovery by Muller that mutations could be induced by X-radiation, the possibilities have been intensively explored. Any high-energy radiation will cause mutations in numbers directly proportional to the total dosage of radiation. Although the possibility of using radiation to induce useful mutations has been extensively investigated, especially in crop plants, almost all of the experimentally produced mutations have been deleterious. A brilliant exception is provided by the work of Demerec on *Penicillium notatum,* the mold which produces penicillin. Previous to Demerec's experiments, the best available cultures of *Penicillium* produced about 70,000 units of the drug for every pint of culture. He X-rayed large numbers of the organisms, and carefully tested penicillin production of pure cultures derived from X-rayed mold. Most of the X-rayed stock was of no special value, but one highly productive strain, yielding 280,000 units of penicillin per pint of culture, was obtained. It is from this stock that most commercial penicillin is now produced. While the possibility that other valuable mutations may be produced by radiation cannot be entirely ruled out, still it is apparent that radiation is, on the whole, a genetic and evolutionary hazard. In a society in which radiation is playing an increasing role in medicine, experimental science, and even industry, it is impossible to estimate what the effects of inadvertent radiation of the gonads may be, both for man and for his associated plants and animals.

Such considerations led Muller to publish on the cover of the September, 1947, issue of the *Journal of Heredity* the warning that ". . . like most species, we are already encumbered by countless undesirable mutations, from which no individual is immune. In this situation we can, however, draw the practical lesson, from the fact of the great majority of mutations being undesirable, that their further random production in ourselves should so far as possible be rigorously avoided. As we can infer with certainty from experiments on lower organisms that all high-energy radiation must produce mutations in man, it becomes an obligation for radiologists —though one far too little observed as yet in most countries—to insist that the simple precautions are taken which are necessary for shielding the gonads, whenever people are exposed to such radiation, either in industry or in medical practice. And with the coming increasing use of atomic energy, even for peacetime purposes, the problem will become very important of insuring that the human germ plasm—the all-important material of which we are the temporary custodians—is effectively protected from this additional and potent source of permanent contamination."

The Future of Man. The physical future of man is, of course, the least predictable of all of the topics discussed in this chapter. No new attempt at prediction will be made, but rather the predictions of bolder authors will be presented. Speculation has taken two broad alternative courses: on the one hand, it has been assumed that the physical evolution of man has reached an end point, but that much future evolution may be expected in terms of the development of more efficient and peacable societies; others have accepted the probability of further physical change, and they have attempted to guess the character of these changes from past trends and apparent selective forces now operative upon man. Some excellent papers have been published upon the basis of the first alternative, but, as these are primarily sociological in character, they are outside the scope of the present work.

Since the beginning of recorded history, predictions of the imminent degeneration of mankind have been frequent. In our time, these are often based upon the premise that man, through his benevolent institutions, has arrested the action of natural selection. Mankind is burdened with the unfit, and their genes, which would be eliminated by natural selection in other species, are perpetuated. Hence biological disaster is inevitable. This idea has not gone unchallenged. Many selective agencies are currently operative in man. An example may be mentioned. We live in a high-speed, mechanized society, a thing unprecedented in history. Hence people who are physically and mentally adapted to live in such a society should, in the long run, have a selective advantage. This should put a premium upon the ability to think quickly and accurately, and to make precise, skillful movements. The effects of selection are obtained not only through the difficulties of providing for a family in the absence of such abilities, but also through the much higher accident rate among those who are maladjusted to modern life. Accidents may "just happen," but every careful study of accidents has indicated that they are caused, and they may properly be regarded as a highly effective source of selection pressure in modern society. Many other examples could be cited. The case was summarized by Dobzhansky * as follows: "The eugenical Jeremiahs keep constantly before our eyes the nightmare of human populations accumulating recessive genes that produce pathological effects when homozygous. These prophets of doom seem unaware ... that wild species ... fare no better ... than man ... , yet life has not come to an end on this planet. The eschatological cries proclaiming the failure of natural selection to operate in human populations have more to do with political beliefs than with scientific findings."

A few writers have treated the future of man without Jeremiads.

---

* Dobzhansky, Th., "Genetics and the Origin of Species," 1st Ed., Columbia University Press, 1937.

R. C. Andrews has looked into the distant future to visualize the outcome of the selection in a mechanized world. He envisions man as a mental giant who will become physically completely dependent upon his machines. The cranial capacity he expects to be enormously enlarged, while the face and all of the rest of the body he expects to be reduced. He expects all hair to be lost, and possibly all of the teeth. In short, Andrews envisions a future man who could not live except in a push-button civilization.

Hrdlicka has published a prediction from a much sounder point of view. He begins with the principle that prediction is in general unsafe, and that it becomes more so as the span of time covered by the prediction becomes greater. Hence he limits his prediction of future trends of human evolution to a maximum of 5000 years, and he bases his prediction upon trends observable in human remains of the past 5000 years. These predictions may be briefly summarized. He expects no increase in the size of the brain, but he does expect an increase in the efficiency of its organization. The sensory organs will become more efficient in response to selection in a high-speed society. The form of the skull will be more nearly globular, a trend which has been quite pronounced in the recent past, and which has been interpreted as a refinement of the adaptation to erect posture. The bones of the skull will generally become thinner because of the reduced stress from the jaw muscles. The hair will be reduced in quantity and finer in quality. The face he expects to become generally more refined and beautiful (by present standards) because of reduced physical stress and because of sexual selection. The forehead will be larger, and the eyes set more deeply. The nose will be prominent and narrow. The mouth will be smaller, but the chin will be more prominent. The teeth will be smaller and less resistant to disease than they now are. The wisdom teeth will be missing, and possibly others also. Unfortunately, he sees no signs of reduction of the beard.

In stature, he expects the man of the future to be taller, but he anticipates no gigantism. The body will be slender. Breasts of the female will be smaller on the average than they now are. The pelvis will show no significant change. The legs will be longer, but the arms will be shorter. Hands and feet will be narrower than at present, and the little toe will be still further reduced.

Internally, he anticipates continued reduction of the appendix and shortening of the intestines. Physiologically, he expects a higher metabolic rate, faster heart beat and respiratory rate, and higher body temperature. Diseases which he expects to become more frequent include skin diseases; mental disease; diabetes; dental disorders; diseases of age, such as cancer, heart disease, and apoplexy; and disorders of the sense organs. But he offers hope that these may be offset by advances of medical science.

Boyd has predicted the future on the basis of population genetics. Because of current trends of population, he expects the world of the future to be populated predominantly by descendants of the peoples of Asia, Africa, and parts of Central and South America. Europe and North America will contribute very much less. Under pressure of increasing world population, interracial marriages may become the rule, so that racial differences will disappear, and a single human race will emerge. This human race of the future he has characterized on the basis of the relative contributions which may be expected from the various races of today. Like Hrdlicka, he expects the skull to be more nearly globular. In stature, he expects this future race to be about like present-day southern Europeans. The eyes and skin will be brown, and the hair will be straight or slightly wavy. But he expects the man of the future to be much more variable than any of the present races, and he hopes that some extreme variant may be endowed with sufficient originality to cope with the major problems which confront man.

Evolution, a Young Science. Most sciences have only gradually emerged from their predecessors, or from natural philosophy, but the science of evolution is one of a very few for which a fairly definite time of origin is known. Although the concept of origin of species by modification is an ancient one, only since the publication of the "Origin of Species" in 1859 has it had a firm scientific basis, capable of commanding the respect of competent scientists. An oldest date of origin might be set as the fall of 1834, when Darwin visited the Galapagos Islands, while the latest date would be November 24, 1859, when the "Origin of Species" was published.

Once established, the new science developed very rapidly indeed. Scientists of the latter decades of the nineteenth century labored mightily to establish the fact of evolution, especially through studies in comparative anatomy and comparative embryology. However, the other basic evidences of evolution were also developed at this time, as explained above in Part I.

Yet it was not only because of the great labors of many men that the new science prospered. Biology had participated in the great upsurge of science in post-Renaissance times. The names of such men as Spallanzani, Redi, Ray, Wolff, and Harvey serve as reminders of biological participation in this great movement. Yet the vast store of data which these men established was largely chaotic, for there was no unified, theoretical framework within which their diverse contributions could be marshalled.

Linnaeus tried to fill this need with his taxonomic system, and for a time he seemed to succeed in giving biology an appearance of order. In the end, however, he stimulated a great deal more exploration and fact-finding without satisfying the need for a basic theory, for the taxonomic system itself was inexplicable. Thus Darwin found biology a burgeoning chaos of more or less unrelated data without any comprehensive theory to make it a cohesive whole. The theory of the origin of species by means of natural selection filled that need. It gave meaning to the taxonomic system of Linnaeus and to the many other biological sciences which Linnaeus had sought to clarify by his system. Today, almost all biological work is based, directly or indirectly, on recognition of the fact that the plant or animal of today is the most recent product of an historical process. *Time* has become an essential dimension of biology.

Thus the role of evolutionary theory in biology is quite comparable to that of the laws of thermodynamics in physics: it is a basic law from which a major branch of science, biology, derives its comprehensibility. It is for this reason especially that the new science of evolution has developed into so great a body of experimental and theoretical knowledge in little more than a century. We may confidently hope for even greater progress during the second century of evolutionary studies!

REFERENCES

ANDREWS, R. C., 1939. "What We'll Look Like Tomorrow," Colliers, **104**, 12–13, 55–57 (July 8, 1939). Summarized above.

BOYD, W. C., 1950. "Genetics and the Races of Man," Little, Brown & Co., Boston. An introduction to physical anthropology from a genetic viewpoint, this book includes an interesting prediction of the future of man.

DARWIN, CHARLES GALTON, 1953. "The Next Million Years," Doubleday & Co., Inc., New York, N.Y., and Rupert Hart-Davis, London. Darwin's grandson, a distinguished physicist, predicts that man's reproductive powers will outstrip the resources of the world, resulting in tragic impoverishment.

HRDLICKA, A., 1929. "Man's Future in the Light of His Past and Present," *Proc. Am. Philosophical Soc.*, **68**, 1–11. Summarized above.

Index